T0192461

Modern Acoustics and Signal Processing

Series Preface for Modern Acoustics and Signal Processing

In the popular mind, the term "acoustics" refers to the properties of a room or other environment—the acoustics of a room are good or the acoustics are bad. But as understood in the professional acoustical societies of the world, such as the highly influential Acoustical Society of America, the concept of acoustics is much broader. Of course, it is concerned with the acoustical properties of concert halls, classrooms, offices, and factories—a topic generally known as architectural acoustics, but it also is concerned with vibrations and waves too high or too low to be audible. Acousticians employ ultrasound in probing the properties of materials, or in medicine for imaging, diagnosis, therapy, and surgery. Acoustics includes infrasound—the wind driven motions of skyscrapers, the vibrations of the earth, and the macroscopic dynamics of the sun.

Acoustics studies the interaction of waves with structures, from the detection of submarines in the sea to the buffeting of spacecraft. The scope of acoustics ranges from the electronic recording of rock and roll and the control of noise in our environments to the inhomogeneous distribution of matter in the cosmos.

Acoustics extends to the production and reception of speech and to the songs of humans and animals. It is in music, from the generation of sounds by musical instruments to the emotional response of listeners. Along this path, acoustics encounters the complex processing in the auditory nervous system, its anatomy, genetics, and physiology—perception and behavior of living things.

Acoustics is a practical science, and modern acoustics is so tightly coupled to digital signal processing that the two fields have become inseparable. Signal processing is not only an indispensable tool for synthesis and analysis, it informs many of our most fundamental models for how acoustical communication systems work.

Given the importance of acoustics to modern science, industry, and human welfare Springer presents this series of scientific literature, entitled Modern Acoustics and Signal Processing. This series of monographs and reference books is intended to cover all areas of today's acoustics as an interdisciplinary field. We expect that scientists, engineers, and graduate students will find the books in this series useful in their research, teaching and studies.

William M. Hartmann
Series Editor-in-Chief

More information about this series at http://www.springer.com/series/3754

Federico Miyara

Software-Based Acoustical Measurements

Springer

Federico Miyara
Laboratorio de Acústica y Electroacústica
Universidad Nacional de Rosario
Rosario
Argentina

ISSN 2364-4915 ISSN 2364-4923 (electronic)
Modern Acoustics and Signal Processing
ISBN 978-3-319-85768-8 ISBN 978-3-319-55871-4 (eBook)
DOI 10.1007/978-3-319-55871-4

Softcover reprint of the hardcover 1st edition 2017

Printed on acid-free paper

This Springer imprint is published by Springer Nature
The registered company is Springer International Publishing AG
The registered company address is: Gewerbestrasse 11, 6330 Cham, Switzerland

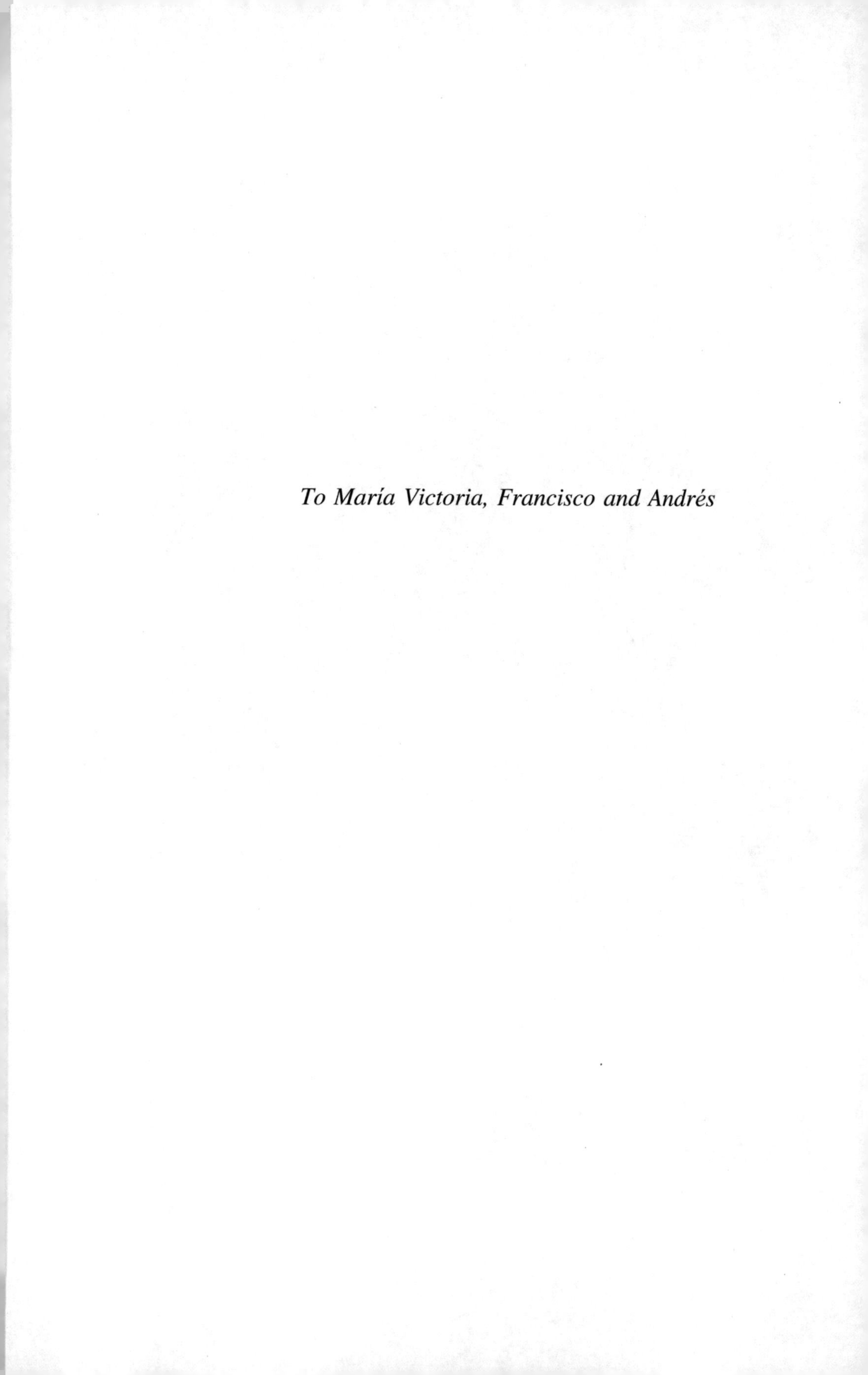

To María Victoria, Francisco and Andrés

Preface

Acoustic metrology has long presented the problem that sound measuring instruments are very expensive. This is due to two types of components: the transducer, i.e., the microphone, and the electronics required to perform signal conditioning and processing (filtering, analysis, comparisons) and display the results.

At the beginning, both the microphone and the electronics were intrinsically expensive. The microphone, because it was a sophisticated electromechanic device which required a complex manufacturing process and high-quality materials to ensure robustness, flat frequency response, low distortion, and stability over time. The electronics, because of the need of highly trained labor as well as expensive high-quality components.

Nowadays, instrumentation grade microphones are still expensive, but the cost of electronics has dropped several orders of magnitude. So one would expect that acoustical instrumentation prices had dropped as well. However, they tend to rise! For instance, a decade ago it was possible to purchase a first-brand sound level meter for about $2000. Today, the cheapest instrument that replaces that one is twice as expensive, even considering that the international inflation of the dollar has been only 30 % during the same time span, and the manufacturing cost has been further reduced by offshoring in countries like China and by the replacement of many discrete components by a small count of custom integrated circuits. In contrast, the computer industry exhibits the opposite trend: the price of what at a given time is considered to be a state-of-the-art computer not only does not increase but is actually falling below the dollar inflation.

Measurement instrument manufacturers try to imply that the price rise pays for a lot of exciting new features that are now possible through the use of sophisticated microprocessing units. However, the real cost of hardware and software is a very small fraction of the retail price. Moreover, they have adopted a commercial policy from the software industry, which is to license different software (firmware) modules that run on a single hardware platform. From the manufacturer standpoint, this reduces even further the manufacturing cost since the same hardware can be used to provide the functionality of several instruments as well as the promise of future updates (which certainly are not cheap). In some cases, firmware modules are preinstalled and all that is needed is to buy the code that unlocks the desired function.

In this scenario, the idea to exploit the potential of personal computers arises. Any user with some programming experience has at their fingertips the possibility to replace or emulate the signal processing that takes place inside a typical advanced sound level meter or analyzer. A sine qua non prerequisite is to have the signal available in digital format. This is accomplished by means of an external digital audio recorder which records the calibrated audio output signal that almost any sound level meter provides. Once digitized, the signal may be subject to many different processes, from computing any of the classic parameters and indicators to any possibility deserving to be investigated that our imagination can think of.

The traditional measuring paradigm is, hence, starting to change. Instead of recording a reading as an abstract translation of some property of a given phenomenon with no possibility of further verification, we record the very object to be measured. It is true that the original object is a sound field with complex spatial, temporal, and spectral distribution, and that by capturing it by means of a single microphone a severe abstraction process has been already performed. But it is the same abstraction process that the sound level meter carries out prior to the actual measurement!

The proposal put forward in this book is to examine traditional measurement processes and to present alternative approaches to be implemented using mathematical software running on a general purpose personal computer. Compatibility with International Standards in the field of acoustical measurements is also analyzed to ensure that the algorithms to be used comply with relevant standards. Several examples of parameters that are usually absent even in sophisticated commercial instruments are also given. In all cases, the proposal involves the minimum hardware and software investment, which is why all the examples use free, open source, and multiplatform software.

While every effort has been made to render the book reasonably self-contained, some prerequisites should be fulfilled to profit the most from its reading: (1) A fluent working knowledge of algebra and calculus of several variables, including the complex variable and the basics of differential equations. (2) A fair acquaintance with the theory of signals and systems, including the Laplace and Fourier transforms, the concept of spectrum and basic filter theory. (3) At least an elementary knowledge of Acoustics. (4) A good command of computers. (5) Some programming experience. (6) A grasp of elementary probability and statistics (though there is an appendix covering the basics).

Chapter 1 introduces the general concepts of acoustical measurements, such as equivalent levels, time and frequency weighting, statistical descriptors, and spectrum analysis, with emphasis on the process of measurement stabilization. Chapter 2 covers the most distinct aspect of measurement quality: uncertainty. While measurement uncertainty is extensively treated in many books and standards, the particular case of acoustical measurements is often neglected or presented in a fragmentary way. An in-depth treatment is provided here. This chapter may be skipped in a first reading without loss of continuity. Chapter 3 covers many aspects of digital recording, from sampling and analog/digital/analog conversion to audio file formats, recording media, and long-term preservation. This chapter can also be skipped or presented in a different order if starting earlier to program algorithms is the priority and the reader has a practical acquaintance with digital recording. Chapter 4 is an introduction to digital audio editing with special reference to its applications to measurement postprocessing. Chapter 5 covers several aspects of transducers that are relevant to their use in a measurement system, such as frequency response and directivity. Chapter 6 introduces digital signal processing, covering the main standard techniques such as convolution, discrete systems, difference equations, Z transforms, and stability. This chapter can be skipped if the reader is already acquainted with the basic theory of digital signal processing. Chapters 7 and 8 cover specific techniques and algorithms used in acoustical measurements. Finally, Chap. 9 presents the test methods to be applied to digital recorders to ensure the hardware complies with the relevant standards.

Each chapter includes a problem set in which the main ideas are applied to practical situations. In several problems, additional or advanced methods related to the subject of the chapter are introduced.

Finally, the book contains sixteen appendices covering supplementary material, derivations, and basic theory or examples.

Because of space restrictions, it is not possible to include here but short excerpts of some programs implementing the general ideas that are studied herein. That is why full-length examples are to be included in the Web site of the Acoustics and Electroacoustics Laboratory of the National University of Rosario. They will be available for download from http://www.fceia.unr.edu.ar/acustica/measurements.

Rosario, Argentina Federico Miyara
March 2016

Acknowledgements

Special thanks to my wife, María Victoria Gómez, and to my sons, Francisco and Andrés, for their permanent support to my crazy projects. To my parents, José Miyara and Julia Elvira Verdeja, my gratitude for the best legacy they could have given me: the intellectual stimulus and the awareness of the importance of knowledge, two necessary conditions so that someday I could write this book. Thanks to my fellows, colleagues, scholars, and friends from the Noise Group and the Acoustics and Electroacoustics Laboratory (National University of Rosario): Susana Cabanellas, Vivian Pasch, Marta Yanitelli, Patricia Mosconi, Juan Carlos Rall, Jorge Vazquez, Joan Puigdomènech, Pablo Miechi, David Giardini, Fernando Marengo Rodriguez, Ernesto Accolti, Ezequiel Mignini, Federico Bergero, Mauro Treffiló, Ezequiel Raynaudo, and Luciano Boggino, with whom I have shared many metrological adventures, including the development of measurement software, all of which have been the basis for the methods discussed herein. Thanks to Marina Fernández de Luco for her frequent help with Statistics. I am also indebted to the Agencia Nacional de Promoción Científica y Tecnológica, a national research program funding organization that supported several research projects involving the development of techniques that inspired parts of this book. My gratitude to the Acoustical Society of Argentina (Asociación de Acústicos Argentinos), AdAA, and particularly to its president, Nilda Vechiatti, for giving me the opportunity to publish the original Spanish version of this book under its imprint. Finally, thanks to my publishing editor at Springer, Sara Kate Heukerott, for her support in the process of getting this book published and to the reviewer William Hartman, whose comments have helped me to give the manuscript a final polish, and to the Production Editor Govardhana Manoharan for the final typesetting of the manuscript.

Contents

1 Introduction 1
1.1 Acoustical Measurements 1
1.2 Instantaneous and Effective (RMS) Value 2
1.3 Sound Pressure Level 4
1.4 Equivalent Level 4
1.4.1 Stabilization Time for L_{eq} 6
1.5 Weighted Levels 8
1.6 Exceedance Levels 13
1.6.1 Stabilization Time for Exceedance Levels 14
1.7 Spectrum Analysis 17
1.7.1 Constant Percentage Spectrum Analyzer 18
1.7.2 Line Spectrum Analyzer 22
1.7.3 Line Spectrum and Time-Frequency Uncertainty 23
1.7.4 Line Spectrum and Power Spectral Density 24
1.7.5 Band Filters 27
1.7.6 Transient Response of Band Filters 31
1.7.7 Parseval's Identity 35
1.7.8 Effect of Spectrum Tolerance on Parseval's Identity 37
1.7.9 Weighted Levels Computed from One-Third Octave Bands 39
References 41

2 Uncertainty 43
2.1 Introduction 43
2.2 Resolution, Precision, Accuracy 44
2.3 Measurement Method and Procedure 45
2.3.1 Direct and Indirect Measurement Methods 46
2.4 Measurement Model 46
2.5 Uncertainty 49

2.5.1 Type A Uncertainty . . . 50
2.5.2 Type B Uncertainty . . . 51
2.5.3 Expanded Uncertainty . . . 52
2.5.4 Combined Standard Uncertainty . . . 53
2.6 Examples . . . 55
2.6.1 Relationship Between Uncertainties in Level and Pressure . . . 55
2.6.2 Uncertainty in the Correction of Environmental Conditions . . . 56
2.6.3 Uncertainty in the Calculation of the Equivalent Level . . . 56
2.6.4 Uncertainty in A-Weighting Computed from Octave Spectrum . . . 58
2.6.5 Uncertainty in the Measurement of Sound Transmission Loss . . . 60
2.7 Uncertainty and Resolution . . . 61
2.8 Uncertainty and Systematic Error . . . 62
2.8.1 Additive Systematic Error . . . 62
2.8.2 Multiplicative Systematic Error . . . 64
2.8.3 Nonlinear Systematic Error . . . 68
2.8.4 Uncorrected Systematic Errors . . . 71
2.9 Chain Calculation of Uncertainty . . . 72
References . . . 75

3 Digital Recording . . . 77
3.1 Introduction . . . 77
3.2 Digital Audio . . . 77
3.3 Sampling . . . 78
3.4 Digitization . . . 83
3.5 Signal-to-Noise Ratio . . . 84
3.6 Low-Level Distortion . . . 85
3.7 Dither . . . 86
3.8 Jitter . . . 89
3.9 D/A and A/D Conversion . . . 91
3.9.1 Digital/Analog Conversion . . . 91
3.9.2 Analog/Digital Conversion . . . 92
3.10 Pulse Code Modulation (PCM) . . . 94
3.11 Differential Pulse Code Modulation (DPCM) . . . 95
3.12 Delta Modulation (DM) . . . 98
3.13 Sigma-Delta Modulation (SDM) . . . 100
3.14 Digital Recording Devices . . . 104
3.15 Sound File Formats . . . 106
3.15.1 WAV Format . . . 108
3.15.2 FLAC Format . . . 115

3.16 Project Files 123
3.16.1 AUP Format 124
3.16.2 AU Format 130
3.16.3 AUF Format 131
3.16.4 External Access to Audio Data 131
3.17 Recording and Storage Media 132
3.17.1 Hard Disk 132
3.17.2 Flash Memory 136
3.17.3 Optical Discs 138
3.17.4 Digital Audio Tape (DAT) 152
3.18 File Systems 152
3.18.1 FAT 32 153
3.18.2 NTFS 155
3.18.3 EXT4 157
3.18.4 HFS+ 158
3.18.5 UDF 159
3.19 Long-Term Preservation 159
3.20 Conclusion 162
References 164

4 Digital Audio Editing 167
4.1 Introduction 167
4.2 Audacity 168
4.2.1 Opening an Existing File 169
4.2.2 Recording Sounds 171
4.2.3 Generating Signals 171
4.2.4 Adding New Tracks 172
4.2.5 Saving a Project 172
4.2.6 Selection 173
4.2.7 Labels 174
4.2.8 Selection in the Presence of Labels 175
4.2.9 Calibration Tone 175
4.2.10 FFT Filters 178
4.2.11 Spectrogram and Spectrum Analysis 181
4.2.12 Noise Reduction 182
References 186

5 Transducers 187
5.1 Microphones 187
5.2 Polarization 189
5.3 Preamplifier 190
5.4 Sound Fields 190
5.4.1 Free Field 191
5.4.2 Diffuse Field 191

5.4.3 Pressure Field 191
5.4.4 Stationary Field 191
5.5 Microphones and Sound Fields 192
5.6 Frequency Response 193
5.7 Directional Response Pattern 196
5.8 Noise 198
5.9 Distortion 199
5.10 Micromachined Microphones 200
5.10.1 Frequency Response 201
5.11 Audiometric Earphones 201
5.12 Omnidirectional Sources 206
References 209

6 Digital Signal Processing 211
6.1 Discrete Signals 211
6.2 Discrete Impulse 211
6.3 A Signal as Its Convolution with a Discrete Impulse 213
6.4 Discrete Systems 213
6.5 Finite- and Infinite-Impulse-Response Systems 214
6.6 Difference Equation of a Discrete System 215
6.7 Frequency Response of a Discrete System 217
6.8 z Transform of a Discrete Signal 219
6.9 z Transform of a Difference Equation 220
6.10 z Transform of a Convolution 222
6.11 z Transform and Frequency Response 222
6.12 Solution of a Difference Equation 223
6.13 Poles and Stability of a Discrete System 225
6.14 Inversion of a Rational z Transform 225
6.15 Continuous- to Discrete-System Bilinear Conversion 226
Reference 230

7 Basic Algorithms for Acoustical Measurements 231
7.1 Introduction 231
7.2 Opening a .wav File 232
7.3 Energy Average and Equivalent Level 232
7.4 Calibration 233
7.5 Energy Envelope 234
7.6 A Weighting 236
7.7 Statistical Analysis 239
References 243

8 Spectrum Analysis 245
8.1 Introduction 245
8.2 Spectrum Analysis Paradigms 245
8.3 Digital Filters 246

8.4 Discrete Fourier Transform (DFT) 251
8.5 Fast Fourier Transform (FFT). 253
8.6 Spectrum Analysis with the FFT 254
8.6.1 Line Spectrum Analysis. 255
8.6.2 The Problem of Frequencies Close to $F_s/2$. 256
8.6.3 The Problem of Subharmonic Frequencies 258
8.6.4 Precise Detection of Pure Tones 259
8.6.5 Windows 260
8.6.6 FFT Band Spectrum Analysis 264
8.6.7 Spectral Density and Spectrum Averaging. 267
8.7 FFT Filters 268
8.8 Some Applications of the FFT 273
8.8.1 Determination of Tonality 273
8.8.2 FFT Convolution. 275
8.8.3 FFT Correlation. 278
8.8.4 Critical Band Filters 279
8.8.5 Determination of Transfer Functions 280
8.9 Contrasting Algorithms for Use in Measurements 280
8.9.1 International Standard IEC 61260 281
8.9.2 Contrasting Algorithms 282
8.9.3 Procedure 283
8.9.4 Sound Card Calibration 283
8.9.5 Verification of the Analyzer Response 284
8.9.6 Contrasting Different Algorithms. 285
8.9.7 Results. 286
8.10 Conclusion 291
References. 294

9 Testing Digital Recorders 297
9.1 Introduction 297
9.2 Specifications of the Digital Recorder. 298
9.3 Tests. 298
9.3.1 Frequency Response 298
9.3.2 Noise. 300
9.3.3 Linearity 301
9.3.4 Transient Response 302
9.3.5 Uncertainty 303
9.3.6 Conclusion 304
References. 305

Further Readings 307

Appendix A: Glossary and Definitions on Metrology. 309

Appendix B: Fundamentals of Statistics 321

Appendix C: Statistical Dispersion of the RMS Value of a Stationary Noise as a Function of the Integrating Time 337

Appendix D: Statistical Dispersion of the RMS Value of a Nonstationary Noise as a Function of the Integrating Time 341

Appendix E: Statistical Dispersion of Percentiles 345

Appendix F: Envelope of a Filtered Noise 347

Appendix G: Transient Response of a Third-Order Bandpass Filter . . . 351

Appendix H: Combined Uncertainty 353

Appendix I: Example of Uncertainty Calculation in the Case of a Nonlinear Systematic Error 357

Appendix J: The Sampling Theorem 363

Appendix K: Structure of a FLAC File 371

Appendix L: Document Type Definition (DTD) for Audacity Project Files 385

Appendix M: Brief Description of Scilab 395

Appendix N: Fast Fourier Transform (FFT) 407

Appendix O: Parseval's Identity and Symmetry of the DFT 411

Appendix P: Spectrum of Window Functions 415

Index 421

About the Author

Federico Miyara graduated as an Electronic Engineer from the Universidad Nacional de Rosario (UNR) in 1984. He then joined the faculty of his alma mater, first as an assistant teacher and later as an associate professor. There, he founded the Laboratory of Acoustics and Electroacoustics and has served as its director for more than 20 years. Over the years, he has led several postgraduate courses at the UNR, as well as at other universities in Argentina, Spain, Chile, Uruguay, and Bolivia.

He has conducted research in the field of noise assessment, particularly in noise mapping and in noise at the work place, as well as in psychoacoustics, musical acoustics, and architectural acoustics. His publications include four previous books on the subject of Acoustics, including the original Spanish-language version of this book. As a pianist and composer, the author attributes his initial interest in Acoustics to his curiosity about the science of music.

Chapter 1
Introduction

1.1 Acoustical Measurements

Unlike the measurement of physical quantities that are essentially time-invariant, such as the length or the mass of an object, the aim of acoustical measurements is to get relevant information from a signal that changes over time, and most often corresponds to an unrepeatable phenomenon.[1]

While the time history of an acoustic signal is of interest in cases such as transient analysis (IEC 3382-1: 2009), it is usually sufficient to measure global parameters that are representative of the phenomenon under study. For the sake of economy, most often one measures during a relatively short time and from the gathered information one attempts to find characteristic values that represent a much larger time span.

On the other hand, since the sound field varies from one spatial position to another, the location and even the orientation of the microphone has a distinct influence on the result. Sometimes the sensor, the instrument and its accessories, as well as the very presence of the operator, can alter significantly the signal to be measured.

As a consequence, measurement uncertainties involve the instrument, the operator, the opportunity, the sensor position, and even the nature of the phenomenon to investigate.

---

[1]Actually, any physical quantity to be measured changes over time. An object's length varies according to random temperature changes, mass varies because of particle deposition, oxidation, sublimation and so on. The difference is, rather, in the underlying model that is relevant for a given situation. So, in most cases an object's length is considered as a constant, unlike sound pressure that is essentially variable.

F. Miyara, *Software-Based Acoustical Measurements*, Modern Acoustics and Signal Processing, DOI 10.1007/978-3-319-55871-4_1

1.2 Instantaneous and Effective (RMS) Value

The acoustic quantity that is easiest to measure is sound pressure p. In many applications, the energy content of a signal is more relevant that its instantaneous value $p(t)$, since two sounds that cause the same effects might have a substantially different time history. As regards the effects of sound, the energy average or *effective value* or *root mean square value* (*RMS*) is of interest. It depends on an averaging time T that depends on the application and may range from a few seconds to years. It is defined as

$$P_{\mathrm{ef},T} = \sqrt{\frac{1}{T}\int_0^T p^2(t)\,\mathrm{d}t} \tag{1.1}$$

As an example, Fig. 1.1 shows the effective value for a sine wave.

Sometimes it is more convenient to have a moving energy average, i.e., one that is continuously updated:

$$P_{\mathrm{ef},T}(t) = \sqrt{\frac{1}{T}\int_t^{t+T} p^2(\theta)\mathrm{d}\theta} \tag{1.2}$$

Equation (1.2) gives the *energy envelope* of the signal. In this case, the selection of T depends on how rapidly the signal varies, which in turn depends on the duration of individual events that constitute the signal. For instance, for traffic noise T should be rather large, but for a noise which is formed by a series of blows T should be smaller. The energy envelope can be obtained replacing the integration by a low-pass filter with time constant τ. In such a case we have:

$$P_{\tau}(t) = \sqrt{\frac{1}{\tau}\int_0^t p^2(\theta)\mathrm{e}^{-\frac{t-\theta}{\tau}}\mathrm{d}\theta} \tag{1.3}$$

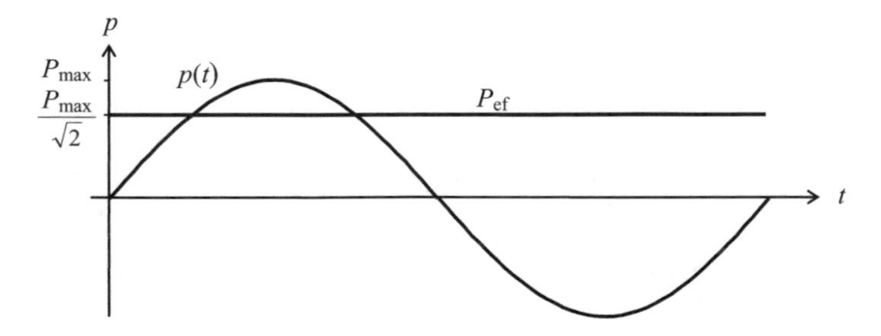

Fig. 1.1 A sine wave and its effective value P_{ef}, extended to a single period

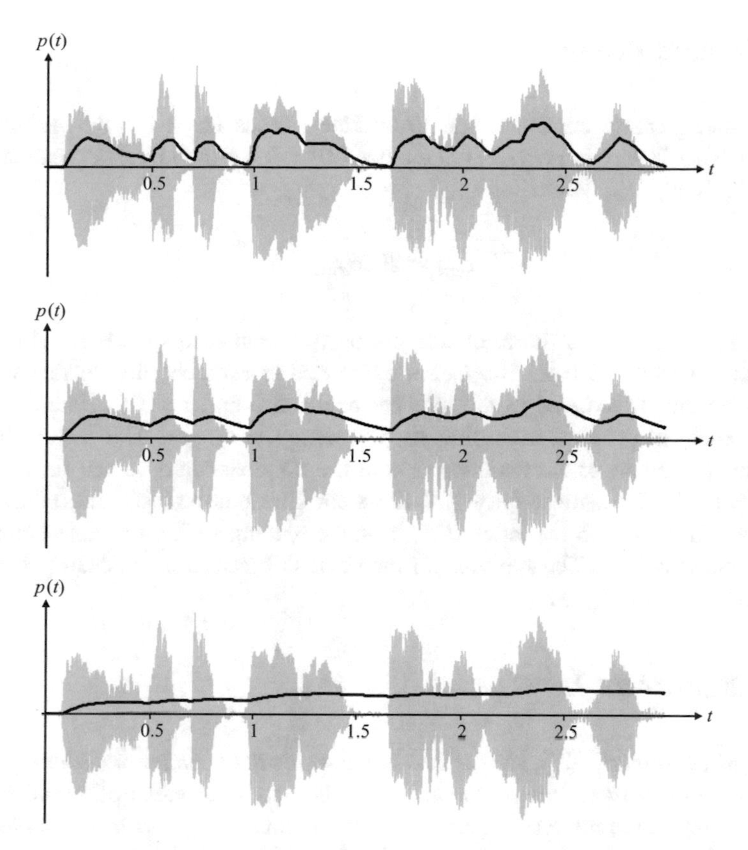

Fig. 1.2 An example of signal (an utterance) and its energy envelope for three different time constants. From *top* to *bottom*, $\tau = 35$ ms; $\tau = 125$ ms; and $\tau = 1$ s. For smaller time constants more time details are preserved

This is an exponential time average. As can be noticed, the values of $p^2(\theta)$ are maximally weighted for $\theta \cong t$. In other words, priority is given to recent values of the signal. The result of applying this time average is shown in Fig. 1.2 for three different time constants τ. The time constant is referred to as *time weighting*.

The envelopes obtained from (1.2) and (1.3) are not identical, but are very similar. While Eq. (1.3) seems to be more complicated, in practice the algorithm used to get $P_\tau(t)$ is computationally more efficient.[2]

[2]This algorithm does not consist in computing the integral numerically but to apply a digital filter that simulates a first order analog filter. See Sect. 7.5.

1.3 Sound Pressure Level

The sound pressure corresponding to audible sounds has a large dynamic range (10^6:1). That is why in general it is replaced by a logarithmic version called *sound pressure level*, $L_{p,\tau}$, defined as

$$L_{p,\tau} = 20 \log_{10} \frac{P_\tau}{P_{ref}} \tag{1.4}$$

where P_{ref} = 20 μPa is the reference pressure, which corresponds roughly to the hearing threshold at 1000 Hz. $L_{p,\tau}$ is expressed in decibels, dB. In this way, the dynamic range is reduced to the interval between 0 and 120 dB. Sound pressure level varies, in general, according to the energy envelope. The larger the time constant τ, the slower the variation. There are three standard values for time constant τ: 1 s for S response (slow), 0.125 s for F response (fast) and 0.035 s for I response (Impulse). In the latter case, to make reading easier, the time constant of falling phase is 1.5 s. The symbols for the results of measurements using these time constants are $L_{p,S}$, $L_{p,F}$, $L_{p,I}$.

1.4 Equivalent Level

The *equivalent level*, $L_{eq,T}$, also referred to as *equivalent continuous sound pressure level* (or *equivalent continuous sound level*) is a constant value of sound pressure level during a time interval T that has the same total energy as the variable sound pressure level. This value is precisely the logarithmic expression of the effective pressure during the same time interval, obtained by integration of the squared sound pressure (which is proportional to acoustic power). This is expressed in decibels:

$$L_{eq,T} = 20 \log \frac{\sqrt{\frac{1}{T}\int_0^T p^2(t)\mathrm{d}t}}{P_{ref}} = 10 \log \frac{\frac{1}{T}\int_0^T p^2(t)\mathrm{d}t}{P_{ref}^2}. \tag{1.5}$$

The equivalent level depends on the measurement interval T, hence it is possible to plot it as a function of integration time. For instance, in the case of a sine wave of angular frequency ω, we have

$$L_{eq,T} = L_{eq,\infty} + 10 \log\left(1 - \frac{1}{2\omega T} \sin 2\omega T\right). \tag{1.6}$$

This behavior is shown in Fig. 1.3. In the case of white noise we have a similar behavior, as shown in Fig. 1.4.

We can see from these examples that as the integrating T time increases, the equivalent level gets closer to the long term equivalent level (i.e., that one which would result from integrating during infinite time or, in practice, a very long time).

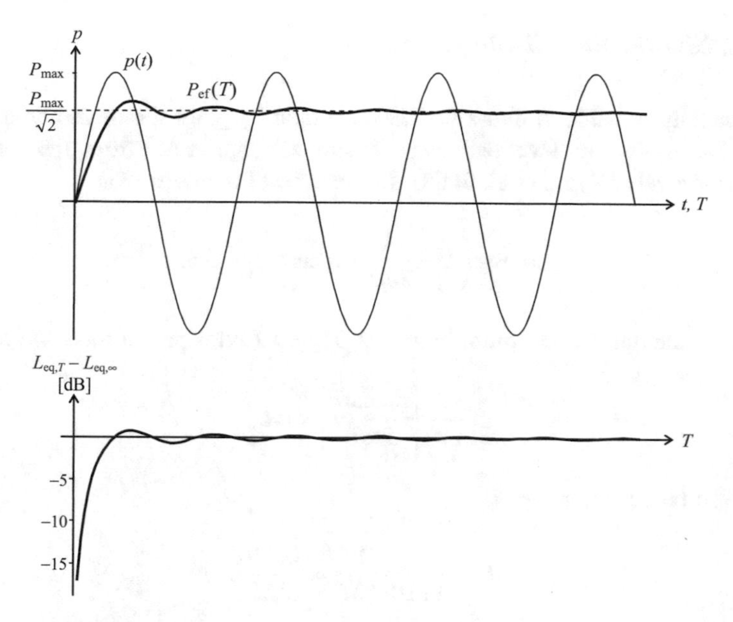

Fig. 1.3 *Top* effective value P_{ef} of a sine wave as a function of integrating time T. *Bottom* its logarithmic version, the equivalent level $L_{eq,T}$, relative to the long term equivalent level $L_{eq,\infty}$

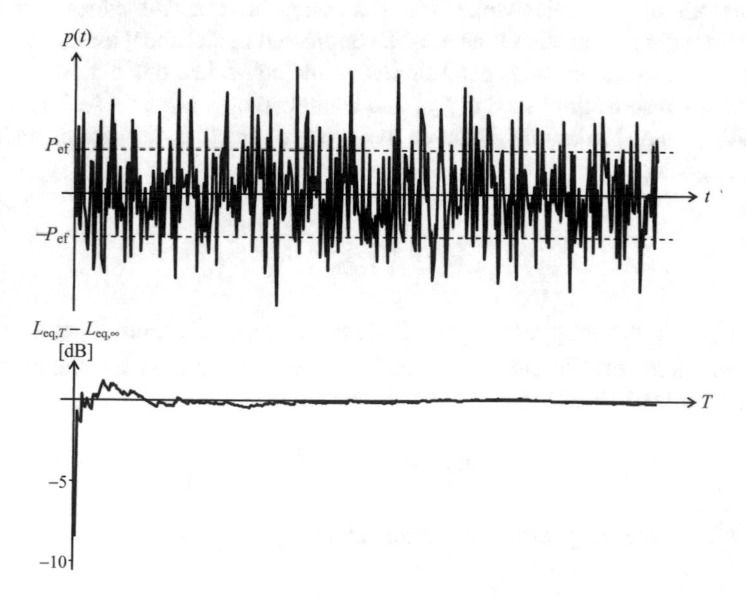

Fig. 1.4 *Top* white noise signal. *Bottom* its equivalent level $L_{eq,T}$ relative to the long-term equivalent level $L_{eq,\infty}$ as a function of T

1.4.1 Stabilization Time for L_{eq}

It is generally possible to find a *stabilization time* $T_{\mathrm{e},\Delta L}$ for the measurement, after which the equivalent level will stay within a range $\pm\Delta L$ from the long-term equivalent level (Miyara et al. 2008). In the case of a sine wave,

$$\left|10\log\left(1-\frac{1}{2\omega T}\sin 2\omega T\right)\right|<\Delta L. \tag{1.7}$$

Approximating the logarithm by a first degree Taylor polynomial, we have

$$\frac{10}{\ln 10}\left|\frac{\sin 2\omega T}{2\omega T}\right|<\Delta L,$$

which can be accomplished if

$$T>\frac{5}{\ln 10}\frac{1}{\omega\Delta L}=\frac{0.346}{f\Delta L}. \tag{1.8}$$

For instance, for a 100 Hz tone the stabilization time for ± 0.1 dB turns to be 34.6 ms.

In the case of a Gaussian white noise (i.e., one whose instant values have normal distribution) the stabilization time must be expressed in statistical terms. In a typical situation, we can suppose it is band-limited white noise, i.e., one that has a constant spectrum up to some limit such as $f_{\max}$ and from then on it is zero.[3] As it is shown in Appendix 3, the standard deviation of the squared effective pressure extended to a time T is given by

$$\sigma_{P^2_{\mathrm{ef},T}}\cong\sqrt{\frac{1}{f_{\max}T}}P^2_{\mathrm{ef},\infty}, \tag{1.9}$$

where $P^2_{\mathrm{ef},\infty}$ is the long-term squared effective pressure. Approximating the statistical amplitude distribution as a normal one,[4] we can bound the squared effective value with a 99.7 % confidence level to the interval

$$P^2_{\mathrm{ef},\infty}-3\sigma_{P^2_{\mathrm{ef},T}}<P^2_{\mathrm{ef},T}<P^2_{\mathrm{ef},\infty}+3\sigma_{P^2_{\mathrm{ef},T}}, \tag{1.10}$$

where $P^2_{\mathrm{ef},T}$ is the long-term value. Equivalently,

[3]There is no ideal white noise in Nature, since it should have an infinite bandwidth. Probably the closest one is electric resistance thermal noise, whose bandwidth spans up to about 100 GHz.

[4]It cannot be strictly normal, since the squared pressure is never negative, but if the standard deviation is small, in the neighbourhood of the mean value it is approximately normal.

$$1-3\frac{\sigma_{P^2_{\mathrm{ef},T}}}{P^2_{\mathrm{ef},\infty}}<\frac{P^2_{\mathrm{ef},T}}{P^2_{\mathrm{ef},\infty}}<1+3\frac{\sigma_{P^2_{\mathrm{ef},T}}}{P^2_{\mathrm{ef},\infty}}. \tag{1.11}$$

Let ΔL be the error between the equivalent level reached after integrating during a time interval T and the long-term equivalent level, i.e.,

$$\Delta L = L_{\mathrm{eq},T} - L_{\mathrm{eq},\infty}. \tag{1.12}$$

Then, from (1.11) and (1.9) we have

$$10\log\left(1-\frac{3}{\sqrt{f_{\max}T}}\right)<\Delta L<10\log\left(1+\frac{3}{\sqrt{f_{\max}T}}\right). \tag{1.13}$$

Using the first degree Taylor expansion of the logarithm, which is adequate for large T, we can rewrite the last equation as

$$|\Delta L|<\frac{10}{\ln 10}\frac{3}{\sqrt{f_{\max}T}}. \tag{1.14}$$

We can accept that $|\Delta L| < \Delta L_{\max}$ with a 99.7 % confidence level if we adopt an averaging interval T such that

$$T>\left(\frac{30}{\ln 10}\right)^2\frac{1}{f_{\max}\Delta L^2_{\max}}=\frac{169.8}{f_{\max}\Delta L^2_{\max}}. \tag{1.15}$$

For instance, if $f_{\max}$ = 22 050 Hz and $\Delta L_{\max}$ = 0.1 dB, then the stabilization time turns to be $T > 0.77$ s. This means that if we integrate during 0.77 s the equivalent level will differ from the long-term equivalent level by no more than 0.1 dB with a confidence level of 99.7 %.

When the signal is not stationary the stabilization time may be significantly larger, particularly when the signal is the superposition of a series of isolated and infrequent events. One example is traffic noise. It is shown in Appendix 4 that the stabilization time with a maximum error $\Delta L_{\max}$ and a confidence level of 99.7 % is given by

$$T_{\mathrm{stabil}}=\frac{9}{Q}\left(\frac{\sigma_{L_{\mathrm{eq},k}}}{\Delta L_{\max}}\right)^2, \tag{1.16}$$

where Q is the number of vehicles per unit time, and $\sigma_{L_{\mathrm{eq},k}}$ is the standard deviation of the equivalent level of different individual vehicles.

For instance, if $\sigma_{L_{\mathrm{eq},k}}$ = 3 dB, $\Delta L_{\max}$ = 0.1 dB and traffic flow is 1200 vehicles per hour, then

$$T_{\text{estab}} = \frac{9}{1200/3600}\left(\frac{3}{0.1}\right)^2 = 24\,300 \text{ s}. \tag{1.17}$$

As can be observed, the stabilization time with a 99.7 % confidence level is much longer than in the case of stationary random noise. The stabilization time is substantially reduced if the acceptable error is larger, for instance 1 dB instead of 0.1 dB.

The essential concept underlying all cases is that stabilization is accomplished by considering a large number of events, so that variability is reduced. In the case of a sine wave, the events are the individual cycles. Therefore, the larger the frequency, the shorter the time necessary to attain the required number of events. In the case of a band-limited white noise, the events are the discrete samples. Since the sample rate is high, stabilization is reached after a very short time interval. In the case of traffic noise, the events are individual vehicles passing by the observer. Since they are rather sparse (in our example there is one every 3 s), it is necessary a very long time to reach a sufficient number of events.

1.5 Weighted Levels

Most acoustic criteria connected with sound perception or the effects of noise in humans (hearing loss, annoyance, effects on performance) are expressed using some sort of frequency weighting. The purpose is to take into account that hearing sensitivity, as well as hearing risk, depends on frequency. To this end, the signal is fed to a filter with a suitable frequency response, and the filter output is measured.

The first frequency weighting contours that were introduced were those called *A*, *B* and *C*. While they were originally intended for taking into account hearing response at levels close to 40, 70 and 100 dB respectively, later *A* and *C* weightings would be used to assess noises of any level (particularly, high levels). The reason is that much research was carried out using the instruments that were more common in those times, i.e., those with frequency *A* and *C* weighting. Investigations have shown that *A*-weighted sound levels exhibit a reasonable correlation with effects such as hearing impairment and annoyance.

Nowadays, there is a strong discussion on this topic, since subjective effects of noise seem to depend on many other features of sound that span from spectral profile and amplitude fluctuation to semantic content.

Several new weightings were since introduced which are applicable to particular cases. For instance, *D* weighting has been proposed to evaluate air transportation noise, and *G* weighting for low frequency noise.

Figure 1.5 plots the frequency response curves that correspond to *A*, *B*, *C* and *D* weighting, and Fig. 1.6 plots *G* weighting. Equations (1.18)–(1.22) give the analytic expressions, and Tables 1.1 and 1.2 provide numeric values of the frequency responses for the one-third octave frequency series.

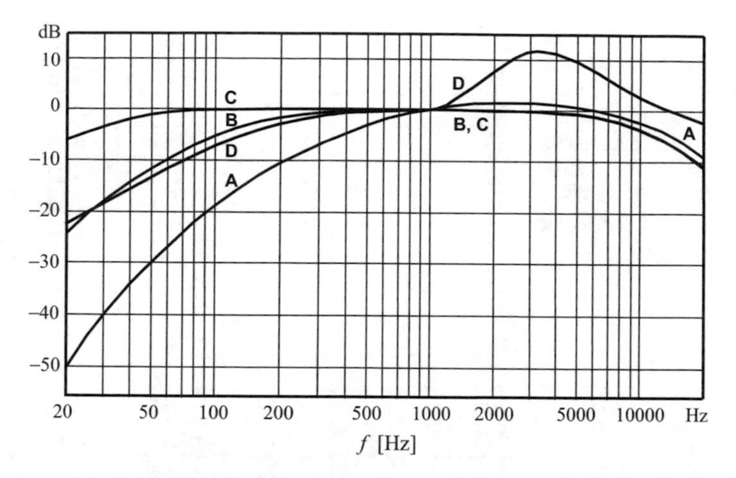

Fig. 1.5 *A*, *B*, *C* and *D* frequency weighting response. For a given frequency, the ordinate represents the additive correction to the sound pressure level of a tone of that frequency to get its weighted sound level. At 1 kHz all weightings are nominally equal to 0 dB

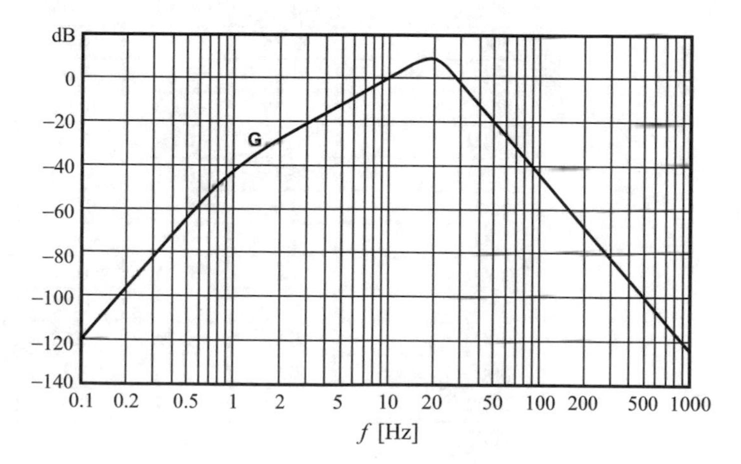

Fig. 1.6 *G* weighting frequency response to be used to measure very low frequency noise. For a given frequency, the ordinate represents the additive correction to the sound pressure level of a tone of that frequency to get its *G* weighted sound level. At 10 Hz the correction is 0 dB

$$A(f) = 20\log\frac{1.2589049 \times 12\,194^2 f^4}{(f^2+20.6^2)\sqrt{f^2+107.7^2}\sqrt{f^2+737.9^2}(f^2+12\,194^2)} \tag{1.18}$$

$$B(f) = 20\log\frac{1.0197249 \times 12\,194^2 f^3}{(f^2+20.6^2)\sqrt{f^2+158.5^2}(f^2+12\,194^2)} \tag{1.19}$$

Table 1.1 Numerical values of the additive corrections in dB corresponding to A, B, C and D weightings for the standard 1/3 octave frequency series (according to equations from Standards IEC 61672-1 (2002), IEC 61651 (1979) and IEC 60537 (1976))

Frequency (Hz)	A (dB)	B (dB)	C (dB)	D (dB)
10	−70.44	−38.24	−14.33	−26.63
12.5	−63.59	−33.32	−11.34	−24.69
16	−56.43	−28.28	−8.43	−22.56
20	−50.39	−24.16	−6.22	−20.63
25	−44.82	−20.48	−4.44	−18.70
31.5	−39.53	−17.13	−3.03	−16.72
40	−34.54	−14.10	−1.98	−14.68
50	−30.28	−11.63	−1.30	−12.79
63	−26.22	−9.36	−0.82	−10.87
80	−22.40	−7.31	−0.50	−8.94
100	−19.15	−5.65	−0.30	−7.20
125	−16.19	−4.23	−0.17	−5.57
160	−13.24	−2.94	−0.08	−3.92
200	−10.85	−2.04	−0.03	−2.63
250	−8.68	−1.36	0.00	−1.59
315	−6.64	−0.85	0.02	−0.81
400	−4.77	−0.50	0.03	−0.36
500	−3.25	−0.28	0.03	−0.28
630	−1.91	−0.13	0.03	−0.46
800	−0.79	−0.04	0.02	−0.61
1000	0.00	0.00	0.00	0.00
1250	0.58	0.01	−0.03	1.92
1600	0.99	−0.02	−0.09	5.05
2000	1.20	−0.09	−0.17	7.95
2500	1.27	−0.21	−0.30	10.32
3150	1.20	−0.40	−0.50	11.54
4000	0.96	−0.72	−0.83	11.10
5000	0.56	−1.18	−1.29	9.61
6300	−0.11	−1.89	−1.99	7.63
8000	−1.14	−2.94	−3.04	5.46
10 000	−2.49	−4.30	−4.40	3.44
12 500	−4.25	−6.07	−6.17	1.43
16 000	−6.70	−8.52	−8.63	−0.77
20 000	−9.34	−11.17	−11.27	−2.74

$$C(f) = 20\log\frac{1.0071525 \times 12\,194^2 f^2}{(f^2 + 20.6^2)(f^2 + 12\,194^2)} \tag{1.20}$$

Table 1.2 Numerical values of the additive correction corresponding to G weighting for the standard 1/3 octave frequency series (according to poles and zeros from Standard ISO 7196 (1995))

Frequency (Hz)	G (dB)
0.25	–88.18
0.315	–80.17
0.4	–71.94
0.5	–64.34
0.63	–56.69
0.8	–49.24
1	–43.01
1.25	–37.61
1.6	–32.45
2	–28.22
2.5	–24.19
3.15	–20.11
4	–15.93
5	–12.05
6.3	–8.03
8	–3.88
10	0.00
12.5	3.86
16	7.87
20	8.97
25	3.88
31.5	–3.89
40	–12.17
50	–19.92
63	–27.95
80	–36.25
100	–44.00
125	–51.76
160	–60.33
200	–68.09
250	–75.84
315	–83.88

$$D(f) = 20\log\frac{14\,499.7f\sqrt{(f^2-1018.78^2)^2+1039.6^2f^2}}{\sqrt{f^2+282.71^2}\sqrt{f^2+1160^2}\sqrt{(f^2-3126.4^2)^2+3424.1^2f^2}} \tag{1.21}$$

$$G(f) = 20\log(630\,673.8G_1\ G_2\ G_3\ G_4) \tag{1.22}$$

where

$$\begin{aligned} G_1 &= \frac{f}{\sqrt{(f^2 - 1)^2 + 2f^2}} \\ G_2 &= \frac{f}{\sqrt{(f^2 - 397.96)^2 + 1485.3f^2}} \\ G_3 &= \frac{f}{\sqrt{(f^2 - 3981.8)^2 + 796.37f^2}} \\ G_4 &= \frac{f}{\sqrt{(f^2 - 397.96)^2 + 106.5f^2}} \end{aligned} \quad (1.23)$$

A and *C* weightings are specified in International Standard IEC 61672-1 (2002). *B* weighting is given in the obsolete Standard IEC 61651 (1979) and *D* weighting in Standard IEC 60537 (1976), also obsolete. *G* weighting is specified in ISO 7196.

These frequency weightings can be used in two ways. First, applying an analog or digital filter with the corresponding frequency response to the signal. Second, adding the numerical values from Tables 1.1 and 1.2 to the one-third octave band spectrum of the signal expressed in dB (see Sect. 1.7.9).

Of these frequency weightings, the most widely used is the *A* weighting. It is applied in hearing risk and annoyance criteria, that is why it is frequently adopted in noise legislation such as occupational safety and health acts and community noise ordinances. However, its validity has long been challenged, and it is currently under discussion by many researchers (Kogan 2004). It is significant that several standards apply corrections to *A*-weighted measurements (i.e., a correction over a correction!) to account for the influence of several spectral, temporal or semantic features.

C weighting is usually applied to assess noises with low frequency content. It is common practice to compute the difference between *C*-weighted and *A*-weighted sound levels. If that difference is 20 dB or more it is considered that the noise has strong low frequency content. *C* weighting is also used to measure peak values of impulse noise ($L_{C,\text{peak}}$). This is because its transient response for very short noises is more accurate than in the case of *A* weighting. Figure 1.7 shows the transient response of *A* and *C* filters when excited with a step function and a very short pulse. As can be noted, the *C*-weighted peak is closer to the real signal peak.

B weighting is no longer used, and current standards (such as IEC 61672-1 2002) do not even specify it.

D weighting is also obsolete, but it can sometimes be used as an alternative, simplified method to assess the effective perceived noise level, L_{EPN}, (an indicator used in airport noise assessment) of airplane flyover (Yanitelli et al. 2001).

Finally, *G* weighting has applications for the assessment of noise with very strong low-frequency content. Since its frequency response covers a spectral region

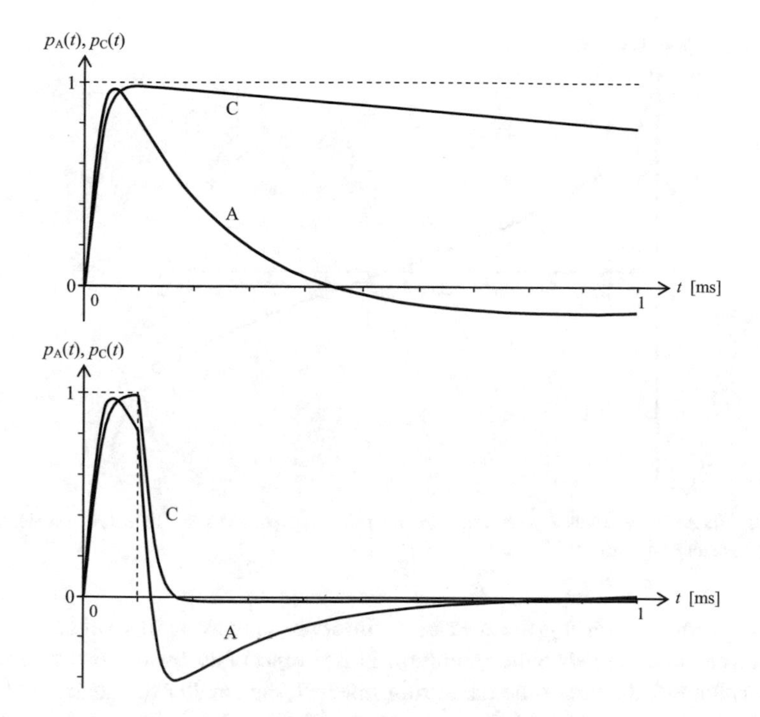

Fig. 1.7 *Top* transient response of weighting filters *A* and *C* with a unit step input. *Bottom* transient response with a short pulse input. In both cases, the *dashed lines* represent the input signal

that is virtually excluded from *A*-weighted measurements, we can say that both responses are complementary.

1.6 Exceedance Levels

Given a time constant τ and a frequency weighting *X* (where *X* can be *A*, *B*, *C*, *D*, *Z* or any other) we define the *exceedance level* L_{XN} as the value that is exceeded an *N* % of the measuring time. For example, in a very simple hypothetical case in which the sound level would increase linearly from 60 to 70 dBA during a measuring interval of 10 min, we can say that $L_{A,10}$ is 69 dBA, since that level is exceeded only during the last 10 % of the measuring period.

NOTE: Sometimes these values are incorrectly called *percentile levels*. Indeed, the *p*th percentile of a random variable is the value that *exceeds* a *p* % of the population, instead of *being exceeded*.

In Fig. 1.8 is shown the example of exceedance levels of a noise whose level is normally distributed with standard deviation $\sigma = 2$ dB.

Exceedance levels are very useful for the application of certain criteria. For instance, in order to determine the background noise one could try to find the

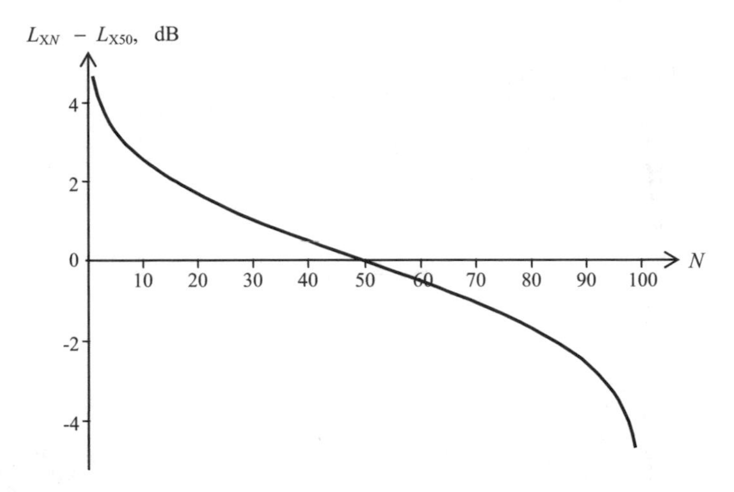

Fig. 1.8 Exceedance levels L_{XN} referred to the median L_{X50} for a noise whose levels L_X have normal distribution with σ = 2 dB.

absolute minimum during the measuring interval. However, this value is not representative, since the absolute minimum is not repeatable from a measurement to another. Indeed, the larger the measuring interval, the smaller the detected absolute minimum. Instead, we could define background noise level as L_{A90}, i.e., the level that is exceeded 90 % of measuring time. Unlike the absolute minimum, when we are in the presence of a stationary phenomenon[5] this value tends to stabilize, as we shall see in the following section. It is also possible to classify background noise according to its origin. Indeed, L_{A90} may be considered as the background noise of nearby sources, while L_{A95} would represent, more likely, noise from the combination of many distant sources.

In a similar fashion, the absolute maximum is not a representative indicator, unlike L_{A10} or L_{A5}, which have a tendency to stabilize for long measuring intervals.

1.6.1 Stabilization Time for Exceedance Levels

Let us analyze now the stabilization time for the exceedance levels. To this end, suppose that the level is sampled with sample rate F_s, so that a measuring interval T is equivalent to a number of samples:

$$n = F_s T. \tag{1.24}$$

[5]The evolution of urban noise across a whole day is an example of a non stationary phenomenon. During rush hours L_{A90} will be significantly larger than during the small hours.

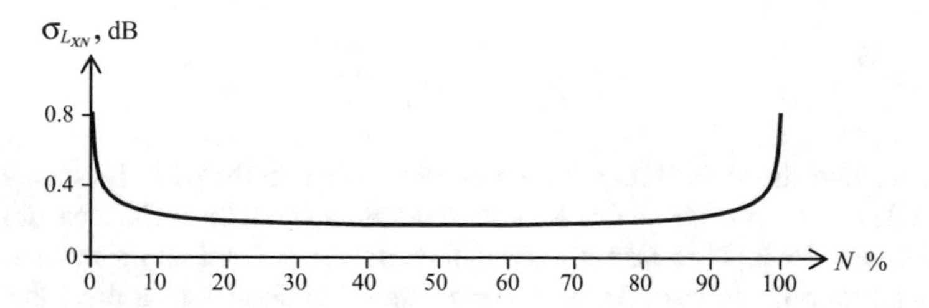

Fig. 1.9 Standard deviation in dB of the exceedance levels L_N for a noise whose level is normally distributed with $\sigma = 2$ dB

Conceptually, the stabilization of the exceedance level L_{XN} means a reduction of the statistical dispersion below an arbitrarily small limit with a given confidence level. For example, we could accept that L_{XN} has stabilized when the absolute error with respect to the final (long-term) value falls below 1 dB with a probability of 95 %. This type of problems are addressed, in general, analyzing the effect of the sample count on the standard deviation of the relevant variable, in this case, L_{XN}.

There is a result of this type available for percentiles, as detailed in Appendix 5. Bear in mind that the exceedance level L_{XN} is equivalent to the percentile $p = 100 - N$ of level L_X. We can then apply Eq. (A5.7) of Appendix 5, to get

$$\sigma_{L_{XN}} = \sqrt{\frac{\left(1-\frac{N}{100}\right)\frac{N}{100}}{n}\frac{1}{f(L_{XN})}}, \tag{1.25}$$

where f is the probability density function of the statistical distribution of L_X. Figure 1.9 shows an example for a level L_X with normal distribution. Close to the extreme values ($N = 0$ and $N = 100$), while the numerator tends to 0, the probability density function tends to 0 much faster; therefore, the standard deviation is very high.

If we assume that the distribution is approximately normal,[6] we know that about 99.7 % of all possible values of L_{XN} will fall inside an interval $\pm 3\sigma_{L_{XN}}$ from the mean value, so if we accept an error ΔL_{XN} the number of samples to take should be

$$n \geq \frac{9}{(\Delta L_{XN})^2}\frac{\left(1-\frac{N}{100}\right)\frac{N}{100}}{f(L_{XN})^2}, \tag{1.26}$$

Considering (1.24), this means that the stabilization time with an error ΔL_{XN} and a 99.7 % confidence level must be[7]

[6]The distribution in not necessarily normal, but it tends to be normal when the noise to be measured is the result of many independent events.

[7]Even if the distribution is not normal, Eq. (1.27) yields the time needed to cover $\pm 3\sigma$, though depending on the distribution the confidence level may differ from 99.7 %.

$$T \geq \frac{1}{F_s} \frac{9}{(\Delta L_{XN})^2} \frac{\left(1 - \frac{N}{100}\right)\frac{N}{100}}{f(L_{XN})^2}, \tag{1.27}$$

We see that the stabilization time increases reciprocally with the square of the acceptable error. This means that in order to attain a small error the measuring time must be very long. It is also observed that if we increase the sample rate, the measuring time is reduced. Note, however, that this applies to a level that varies randomly and is stationary during the measuring time. If the noise is not stationary, for instance the noise of a single vehicle passing by the observer, the preceding inequality might yield an incorrect selection of T. We might believe that we can reduce the stabilization time by increasing the sample rate F_s. However, this is not correct since in that case we might be considering a portion of the signal that is not representative of the whole phenomenon.

If the level distribution is normal with standard deviation $\sigma_{L_{XN}}$, then we can apply Eq. (A5.6) from Appendix 5. Considering that the percentile which corresponds to the exceedance level N is $p = 100 - N$,

$$L_{XN} = \sqrt{2}\sigma_{L_X} \mathrm{erf}^{-1}\left(1 - \frac{2N}{100}\right) + \mu_{L_X}, \tag{1.28}$$

where erf is the *error function*,[8] defined by Eq. (A5.5) from Appendix 5. The probability density function at this value of the argument is

$$f(L_{XN}) = \frac{1}{\sqrt{2\pi}\sigma_{L_X}} \mathrm{e}^{-\left(\mathrm{erf}^{-1}\left(1-\frac{2N}{100}\right)\right)^2}. \tag{1.29}$$

As an example, let us compute the stabilization time for L_{A90}, which is usually considered as the background noise level, with a 0.5 dB error and a 99.7 % confidence level, for an A-weighted sound level with normal distribution and $\sigma_{L_{AN}} = 2$ dBA. From (1.29) we have

$$f(L_{A90}) = \frac{1}{\sqrt{2\pi}\,\sigma_{L_X}} \mathrm{e}^{-\left(\mathrm{erf}^{-1}(-0.8)\right)^2} = \frac{0.1755}{\sigma_{L_X}}. \tag{1.30}$$

Replacing $\sigma_{L_{AN}} = 2$ dBA,

$$f(L_{A90}) = 0.087749. \tag{1.31}$$

Finally, we replace in (1.27), to get

[8]The error function and its inverse may be computed numerically. Mathematical software packages contain these functions as primitives.

$$T \geq \frac{1}{44\,100} \frac{9}{(0.5)^2} \frac{0.1 \times 0.9}{0.087749^2} \text{ s} = 9.5 \text{ ms}. \tag{1.32}$$

It should be noted that such a short stabilization time is difficult to attain in practice since the level variation is not truly random, i.e., it is not true that consecutive samples are uncorrelated. This is because L_A is obtained after filtering and getting the energy envelope by means of Eq. (1.3). The resulting signal is slowly variable, so when it is sampled at a high sample rate such as the one in the example (F_s = 44 100 Hz), consecutive samples are strongly correlated. Conceptually, while we are taking a lot of samples, many of them represent basically the same information repeated over and over again.

If, for instance, we use fast (F) time weighting (τ = 125 ms), when sampling the envelope at a sample rate of 44 100 Hz it is very unusual to get a difference between successive samples greater than 0.02 dB. One way to take this into account is to replace in (1.27) the actual sample rate by a much smaller sample rate of about 10 f_c, where f_c is the cutoff frequency of the lowpass filter. In the case of F time weighting it is

$$f_c = \frac{1}{2\pi\tau_F} = \frac{1}{2\pi \times 0.125 \text{ s}} = 1.2732 \text{ Hz}.$$

Using the criterion introduced above the stabilization time is about 33 s.

This example corresponds to white noise. In the case of noises such as traffic noise, individual events have a much slower envelope, hence they are not stationary except if we consider much longer measuring intervals.

In this case we can replace the sample rate by traffic flow Q in vehicles per unit time. In the example of Sect. 1.4.1 where Q = 600 veh/h, i.e., 1/6 veh/s, the stabilization time turns to be 2525 s. If the acceptable error ΔL_{A90} is increased to 1 dB, the stabilization time is reduced to 1/4 of the original one, i.e., about 11 min.

Finally, when N gets close to 0 or to 100, as it has been already mentioned, the stabilization time is very large. When we finally reach N = 0 or N = 100, i.e., when we reach respectively the maximum and the minimum of L_X, the stabilization time rises to infinite. In other words, no finite time is able to capture the extreme values of the variable. This is the reason for which the maximum and minimum of a measurement are not representative.

1.7 Spectrum Analysis

Often, weighted or unweighted sound pressure level is not sufficient to describe all relevant aspects of an acoustic situation. For example, if only the sound level near the external surface of a façade is known, there is no way to determine the sound level inside since the façade's sound insulation depends on frequency. The same outdoor level may give rise to different indoor levels depending on whether it has predominantly low frequency or high frequency. For instance, a 1600 Hz, 70 dB tone has an *A*-weighted sound level of 71 dBA, the same as a 100 Hz, 90 dB tone. However, after going through the façade the first tone will be attenuated about 30 dB, while the second one, only 15 dB. Indoor level will be 76 dB in the first case and 41 dB in the second.

Another typical example is sound source identification. In many cases the noise generation mechanism involves periodic phenomena (rotation, vibration, resonance) with well-defined frequencics. Those frequencies remain unaltered along the propagation path, allowing their detection at remote locations. However, sound pressure level measurements do not provide enough information to be able to detect those characteristic frequencies.

In such cases it is convenient to perform a *spectrum analysis*, i.e., a decomposition of sound into the different simultaneous frequencies that it contains. This is accomplished with a *spectrum analyzer*. Many modern sound meters offer the possibility of optional software or hardware modules that add spectrum analyzing functionality. As we shall see further on, if we have a digital audio recording of the sound, it is possible to carry out the spectrum analysis in any personal computer.

There are three types of spectrum analyzers: *constant bandwidth*, *constant percentage*, and *line spectrum analyzers*. Constant bandwidth have typically a single band whose center frequency and its bandwidth can be independently adjusted. They are used mainly in radiofrequency tests and very seldom in audio applications, so we will not study them. Constant percentage analyzers have a number of bands that cover the whole audio spectrum (20 Hz to 20 000 Hz) in such a way that each band's bandwidth is a fixed fraction or percentage of the center frequency. Line analyzers present a large number (hundreds or thousands) of spectral lines, each one representing a very narrow constant bandwidth.

1.7.1 Constant Percentage Spectrum Analyzer

Constant percentage spectrum analyzers are generally specified as *octave fraction analyzers*, for instance *octave band analyzer* or *one-third octave band analyzer* (*1/3 octave analyzer*). Calling α the octave fraction, we have

$$f_{\mathrm{u},k} = 2^{\alpha/2} f_{\mathrm{o},k} \tag{1.33}$$

$$f_{\mathrm{l},k} = 2^{-\alpha/2} f_{\mathrm{o},k} \tag{1.34}$$

where $f_{\mathrm{o},k}$, $f_{\mathrm{l},k}$, $f_{\mathrm{u},k}$ are, respectively, the center, lower and upper frequencies of the k-th band.[9] It is easy to see that

[9]The ISO 61260 standard also accepts (and prefers) replacing 2 by $G = 10^{0.3} = 1.995262315$. This choice allows that ten one-third octave bands are exactly equal to one decade.

$$f_{u,k} = 2^{\alpha} f_{l,k} \tag{1.35}$$

and

$$f_{u,k} f_{l,k} = f_{o,k}^2. \tag{1.36}$$

There is also a band adjacency condition by which the lower frequency limit of a band must equal the upper frequency limit of the preceding band:

$$f_{l,k} = f_{u,k-1}. \tag{1.37}$$

This, in turn, implies that

$$f_{o,k} = 2^{\alpha} f_{o,k-1}, \tag{1.38}$$

so the ratio between successive center frequencies is the same as the ratio of the upper to the lower limit of a band. Figure 1.10 shows three octave bands presented in three different ways.

As can be seen, in a linear frequency scale, each octave band is twice as wide as the preceding one. In a logarithmic frequency scale, all bands seem to have the same size. One reason to chose this kind of bands is that human hearing system perceives frequency almost logarithmically in a wide range.

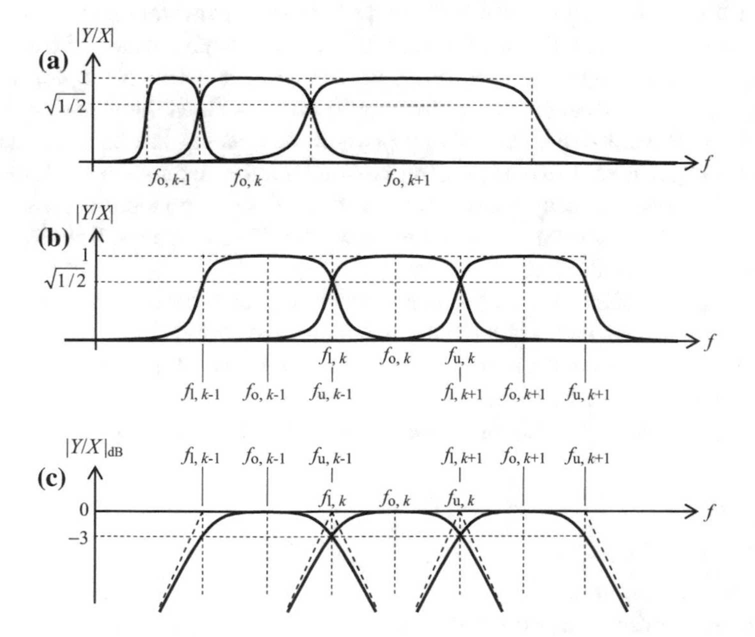

Fig. 1.10 Three consecutive bands of a constant percentage spectrum analyzer, represented: **a** with a linear frequency axis; **b** with a logarithmic frequency axis; **c** with a logarithmic frequency axis and dB amplitude axis

Table 1.3 Characteristic parameters of bandpass filters for several octave fractions

α	f_u/f_l	B
1	2	0.707
$^1/_2$	1.414	0.348
$^1/_3$	1.260	0.232
$^1/_6$	1.122	0.116
$^1/_{10}$	1.072	0.0693
$^1/_{12}$	1.059	0.0578
$^1/_{24}$	1.029	0.0289

Formula (1.38) implies that if we adopt a generating frequency it is possible to find all successive center frequencies. From these, Eqs. (1.33) and (1.34) allow to compute the lower and upper frequencies of each band. As 1000 Hz is approximately in the middle of the audible range (in a logarithmic scale) it has been adopted as the generating frequency.

It is customary to associate bandpass filters with a *relative bandwidth B* defined as the ratio between absolute bandwidth and center frequency:

$$B_k = \frac{f_{u,k} - f_{l,k}}{f_{o,k}} = 2^{\alpha/2} - 2^{-\alpha/2}. \tag{1.39}$$

As can be seen, the relative bandwidth is constant for a given octave fraction. In Table 1.3 there is a list of characteristic parameters for several octave fractions.

Let us consider now the specific case of one-third octave bands, since they are among the most frequently used ones. In this case, $\alpha = 1/3$. If we apply (1.38) strictly, the resulting exact frequencies exhibit many decimal digits and so they are difficult to remember. For the sole purpose of designating the bands, a series of nominal (approximate) frequencies has been adopted in international standards. It attempts to respect exactly decade ratios and as close as possible octave ratios.[10] Nominal values are shown in Table 1.4 along with the exact ones. Relative error is smaller than 1.6 % in all cases.

Although constant percentage analyzers have long been implemented with analog electric filter banks,[11] all modern designs use digital filters. This simplifies the design and prevents the need for parameter adjustment. It also greatly improves long-term stability of filter specifications such as center frequency or bandwidth since they remain unaltered during the whole lifespan of the product.

[10]This series of values is known as Renard series R10. Renard series are designed to distribute uniformly in a logarithmic scale a number of values (in this case, 10) in a decade.

[11]A filter bank is a set of filters that work simultaneously on the same signal, providing several outputs.

Table 1.4 Standard nominal center frequencies to be used in one-third octave analyzers in the audible range

Nominal center frequency (Hz)	Lower frequency (Hz)	Exact center frequency (Hz)	Lower frequency (Hz)
20	17.54	19.69	22.10
25	22.10	24.80	27.84
*31.5	27.84	31.25	35.08
40	35.08	39.37	44.19
50	44.19	49.61	55.68
*63	55.68	62.50	70.15
80	70.15	78.75	88.39
100	88.39	99.21	111.36
*125	111.36	125.00	140.31
160	140.31	157.49	176.78
200	176.78	198.43	222.72
*250	222.72	250.00	280.62
315	280.62	314.98	353.55
400	353.55	396.85	445.45
*500	445.45	500.00	561.23
630	561.23	629.96	707.11
800	707.11	793.70	890.90
*1000	890.90	1000.00	1122.50
1250	1122.50	1259.90	1414.20
1600	1414.20	1587.40	1781.80
*2000	1781.80	2000.00	2244.90
2500	2244.90	2519.80	2828.40
3150	2828.40	3174.80	3563.60
*4000	3563.60	4000.00	4489.80
5000	4489.80	5039.70	5656.90
6300	5656.90	6349.60	7127.20
*8000	7127.20	8000.00	8979.70
10 000	8979.70	10 079.00	11 314.00
12 500	11 314.00	12 699.00	14 254.00
*16 000	14 254.00	16 000.00	17 959.00
20 000	17 959.63	20 159.00	22 627.00

Those marked with an asterisk (*) are also center frequencies of octave analyzers

Constant percentage spectrum analysis is very useful when applied to wide band signals, particularly those without strong tones, since it allows the assessment of the effect of sound barriers, walls, sound absorption, long distance outdoor propagation, etc. The acoustic properties of many acoustical materials are specified in octave bands or one-third octave bands.

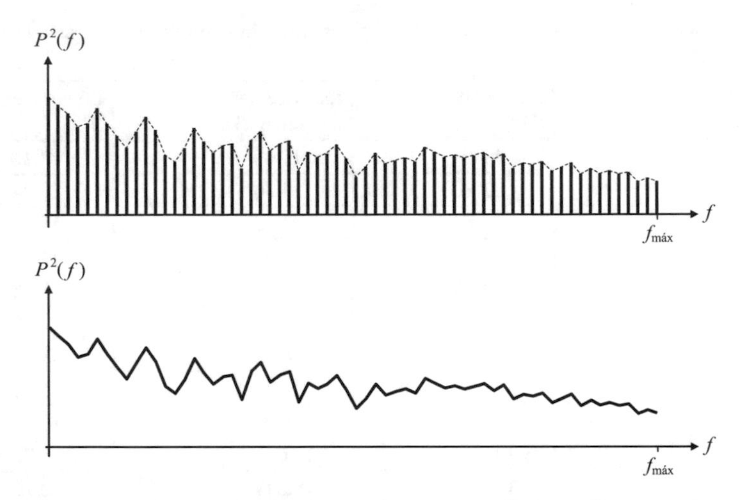

Fig. 1.11 *Top* a sample line spectrum (in this case, 64 spectral lines). The magnitude of each line represents the energy contained in a narrow spectral band. *Bottom* a frequent way of representing the same information by means of the envelope of the spectral lines. It is useful when N is large, since in that case the lines cannot be visually separated

1.7.2 Line Spectrum Analyzer

This type of analyzers became popular with the introduction of digital signal processing. They are based on the Fast Fourier Transform, FFT. Although we will study this algorithm in more detail later, we will advance now some basic properties.

In general, the number N of spectral lines is an integer power of 2, such as 2048. They are uniformly distributed in the range from 0 (DC component)[12] to a given maximum frequency f_{max},[13] slightly above the audible range, for instance 22 050 Hz (see Fig. 1.11). These frequencies are given by

$$f_k = k\frac{f_{max}}{N}. \tag{1.40}$$

This means that the frequency resolution is equal to f_{max} divided by the number of lines:

[12]In Acoustics, the DC (constant) component should be always 0, since sound pressure is the difference between the instantaneous pressure and static pressure. However, the electronics involved in signal conditioning adds some offset that may appear as a DC component in the spectrum.

[13]The maximum frequency is given, as we shall see when we address the study of digital audio, by one half the sample rate.

$$\Delta f = \frac{f_{max}}{N}. \tag{1.41}$$

In the specific case of maximum frequency f_{max} = 22 050 Hz and N = 2048 spectral lines, we have:

$$\Delta f = \frac{22\,050\ \text{Hz}}{2048} = 10.77\ \text{Hz}.$$

Each spectral line represents the energy content of a narrow frequency band. However, the actual meaning is a bit more complicated because of a phenomenon known as *spectral leakage*, by which a single frequency (for instance a sine wave) affects several or even all lines (although the influence is small for distant lines). This problem can be dealt with by the use of several forms of *windowing*, as we will see later.

Line spectrum is useful to detect tonal components, which in turn is interesting to detect, identify and quantify specific noise sources. It also allows to perform modal analysis of different devices and components. We can, for instance hit the housing of a machine to find the *normal vibration modes*. These are the ones that in normal operation will be radiated more intensely. Another application is to superpose (on an energy basis) several lines to get the sound pressure level in a non-standard, custom band.

1.7.3 Line Spectrum and Time-Frequency Uncertainty

A fact that must be considered is that the FFT line spectrum requires the analysis of a signal portion of duration

$$T = \frac{N}{f_{max}} \tag{1.42}$$

This implies that the more lines are required, i.e., the finer is the spectral resolution, the larger the duration of the signal portion needed to attain such resolution. This property is known as *time-frequency uncertainty*. Indeed, combining (1.41) and (1.42) we get

$$T\Delta f = 1. \tag{1.43}$$

This is interpreted as follows: if we want to know the spectrum with much detail, we lose information about its time location. Conversely, if we need to know the time location of an event very accurately, we will have to drop spectral resolution. If, for instance, we needed a spectral resolution of 10 Hz, we should analyze a signal portion of 100 ms, but if we can accept a coarser resolution of 100 Hz, it will suffice to analyze only 10 ms.

1.7.4 Line Spectrum and Power Spectral Density

The line spectrum may also be used to estimate the *power spectral density* of a random signal (also abbreviated as *spectral density*), which is defined as the mean square pressure per unit bandwidth. More precisely, it is the limit of the square of the effective pressure of the signal after passing through an ideal bandpass filter of bandwidth Δf, divided by Δf:

$$\overline{p^2}(f) = \lim_{\Delta f \to 0} \frac{P^2_{f,\Delta f,\mathrm{ef}}}{\Delta f}. \tag{1.44}$$

where

$$P^2_{f,\Delta f,\mathrm{ef}} = \frac{1}{T} \int_0^T p^2_{f,\Delta f}(t)\mathrm{d}t. \tag{1.45}$$

and T is given by Eq. (1.42). The power spectral density is expressed in Pa^2/Hz. The filter and its frequency response are shown in Fig. 1.12.

The spectral density represents how signal energy is distributed along the frequency spectrum. Its integral allows to compute the effective pressure in a given band:

$$P^2_{[f_1,f_2],\mathrm{ef}} = \int_{f_1}^{f_2} \overline{p^2}(f)\mathrm{d}f. \tag{1.46}$$

If $f_1 = 0$ and $f_2 = \infty$, we get the *total* effective pressure:

$$P^2_{\mathrm{ef}} = \int_0^{\infty} \overline{p^2}(f)\mathrm{d}f. \tag{1.47}$$

If we take a very narrow bandwidth Δf centered at frequency f, we can approximate:

$$\overline{p^2}(f) \cong \frac{P^2_{f,\Delta f,\mathrm{ef}}}{\Delta f}. \tag{1.48}$$

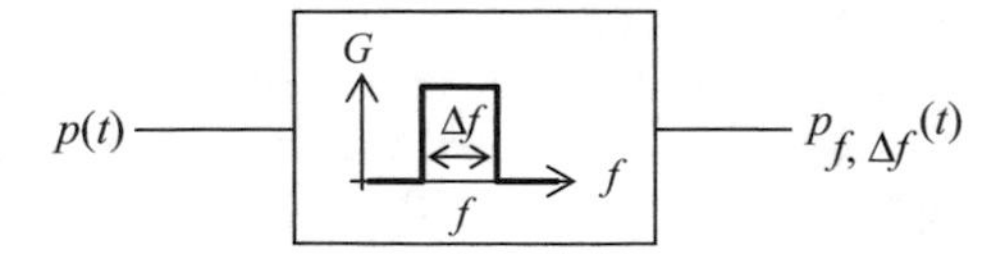

Fig. 1.12 A signal $p(t)$ passing through a bandpass filter with center frequency f and bandwidth Δf

If we choose a number N of spectral lines and $P_{k,\mathrm{ef}}$ is the effective pressure of the k-th line, which corresponds to frequency

$$f_k = k\frac{f_{\max}}{N},$$

then we can estimate the spectral density at f_k as

$$\overline{p^2}(f_k) \cong \frac{P_{k,\mathrm{ef}}^2}{\Delta f} = \frac{P_{k,\mathrm{ef}}^2}{f_{\max}/N}. \tag{1.49}$$

We often want to estimate the long-term average of the spectral density, since any random signal, because of its randomness, may present important spectral fluctuations with time. For instance, if the signal is a long conversation between several people, the short term spectrum will vary not only when different people speak, but also when a single person utters different phonemes.

Equation (1.49) provides only spectral information pertaining to the particular N-sample segment of the signal on which we have computed the line spectrum, which is located around a given time instant (within margins compatible with time-frequency uncertainty). We can estimate the average spectral density taking, instead, M segments of duration T and computing the average, for a specific frequency f_k, of the spectral density of all segments:

$$\overline{p^2}(f_k) \cong \frac{\frac{1}{M}\sum_{m=1}^{M} P_{k,m,\mathrm{ef}}^2}{f_{\max}/N}, \tag{1.50}$$

where $P_{k,m,\mathrm{ef}}^2$ is the mean square pressure that corresponds to frequency f_k and to the mth time segment.

It is interesting to determinate the number M of segments to average in order to meet a given error bound. More precisely, we are looking for M such that the error is under that bound with a probability of, say, 99.7 %. To this end, we need to know the statistical dispersion of the spectral lines. Of course, dispersion will depend on the type of signal. If the signal is a pure tone, there will be no dispersion, whereas if the tone has a strong amplitude modulation, the spectral amplitude will change according to the modulating signal.

In order to illustrate the concept, we shall consider the case of random white noise $n(t)$ limited in band to $f_{\max}$. We can assume that each spectral line will evolve in time as the envelope of the output of a bandpass filter with bandwidth equal to the analysis resolution, as given by (1.41):

$$\Delta f = \frac{f_{\max}}{N}.$$

As mentioned in Appendix 6, such envelope has a Rayleigh statistical distribution, whose mean is

$$\mu_Y = N_{ef}\sqrt{\frac{\pi \Delta f}{2 f_{max}}} = N_{ef}\sqrt{\frac{\pi}{2N}}, \tag{1.51}$$

where N_{ef} is the effective value of $n(t)$ prior to going into the filter. This formula provides the same expected value for all bands that, to be sure, is constant because we are dealing with white noise.

Also, the standard deviation is

$$\sigma_Y = N_{ef}\sqrt{\frac{4-\pi}{2}\frac{\Delta f}{f_{max}}} = N_{ef}\sqrt{\frac{4-\pi}{2N}}. \tag{1.52}$$

This result corresponds to the statistical dispersion of each single spectral line. If we want that the average of M segments has, for instance, an error in dB smaller than a given ΔL with a 99.7 % probability, we must choose M such that ΔL corresponds to three times the standard deviation.

Let Y_{mean} be the result of averaging the values of the same spectral line along M segments. We need that

$$\left| 20 \log \frac{Y_{mean}}{\mu_Y} \right| < \Delta L. \tag{1.53}$$

In terms of the natural logarithm,

$$\left| \ln \frac{Y_{mean}}{\mu_Y} \right| < \frac{\ln 10}{20} \Delta L.$$

If ΔL is small, we can approximate the logarithm linearly,

$$\left| \frac{Y_{mean} - \mu_Y}{\mu_Y} \right| < \frac{\ln 10}{20} \Delta L,$$

i.e.,

$$|Y_{mean} - \mu_Y| < \frac{\ln 10}{20} \mu_Y \Delta L.$$

We know that if M is large enough, Y_{mean} has an almost normal distribution, so 99.7 % of the possible outcomes of the process of averaging M segments will comply with

$$|Y_{\text{mean}} - \mu_Y| < 3\sigma_{Y_{\text{mean}}} = 3\frac{\sigma_Y}{\sqrt{M}}.$$

Then it is sufficient that

$$\frac{\ln 10}{20}\mu_Y \Delta L = 3\frac{\sigma_Y}{\sqrt{M}},$$

from where we get

$$M = \left(\frac{60}{\ln 10}\frac{\sigma_Y}{\mu_Y}\right)^2 \frac{1}{\Delta L}. \tag{1.54}$$

Substituting (1.51) and (1.52) in (1.54) we have

$$M = \frac{4-\pi}{\pi}\left(\frac{60}{\ln 10}\right)^2 \frac{1}{\Delta L} \cong \frac{186}{\Delta L}. \tag{1.55}$$

If, for instance, we want that the spectral density error of our white noise be less than 1 dB with a confidence level of 99.7 %, it will be necessary to average 186 segments. For f_{max} = 22 050 Hz and N = 2048, this means a total measurement time of MN/f_{max} = 17.3 s.

Other type of noise, such as traffic noise, may require to average a much larger number of segments, because of its non-stationary nature, which will increase considerably the measuring time.

1.7.5 Band Filters

Octave and one-third octave filters are specified in International Standard IEC 61260 (1995), where the upper and lower limit of the gain is stated for each frequency. There are three filter classes, depending on how tight is the tolerance. Class 2 is the most permissive (it allows smaller attenuations in the stop band and greater fluctuations respect to the ideal response inside the pass band), and Class 0 is the most stringent one.

Since in a logarithmic frequency scale all filters of the same family have the same response, it suffices to specify a single filter with angular frequency normalized to 1. Figures 1.13 and 1.14 illustrate the limits for the normalized octave filter, and Figs. 1.15 and 1.16, for the normalized one-third octave filter.

At frequencies far away from the center frequency, attenuation should be very high. Such attenuations are sometimes difficult to measure because electric noise could mask the signal. This is particularly true in the case of the less stringent

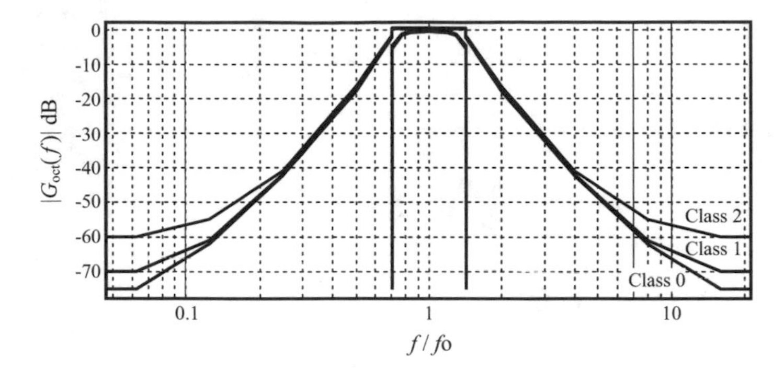

Fig. 1.13 *Upper* and *lower* limits for octave band filters according to International Standard IEC 61260. Global view

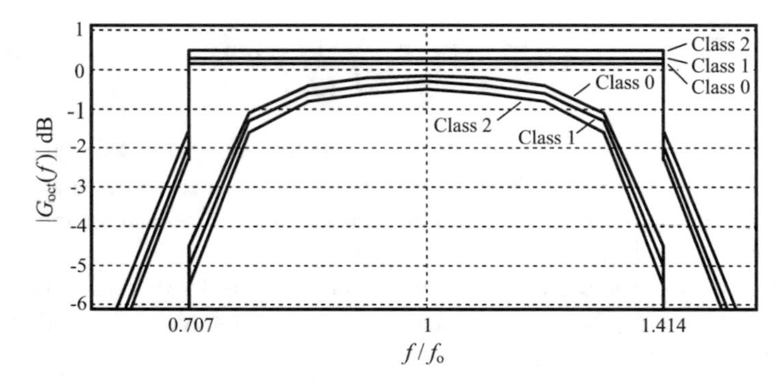

Fig. 1.14 *Upper* and *lower* limits for octave band filters according to International Standard IEC 61260. Pass band details

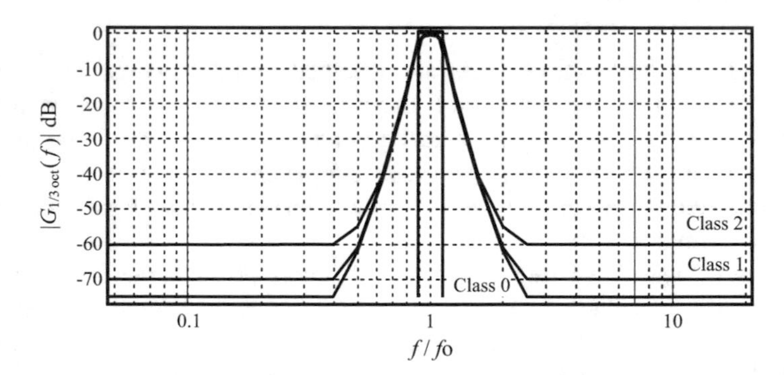

Fig. 1.15 *Upper* and *lower* limits for one-third octave band filters according to International Standard IEC 61260. Global view

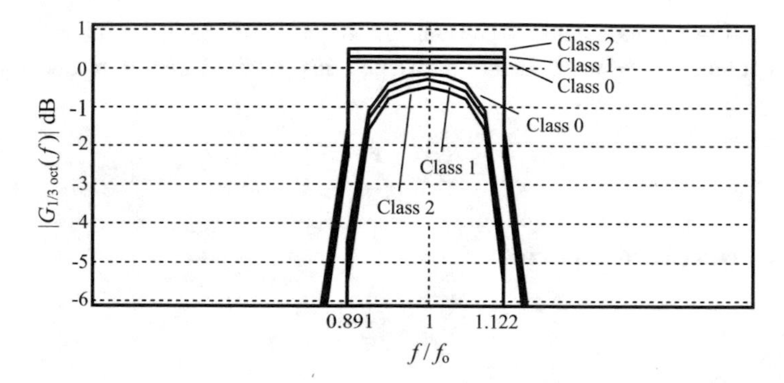

Fig. 1.16 *Upper* and *lower* limits for one-third octave band filters according to International Standard IEC 61260. Pass band details

classes (1 and 2) since they may use cheaper and noisier electronics. That is the reason why the specification demands a smaller attenuation.

Since all filters allow an attenuation of 3 dB at the limit between the pass band and the stop band, they may be implemented with a Butterworth approximation.[14] A third-order Butterworth filter satisfies, in theory, the limits for all classes.[15] However, the actual behavior depends on how stringent are the tolerances of the components.

Traditionally this implementation has been carried out using active filter analog technology. They used three cascaded first-order bandpass cells, each one characterized by a quadratic polynomial.[16] The normalized transfer function of such filters is given by Eqs. (1.56) for octave band and (1.57) for one-third octave band:

$$G_{\mathrm{oct}}(p) = \frac{0.35355p^3}{(p^2+0.45844p+1.8436)(p^2+0.70711p+1)(p^2+0.24867p+0.54242)} \tag{1.56}$$

$$G_{1/3\mathrm{oct}}(p) = \frac{0.12417p^3}{(p^2+0.12735p+1.2221)(p^2+0.23156p+1)(p^2+0.10421p+0.81829)} \tag{1.57}$$

[14]There are several classic approximations for the implementation of filters that must comply with given limits, for instance Butterworth, Chebyshev or Cauer (elliptic). Each one presents pros and cons. For example, Chebyshev has the maximum slope in the attenuation band for a polynomial filter of given degree, at the expense of some ripple in the pass band as well as quite an oscillatory transient response. Butterworth is maximally flat in the pass band and has a tolerable transient response, but its attenuation slope is smaller than Chebyshev's (Miyara 2004).

[15]The order n of a filter is related to the falling slope in the stop band. Specifically, the slope is $\pm 20n$ dB/dec.

[16]As a dynamic system these cells are second-order systems, so they have 2 poles. As a filter they are considered first-order filters, since the slope towards both sides of the center frequency is that of a first-order filter, i.e., ± 20 dB/dec.

where p is the Laplace variable normalized so that the center frequency is $\Omega = 1$. We can unnormalize these functions to get any desired center frequency f_o by replacing

$$p = \frac{s}{2\pi f_o}. \tag{1.58}$$

On the other hand, replacing

$$p = j\frac{f}{f_o} \tag{1.59}$$

we get the frequency response (amplitude and phase) of a band filter centered at f_o.

Each cell's transfer function may be expressed using two characteristic parameters: the *natural angular frequency*. ω_o, and the *quality factor*, Q. For a generic cell,

$$H(s) = \frac{\frac{\omega_o}{Q} s}{s^2 + \frac{\omega_o}{Q} s + \omega_o^2} \tag{1.60}$$

In the case of normalized frequencies,

$$H(p) = \frac{\frac{\Omega_o}{Q} p}{p^2 + \frac{\Omega_o}{Q} p + \Omega_o^2} \tag{1.61}$$

Tables 1.5 and 1.6 give the characteristic parameters for each one of the cells that constitute a normalized octave filter and a one-third octave filter, respectively.

Finally, Figs. 1.17 and 1.18 show the frequency response of octave and one-third octave filters, along with the responses of the three cells forming them.

Table 1.5 Characteristic parameters (natural angular frequency Ω_{ok} and quality factor Q_k) of the three normalized cells of a Butterworth octave band filter

Cell	Ω_{ok}	Q_k
1	1.357783939141098	2.961756551889217
2	1	1.414213562373096
3	0.736494202923462	2.961756551889228

Table 1.6 Characteristic parameters (natural angular frequency Ω_{ok} and quality factor Q_k) of the three normalized cells of a Butterworth one-third octave band filter

Cell	Ω_{ok}	Q_k
1	1.105467721309820	8.680399444876773
2	1	4.318473046963131
3	0.904594481343287	8.680399444876974

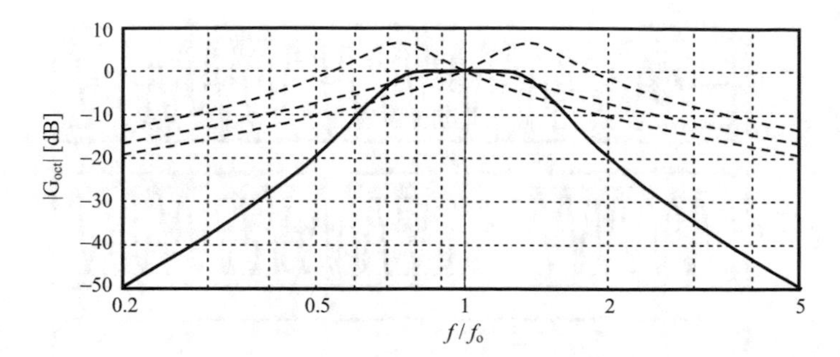

Fig. 1.17 Normalized frequency response of a third-order Butterworth octave filter and its first order factors (in *dashed lines*)

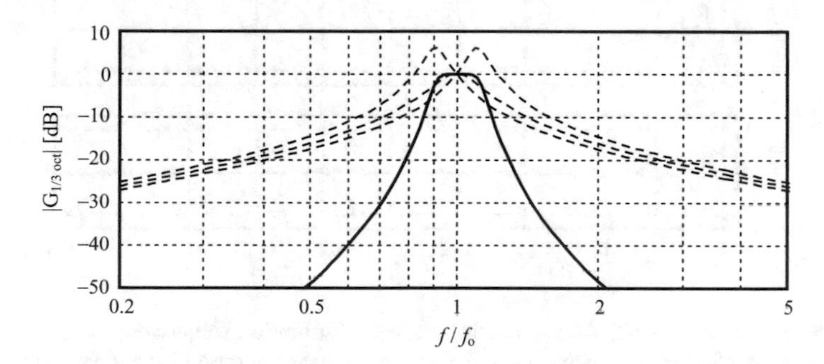

Fig. 1.18 Normalized frequency response of a third-order Butterworth one-third octave filter and its first order factors (in *dashed lines*)

1.7.6 Transient Response of Band Filters

The frequency response of a filter indicates how much the signal is attenuated after going through the filter as a function of frequency. However, the desired output is obtained after some time, i.e., after the *transient* dies away and the output reaches the *steady state*. Indeed, the temporal response to a sine wave of frequency ω starting at $t = 0$ is given by

$$r(t) = A_o \sin(\omega t + \phi) + \sum_{k=1}^{3} A_k e^{-t/\tau_k} \sin(\omega_k t + \psi_k) \tag{1.62}$$

where ω_k is the angular frequency and τ_k the decay time constant of the k-th mode (see Appendix 7 for more details). The first term is the *forced response* or steady state, whereas the second one is the *free response*, *natural response* or transient. Figures 1.19 and 1.20 illustrate the complete response along with its components.

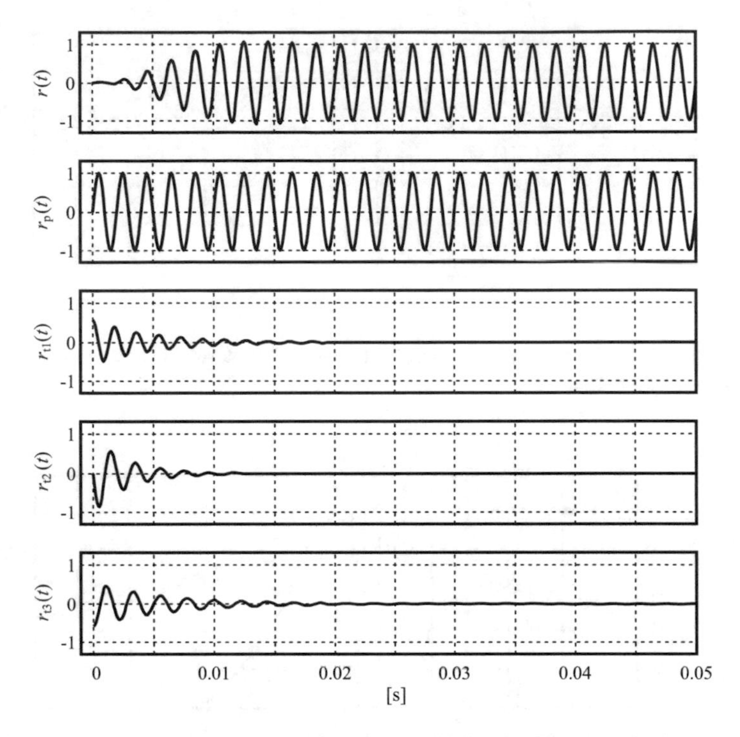

Fig. 1.19 *Top* response $r(t)$ of a one-third octave, third-order Butterworth filter centered at $f_o = 500$ Hz for a 500 Hz sine wave input. *Below* steady-state response $r_p(t)$ and the three transient response components $r_{t1}(t)$, $r_{t2}(t)$ and $r_{t3}(t)$

As can bee seen, the filter output will reach an essentially constant amplitude (equal to that of the forced response) after some time that depends on the time constants τ_k. However, it is not possible to reduce these time constants further than a limit that depends on the filter bandwidth. Indeed, Equation (A7.8) from Appendix 7,

$$\tau_k = \frac{2Q_k}{\omega_{ok}}, \tag{1.63}$$

can be rewritten in terms of the center frequency and normalized frequency:

$$\tau_k = \frac{2Q_k}{\omega_o \Omega_{ok}}. \tag{1.64}$$

We can see that the smaller the center frequency ω_o, the larger the time constants and the transient duration. Therefore, a band filter centered at 31.5 Hz will have a much slower response than one centered at 16 000 Hz.

We can also see that the narrower the pass band, the higher the Q values. Comparing Tables 1.5 and 1.6, we see that for octave bands $Q_{max} = 2.96$ while for one-third octave bands $Q_{max} = 8.68$. Hence, for a given center frequency, a

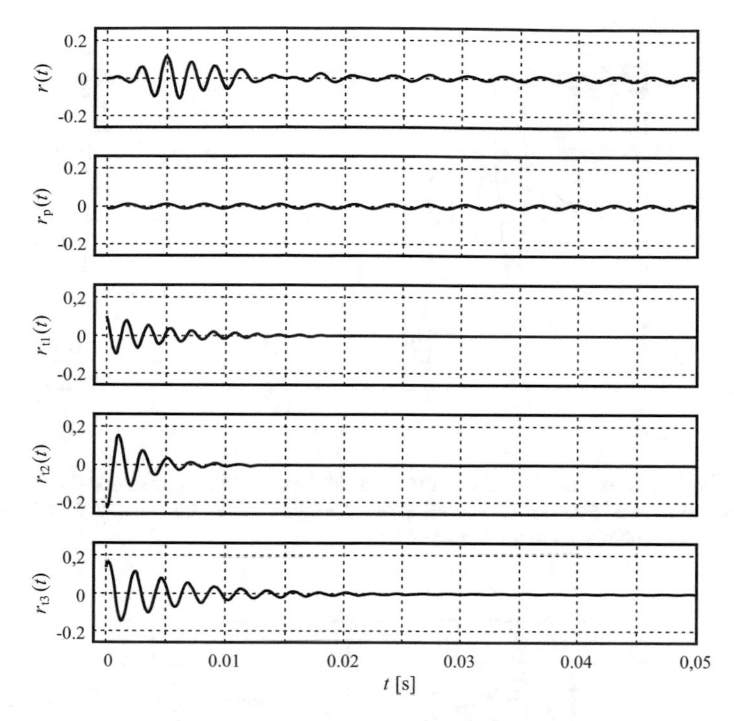

Fig. 1.20 *Top* response $r(t)$ of a one-third octave, third-order Butterworth filter centered at $f_o = 500$ Hz for a 315 Hz sine wave input. *Below* steady-state response $r_p(t)$ and the three transient response components $r_{t1}(t)$, $r_{t2}(t)$ and $r_{t3}(t)$

one-third octave filter will have a slower response (by a factor of three) than an octave filter.

This is another instance of the *time-frequency uncertainty* that we had already seen in connection with the line spectrum (Eq. 1.43). Conceptually, as the analysis band gets narrower, i.e., as we increase the spectral resolution, the time needed to get a stable response gets longer, losing time resolution.[17]

A theoretical analysis of the stabilization time should specify a tolerance margin in dB with respect to the steady state amplitude, and study the time required in order to reach that tolerance, as shown in Fig. 1.21.

This approach is complicated because of the involved relationship between the amplitude of the steady state response and the initial amplitudes of the components of the transient response. As seen in Figs. 1.19 and 1.20, for an input frequency equal to the center frequency, the transient amplitudes are smaller than the steady state, while for an out-of-band frequency, this situation is reversed.

[17]Note that for constant-percentage bands, the smaller the center frequency the smaller the bandwidth. This is why low frequency bands exhibit a longer stabilization time.

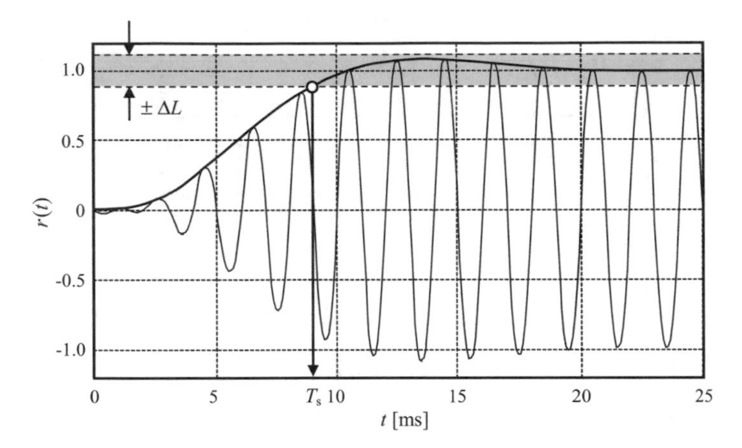

Fig. 1.21 Stabilization time for $\Delta L = 1$ dB for a third-order Butterworth, one-third octave filter with center frequency 500 Hz. The *small empty circle* indicates the point where the envelope enters and stays within the $\pm\Delta L$ tolerance band

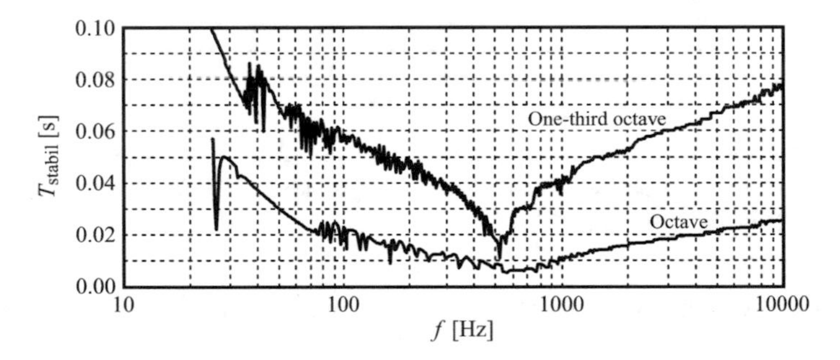

Fig. 1.22 Stabilization time of the response of third-order Butterworth octave and one-third octave filters centered at 500 Hz, for tones of different frequencies. Tolerance was ± 0.15 dB, i.e., that for class 0 filters at center frequency

It is more operational to simulate by software the response for several frequencies and to find the stabilization time with a prescribed tolerance. The results are shown in Figs. 1.22 and 1.23, for ± 0.15 and ± 0.5 dB tolerance, respectively (corresponding to class 0 and class 2 filters).

As expected, the stabilization time is shorter for octave filters than for one-third octave filters. It is also observed that for low frequencies the stabilization time increases. This is not only due to Eq. (1.64) but also to the fact that the final (steady state) amplitude is much smaller, so the transient should fall to a much lower amplitude in order to comply with the tolerance. This also explains the high-frequency behavior, where the steady state amplitude is very small as well.

It is also interesting to note the irregular behavior of the stabilization time with frequency, apart from the general tendency already commented. The reason is that

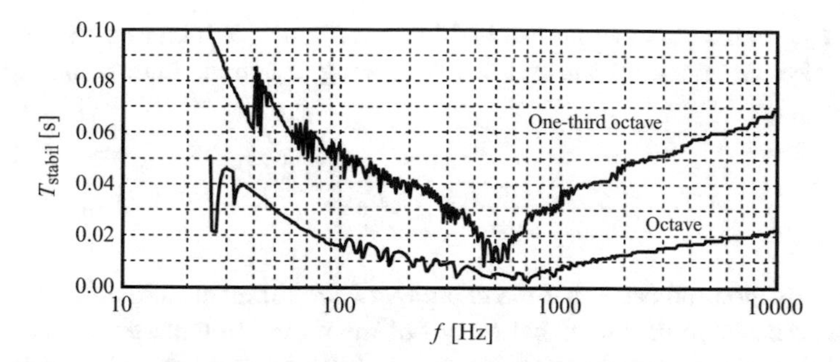

Fig. 1.23 Stabilization time of the response of third-order Butterworth octave and one-third octave filters centered at 500 Hz, for tones of different frequencies. Tolerance was ±0.5 dB, i.e., that for class 2 filters at center frequency

small frequency differences may cause the entry in the tolerance band to jump from one cycle to the next because of slight changes in the amplitude of the steady state or the transients.

Although the plots on Figs. 1.22 and 1.23 were simulated for a center frequency of 500 Hz, they can be applied to any other frequency with a scale correction as given in Eq. (1.64).

NOTE 1: When the tolerance ΔL is very small, it may happen that the envelope of the output signal enters and leaves the tolerance band several times before entering permanently.
NOTE 2: The stabilization time might be redefined introducing a variable tolerance over frequency considering Figs. 1.13, 1.14, 1.15 and 1.16. This would reduce even further the stabilization time.
NOTE 3: Except for particular cases such as pure tone measurement at the laboratory, the out-of-band stabilization time is not very relevant since the steady state response for other bands falls rapidly below background noise.

1.7.7 Parseval's Identity

Let us consider a periodic signal $p(t)$ expressed as its trigonometric series expansion[18]:

$$p(t) = \sum_{n=1}^{\infty} \sqrt{2} P_{n,\mathrm{ef}} \sin(2\pi n f t + \phi_n), \tag{1.65}$$

[18]We have omitted the mean value (or DC value) since in the case of acoustic signals it is not generally relevant. There are some exceptions, such as impedanciometry in otological studies where a positive or negative static pressure is applied to induce a deformation of the tympanic membrane.

where $P_{n,\mathrm{ef}}$ is the effective pressure of the nth harmonic. It can be shown that the global effective pressure satisfies the following equation, known as *Parseval's identity*:

$$P_{\mathrm{ef}}^2 = \sum_{n=1}^{\infty} P_{n,\mathrm{ef}}^2. \tag{1.66}$$

This equation indicates that the energies of the different harmonics can be linearly superposed to get the global energy of the signal. In other terms, the energies (or powers) of the different harmonics are independent and do not mutually interfere. Note that this is not true for signals of the same frequency. For instance, the power of the superposition of two tones of equal frequency and amplitude may vary from 0 to 4 times that of one of them, depending on their relative phase.

Parseval's identity has several generalizations. The first one applies to Fourier Transform,

$$P(\omega) = \int_{-\infty}^{\infty} p(t)\mathrm{e}^{-j\omega t}\mathrm{d}t. \tag{1.67}$$

In this case it is known as Plancherel's identity, analog to Parseval's,

$$\int_{-\infty}^{\infty} p^2(t)\mathrm{d}t = \frac{1}{2\pi}\int_{-\infty}^{\infty} |P(\omega)|^2\mathrm{d}\omega, \tag{1.68}$$

which may also be expressed as

$$\int_{-\infty}^{\infty} p^2(t)\mathrm{d}t = \int_{-\infty}^{\infty} |P(2\pi f)|^2\mathrm{d}f. \tag{1.69}$$

This identity is true for square-integrable signals from $-\infty$ to ∞, i.e., those with finite energy; for instance, short duration signals such as pulses, tone bursts, single events, etc. Note that in these cases the long-term effective value is 0, so the concept of mean power is not useful.

For stationary signals or signals that do not fall rapidly to 0, whose total energy is, hence, infinite, Parseval's identity adopts the form already seen in (1.47),

$$P_{\mathrm{ef}}^2 = \int_0^{\infty} \overline{p^2}(f)\mathrm{d}f, \tag{1.70}$$

where $\overline{p^2}(f)$ is the *spectral density*. In the particular case of a sine wave the spectral density is a *Dirac's delta*.

1.7.8 *Effect of Spectrum Tolerance on Parseval's Identity*

Consider a signal whose spectrum is non-zero between f_{min} and f_{max}. If we split the spectrum into a series of consecutive bands $[f_{lk}, f_{uk}]$, where $f_{l1} = f_{min}$ and $f_{uN} = f_{max}$, and we apply Eq. (1.46) along with the properties of the integral, we will have

$$P_{ef}^2 = \sum_{k=1}^{N} \int_{f_{lk}}^{f_{uk}} \overline{p^2}(f)\mathrm{d}f, \tag{1.71}$$

In other words, the power of disjoint bands is additive. We could apply this property to the band spectrum from where we can conclude that it should satisfy Parseval's identity.

However, the validity of that conclusion is conditioned by the assumption that the bands are really disjoint, i.e., that no band is excited by spectral content from other bands. If we observe the frequency response of Figs. 1.17 and 1.18, or the limits shown in Figs. 1.13, 1.14, 1.15 and 1.16, we can easily conclude that a sine wave outside a band causes a small but distinct response in that band. This phenomenon is generally referred to as *spectral leakage.*

As a consequence, there will be errors in Parseval's identity if it is applied to the octave band levels. Suppose, for instance, a sine wave of frequency $f_o = 1$ kHz to be analyzed by a spectrum analyzer comprising a series of third-order Butterworth filters whose responses at their center frequencies are ideal (unity). The sum of powers of the 31 bands is

$$\sum_{k=1}^{31} P_{k,ef}^2 = P_{ef}^2 \sum_{k=1}^{31} G_k^2(f_o). \tag{1.72}$$

where $G_k(f_o)$ is the gain of the k-th filter at frequency f_o. Since all filters have the same frequency response shifted by an integer number of one-third octave bands, we have

$$G_{k-1}(f) = G_k(2^{1/3}f), \tag{1.73}$$

hence,

$$\sum_{k=1}^{31} P_{k,ef}^2 = P_{ef}^2 \sum_{k=1}^{31} G_{18}^2\left(2^{\frac{k-18}{3}} f_o\right). \tag{1.74}$$

The exponent of 2 was chosen in order to cover the full audio spectrum (20–20 kHz). The calculation was carried out by computer using the theoretical frequency response. The result is

$$\sum_{k=1}^{31} P_{k,\mathrm{ef}}^2 = 1.030013 P_{\mathrm{ef}}^2. \tag{1.75}$$

This implies an excess error of 0.13 dB in Parseval's identity.

In the case of octave bands, we have

$$\sum_{k=1}^{10} P_{k,\mathrm{ef}}^2 = P_{\mathrm{ef}}^2 \sum_{k=1}^{10} G_6^2\left(2^{k-6} f_\mathrm{o}\right). \tag{1.76}$$

Computer calculation yields

$$\sum_{k=1}^{10} P_{k,\mathrm{ef}}^2 = 1.021801 P_{\mathrm{ef}}^2. \tag{1.77}$$

which corresponds to an excess error of 0.09 dB in Parseval's identity.

Suppose, now, the case of a band-limited white noise ranging from f_{min} = 17.8 Hz to f_{max} = 22.4 kHz, the ideal bounds of the one-third octave bands centered at 20 and 20 kHz. If the effective pressure is P_{ef}, then the spectral density, which is constant since it is white noise, can be calculated as:

$$\overline{p_\mathrm{o}^2} = \frac{P_{\mathrm{ef}}^2}{f_{\mathrm{max}} - f_{\mathrm{min}}}. \tag{1.78}$$

The square effective pressure of the spectral segment ideally contained in the k-th band is

$$P_{k,\mathrm{ef,ideal}}^2 = \int_{f_{\mathrm{i},k}}^{f_{\mathrm{s},k}} \overline{p_\mathrm{o}^2}\, \mathrm{d}f = \overline{p_\mathrm{o}^2}\left(f_{\mathrm{s},k} - f_{\mathrm{i},k}\right) \tag{1.79}$$

hence,

$$P_{k,\mathrm{ef,ideal}}^2 = \frac{f_{\mathrm{s},k} - f_{\mathrm{i},k}}{f_{\mathrm{max}} - f_{\mathrm{min}}} P_{\mathrm{ef}}^2 \tag{1.80}$$

These ideal values satisfy Parseval's identity since the upper bound of each band equals the lower bound of the next one, so

$$\sum_{k=1}^{N} \left(f_{\mathrm{s},k} - f_{\mathrm{i},k}\right) = f_{\mathrm{max}} - f_{\mathrm{min}}.$$

However, the value that is actually obtained includes the filter gain. That implies a slight decrease toward the limits of the band, but also an increment because of spectral leakage:

$$P_{k,\mathrm{ef}}^2 = \int_{f_{\min}}^{f_{\max}} \overline{p_{\mathrm{o}}^2} |G_k(f)|^2 \mathrm{d}f = \overline{p_{\mathrm{o}}^2} \int_{f_{\min}}^{f_{\max}} |G_k(f)|^2 \mathrm{d}f \tag{1.81}$$

For instance, assuming that P_{ef} = 1 Pa, the ideal result for the one-third octave band centered at 1 kHz is 0.10160 Pa, while the true result is 0.10397 Pa. This represents an excess error of about 0.2 dB. We can conclude that the spectral leakage prevails against gain reduction close to the band limits.

If we repeat the calculation for all bands and the mean square pressure values are added together, we get an estimated effective pressure $P_{\mathrm{ef,estimate}}$ = 1.0193 Pa, instead of the true value, 1 Pa. This means that the application of Parseval's formula results in an excess error of 0.17 dB.

Therefore, we can conclude that spectral leakage causes errors of up to 0.17 dB in Parseval's formula, both in the case of spectrally narrow signals (sine wave) and of wide band signals (white noise). This constitutes an internal coherence data error, but it is generally acceptable if we consider all uncertainty sources that affect acoustical measurements.

1.7.9 Weighted Levels Computed from One-Third Octave Bands

As mentioned earlier, weighted levels can be obtained from the one-third octave band spectrum. To this end, we first add the numerical dB value of the desired frequency weighting (for example, *A* weighting) to the corresponding band level to get the band's weighted level:

$$L_{\mathrm{p}A,k} = L_{\mathrm{p},k} + A(f_k). \tag{1.82}$$

These values can be computed from the corresponding equations or can be directly obtained from Tables 1.1 and 1.2.

Next we convert each band's weighted level to square pressure and apply Parseval's identity to get the square pressure of the whole weighted signal. Finally we convert back to get the weighted sound pressure level. This process can be summarized as

$$L_{\mathrm{p}A} = 10 \log \left(\sum_{k=1}^{31} 10^{\frac{L_{\mathrm{p},k} + A(f_k)}{10}} \right). \tag{1.83}$$

Problems

Note: While the details of Scilab algorithms will be postponed until Chap. 7, some of the problems for this and next chapters will freely resort to the use of this software (or any other suitable mathematical software of your preference). The basics of Scilab can be found in Appendix 13. Whenever a function **`xxx`** is suggested, it is a good idea typing **`help xxx`** to get usage information.

(1) (a) Find the long-term equivalent level of a square wave of amplitude 1 Pa. (b) How does it differ from the equivalent level extended to a single period? (c) Find the theoretical expression for the equivalent level of the square wave of part (a) extended to an arbitrary time T.

2) (a) Find by direct calculation the long-term equivalent level of the superposition of two tones of amplitudes P_1, P_2, and frequencies f_1, f_2. (b) Using trigonometric identities derive a formula for the equivalent level of this signal extended to an arbitrary time T. (c) Estimate the stabilization time for a given ΔL. (d) Show that as $f_1 - f_2 \rightarrow 0$ the stabilization time $\rightarrow \infty$. Explain this result in terms of beats.

3) (a) Find the long-term equivalent level of a triangle wave of amplitude P and frequency f. (b) Estimate a conservative bound for the stabilization time of the signal in part (a) for a given ΔL. Hint: The squared signal is a series of parabolic segments. Find the average value and remove it. Integrate over 1/4 period and find the maximum. This helps to find a bound for the mean square of the signal.

(4) Find the stabilization time of the equivalent level in the case of a Gaussian noise that has been sampled at 44 100 Hz with a tolerance of 0.5 dB and 95 % confidence level.

(5) Find the stabilization time of the equivalent level for a Gaussian noise that has been filtered by a bandpass filter centered at 1 kHz with a bandwidth of 20 Hz. Hint: The filtered signal resembles a sine wave with an amplitude that varies randomly according to a Rayleigh distribution (see Appendix 6).

(6) Using Scilab (see Appendix 13) generate a Gaussian noise of about 20 s at a sample rate of 44 100 Hz. Then design an algorithm that computes the equivalent level by discrete integration over a variable T and plot the result as a function of T. Hint: Scilab clause **`rand('normal')`** sets the random generator to normal or Gaussian mode; then **`rand(1,n)`** generates a row vector with n samples.

(7) The background noise level is often defined as the 90 % exceedance level (i.e., a level that is exceeded 90 % of the time) of the ambient noise. Find the stabilization time for the measurement of the A-weighted background noise with an F time constant if we can assume it has a normal distribution with a mean of 70 dBA and a standard deviation of 5 dBA

(8) Consider a white noise whose long-term unweighted equivalent level is 60 dB. (a) Find its theoretical line spectrum if it has been sampled at 44 100 Hz and the FFT (Fast Fourier Transform) size is 4096. (b) Find the spectral density. (c) Find the so-called *spectral level*, i.e., the sound pressure level of a 1 Hz

band (since it is *white* the spectral level is independent of the center of the band). (d) Using Scilab apply the A weighting and find the A-weighted equivalent level. Hint: The A-weighted line spectrum is obtained as the product of the original line spectrum (on a frequency by frequency basis) and $A(f)$ at the corresponding frequencies. The total A-weighted RMS value can be obtained applying Parseval's identity.

(9) Repeat problem (8) in the case of a *pink noise* whose spectral density is inversely proportional to the square root of frequency.

(10) Consider a noise whose octave band spectrum falls at a rate of 2 dB/oct and has an overall unweighted level of 65 dB. (a) Find the octave band spectrum. (b) Apply the discrete values of the A-weighting for the center of each band (see Table 1.1) and compute the A-weighted level. (c) Repeat (b) assuming that the spectral density remains constant for each octave band and applying the mathematical formula for the A-weighting. Compare both results.

(11) (a) Applying a convenient scale factor and Figs. 1.22 and 1.23, estimate graphically the stabilization time within a tolerance of 0.5 dB for the 200 Hz one-third octave band when measuring a 50 Hz tone.

(12) When measuring a 1 kHz with a one-third octave analyzer it is observed that the 1 kHz band shows a level of 73 dB, but the other bands present also significant output. (a) Explain the reason. (b) Estimate (you may need Scilab) the level that is measured at the 125 Hz band. (c) Estimate the error you would get if trying to check Parseval's consistency.

References[19]

IEC 61260 (1995) Octave-band and fractional-octave-band filters

IEC 61651 (1979) Acoustics—Sound level meters

IEC 61672-1 (2002) Electroacoustics—sound level meters—part 1: specifications. International Electrotechnical Commission

IEC 60537 (1976) Frequency weighting for measurement of aircraft noise (D-weighting). International Electrotechnical Commission. Withdrawn standard.

IEC 3382-1 (2009) Acoustics. Measurement of room parameters. Part 1: performance spaces

ISO 7196 (1995) Acoustics—Frequency weighting characteristic for infrasound measurements

Kogan P (2004) Análisis de la eficiencia de la ponderación A para evaluar los efectos del ruido en el ser humano. In Spanish. Escuela de Ingeniería Acústica. Universidad Austral de Chile. Valdivia, 2004. http://www.fceia.unr.edu.ar/acustica/biblio/kogan.pdf. Accessed 26 Feb 2016

[19]Internet URL's of the documents included in all reference lists in this book were in effect at the moment of accessing them (approximately from January 2013 to October 2016). However, because of web dynamics, it is possible that they be temporarily or definitively unavailable without notice, or that they be corrected or updated. In some cases the documents may have been relocated, so it may prove useful to do a web search about the file name (the character string after the last slash) to find the updated URL. In other cases the query may directly include the document title.

Miyara F (2004) Filtros Activos. Monografía de la cátedra de Electrónica III, Universidad Nacional de Rosario. Publicación interna B09.00. http://www.fceia.unr.edu.ar/enica3/filtros-t.pdf. Accessed 26 Feb 2016

Miyara F, Miechi P, Pasch V, Cabanellas S, Yanitelli M, Accolti E (2008) Tiempos de estabilización del espectro del ruido de tránsito. VI Congreso Iberoamericano de Acústica FIA2008. Trabajo A033. Rosario, 2008. http://www.fceia.unr.edu.ar/acustica/biblio/A033-Miyara.pdf. Accessed 26 Feb 2016

Yanitelli M, Pasch V, Mosconi P, Cabanellas S, Vazquez J, Rall JC, Miyara F (2001) Manchas acústicas: Ruido de aeropuertos. Cuartas Jornadas Internacionales Multidisciplinarias sobre Violencia Acústica. Rosario, Argentina, 22 al 24 de octubre de 2001. http://www.fceia.unr.edu.ar/biblio/aeropuer.pdf. Accessed 27 Feb 2016

Chapter 2
Uncertainty

2.1 Introduction

Every measurement has the purpose of obtaining the numerical value of a physical quantity. However, this statement that seems to be so clear and obvious requires careful scrutiny. In the first place, the quantity to be measured, called the *measurand*, is not always perfectly defined. In order for the definition to be exhaustive, we should specify all the conditions that could influence the measurand's value. In the acoustic case, for instance:

(1) Atmospheric conditions (atmospheric pressure, temperature, humidity, wind velocity and direction, altitude, presence of particles),
(2) Geometry and materials of the environment (presence of acoustic barriers or screens, sound absorbing or reflecting surfaces, etc.),
(3) Operational conditions of the source (for instance, temperature, load conditions, power supply voltage),
(4) Opportunity (time, kind of process currently taking place, time interval since the beginning of the process, time interval since last maintenance),
(5) Presence of other simultaneous sources,
(6) Location of measuring instrument and its operator.[1]

In the second place, even if all influencing conditions could be identified and specified, it would be physically impossible to warrant they are satisfied during the measurement; among other reasons, because it would be impossible to know the exact values of the associated parameters. This means that in a real-life situation, there is no single value of the measurand.

However, we can conceive the existence of a hypothetical *true value*, i.e. the value that would have the measurand if it would be measured in ideal conditions by

[1]The measuring instrument itself introduces reflection and diffraction phenomena that may change the sound field, which in turn changes the measurand. The same is true for the operator conducting the measurement.

F. Miyara, *Software-Based Acoustical Measurements*, Modern Acoustics and Signal Processing, DOI 10.1007/978-3-319-55871-4_2

means of an ideal instrument. Because of the previous comments, it is not possible to find exactly the true value.[2] We can only expect to get an estimate along with some expression of *uncertainty*, i.e. a way to quantify how far can be the true value from our estimate.

The way to express the uncertainty has changed over time. At the beginning, it was assumed that it was possible to provide some interval around the estimated value where the true value lies. Nowadays, since the publication by the ISO and other organizations of the Guide for the Expression of Uncertainty in Measurements, known by its partial acronym GUM (ISO-GUM 1993; JCGM 2008; Taylor and Kuyayt 1994), the paradigm has shifted towards a statistical expression, specifying the uncertainty as an interval where we can find the true value with a given probability, for instance 95 %.

2.2 Resolution, Precision, Accuracy

Every measurement instrument has a limit for the minimum difference between two readings that are not identical called *resolution*. In an instrument with an analog display (which are now rare), it is given by the distance between two consecutive scale divisions. In a digital instrument, it corresponds to a unit of the least significant digit. In most sound level meters the resolution is 0.1 dB, but some of them allow to configure the resolution as 0.01 dB (as we shall see, this can make sense for equivalent level measurements).

The resolution of a measurement procedure may be better than that of the instrument used to conduct the basic measurements. An example is when the outcome of the measurement is the average of several readings. For instance, if the measurement procedure involves averaging 10 readings of an instrument with a resolution of 0.1 dB, the measurement procedure will have a resolution of 0.01 dB.

When several measurements of the same measurand are conducted in identical conditions[3] (i.e. several *realizations of the measurand* are measured), the readings are not always identical, but they present some statistical dispersion instead. This means that it is more correct to consider the reading as a random variable instead of a single value that is a function of the true value.[4] We define the *precision* of an instrument as the statistical dispersion of readings. It can be expressed as the

[2]According to the GUM, there may be more than one true values. This is because when the specification of the measurand is not exhaustive, there are many different values that agree with the definition.

[3]*Identical conditions* means to control as far as possible all possible variables; for instance, the same physical system, the same instrument, the same operator, the same environmental conditions, and the minimum reasonable time between measurements to prevent drifts in the instrument's characteristics as well as those of the physical system where the measurand manifests itself.

[4]We say "that is a function of" instead of "equal to" because the instrument could be uncalibrated or have some nonlinearity that should be corrected.

standard deviation σ of the readings or as half the width of a confidence interval, where lies a given percentage of the readings (typically, 95 %; for normal distribution, it equals 2σ).

A more controversial concept is that of *accuracy*, defined, according to the GUM, as the "degree of closeness of the agreement between the result of a measurement and a true value of the measurand." Pursuant to the GUM, it is a qualitative concept, so one should not fall into the temptation of assigning a numerical value. However, the GUM itself and other metrological documents use frequently the term *accuracy* as a quantitative concept. For instance, in subclause 3.1.3, the GUM says:

> In practice, the required specification or definition of the measurand is dictated by the required accuracy of measurement.

We can ask how it is possible to specify the "required accuracy" in a non-quantitative form, since a qualitative expression would make it impossible to check whether the specification has been satisfied. In fact, an example in the same subclause reads:

> If the length of a nominally one-metre long steel bar is to be determined to micrometre accuracy, its specification should include the temperature and pressure at which the length is defined (…).

Despite what the GUM asserts, it would be possible to specify the accuracy quantitatively as a dispersion, for instance the standard deviation of the difference between the observed value and the true value (or an equivalent confidence interval).

2.3 Measurement Method and Procedure

Every measurement is carried out following some *measurement procedure* that implements a *measurement method* and arrives at a result directly or through a *measurement model*.

The measurement method describes in a general way the structure of the measuring process. In its simplest version, it reduces to reading the instrument's display. In more elaborate cases, it can involve the use of several transducers, the processing of the magnitudes they generate and the application of corrections, formulas, etc. The measurement method rests on a *measurement principle*, i.e. one or more physical phenomena on which the measurement is based (for instance, the electroacoustic transduction in a condenser microphone) and is embodied in a *measurement procedure* where all the steps of the measurement are described in detail, covering the selection of instruments and samples, a survey of environmental variables or any other circumstance that could influence the measurement outcome, the specification of the measurement conditions, precautions and recommendations to observe during the measurement, as well as formulas and calculations, and the structure and content of the *measurement report*.

For typical and frequent measurements, the measurement procedure is usually stated in national or international standards. In more specific cases, it may be part of internal laboratory documents, or part of the measurement report itself.

Example: The measurement of the acoustic power radiated by a machine in a semi-anechoic chamber is based on the principle of electroacoustic transduction of pressure into electric voltage and in the simple relationship that is held between sound intensity and sound pressure in a free field. The measurement method consists in taking pressure measurements at several points of an imaginary surface around the machine, converting sound pressure into sound intensity, computing a weighted sum of the intensities, and multiplying by the total area of the surface. The measurement procedure indicates the characteristics of the measurement environment, type of instrument to use (for instance, one-third octave spectrum analyzer), environmental conditions, operational conditions of the machine (for instance installation, supply voltage, warming time), coordinates of the measuring points, type, number and duration of measurements, microphone orientation, formulas to use, atmospheric condition corrections, effect of background noise, and uncertainty calculation.

2.3.1 *Direct and Indirect Measurement Methods*

Measurement methods can be classified into *direct* and *indirect*. Direct methods consist in just reading and recording the instrument's display. An example is the measurement of the equivalent level with an integrating sound level meter.

Indirect methods require some transformation of the direct data in order to arrive at the desired result. For instance, starting from the measurement of the sound pressure level in some point of a plane progressive wave (free field), it is possible to estimate the particle velocity by means of the equation.

$$U_{\mathrm{ef}} = \frac{1}{\rho_{\mathrm{o}} c} P_{\mathrm{ref}} 10^{\frac{L_{\mathrm{P}}}{20}}, \tag{2.1}$$

where ρ_{o} is the air density at equilibrium and c is the velocity of sound.

2.4 Measurement Model

Except in very simple cases, in general, the measurement result is not the simple reading of the instrument's display but the outcome of a series of operations that often condense into an explicit or implicit functional relationship between two or more variables, or even a software algorithm.

Calling the measurand Y and the quantities obtained by direct reading or comparison X_k, we can write the explicit form

$$Y = f(X_1, \ldots, X_N), \tag{2.2}$$

or the implicit form,

$$h(Y, X_1, \ldots, X_N) = 0. \tag{2.3}$$

In some cases, it will be possible to directly solve for Y, arriving at an explicit form such as (2.2). In other cases, a numerical solution can be attempted, for instance, applying the Newton–Raphson algorithm.

Example 1 If L_1, …, L_{10} are the octave band levels corresponding to frequencies f_1 = 31.5 Hz, …, f_{10} = 16 kHz, then the A-weighted sound level can be computed by means of the formula.

$$L_{\text{pA}} = 10 \log \sum_{k=1}^{10} 10^{\frac{L_k + A_k}{10}}, \tag{2.4}$$

where A_k is the A-weighting correction for the band centered at f_k.

Example 2 Let $L_{\text{Aeq},\ Tn}$ be the A-weighted equivalent level of a given noise for N intervals of durations T_n covering the total duration T of a desired interval (for instance, a working day). Then we can compute the global equivalent level as

$$L_{\text{Aeq},T} = 10 \log \left(\frac{1}{T} \sum_{n=1}^{N} T_n 10^{\frac{L_{\text{Aeq},T_n}}{10}} \right). \tag{2.5}$$

Example 3 In order to measure the sound transmission loss *STL* of a partition in a sound transmission facility, one-third octave levels are measured at the source room (generated by a proper sound source) as well as at the receiving room (the result of the transmission through the partition and flanking transmission). Reverberation time at each one-third octave band is also measured. If we call these values L_{k1}, L_{k2}, and T_k, we can obtain the sound transmission class for each one-third octave band k as

$$\text{STL}_k = L_{k1} - L_{k2} + 10 \log_{10} \frac{T_k S_{12}}{0.161 V_2}, \tag{2.6}$$

where S_{12} is the partition area and V_2 is the receiver room volume.

Example 4 Suppose we want to measure the sound pressure level at some location and at some environmental reference conditions defined as a temperature of 23 °C, a relative humidity of 50 %, and an atmospheric pressure of 1013.25 hPa, but the actual measurement conditions are 27 °C, 75 %, and 1005.3 hPa. Measurement is conducted using a Brüel and Kjær microphone Type 4189.

It will be necessary to apply some correction to refer the actual measurement to the desired conditions. Since the variations respect the reference conditions are small, their effect is assumed to be linear:

$$L_p(T_1, h_{r1}, P_{a1}) = L_p(T_0, h_{r0}, P_{a0}) + K_T \Delta T + K_{h_r} \Delta h_r + K_{P_a} \Delta P_a \quad (2.7)$$

For a Brüel and Kjær microphone Type 4189, we have (Brüel and Kjær 1995, 1996):

$$\begin{aligned} K_T &= -0.001\,\mathrm{dB}/{}^\circ\mathrm{C} \\ K_{h_r} &= 0.001\,\mathrm{dB}/\% \\ K_{P_a} &= -0.001\,\mathrm{dB}/\mathrm{hPa} \end{aligned}$$

The effect of the environmental conditions turns out to be 0.02895 dB, so the correction to add to the measured value is −0.02895 dB. In most cases this correction can be ignored. One exception may be when performing calibrations by the reciprocity method (Popescu 2007; Brüel and Kjær 1996).

Example 5 Consider the following method to find the resonant frequency f_o of a system. The response Y_k at three frequencies f_k close to the resonant frequency is measured (select the frequencies such that Y_1, $Y_3 < Y_2$). Then a parabolic interpolation[5] is computed and finally the maximum is computed analytically. The associated model is as follows:

$$\begin{aligned} Y_{max} + K(f_1 - f_o)^2 &= Y_1 \\ Y_{max} + K(f_2 - f_o)^2 &= Y_2 \\ Y_{max} + K(f_3 - f_o)^2 &= Y_3 \end{aligned} \quad (2.8)$$

This is a system of three equations with three unknowns, Y_{max}, K, f_o, of which only f_o is relevant to our problem. We can get rid of Y_{max} by subtracting the second equation from the first and the third from the first. Then we can remove K by dividing the resulting equations. This yields

$$\frac{(f_1 - f_o)^2 - (f_2 - f_o)^2}{(f_1 - f_o)^2 - (f_3 - f_o)^2} = \frac{Y_1 - Y_2}{Y_1 - Y_3}, \quad (2.9)$$

where we have reduced the problem to a single equation with a single unknown, f_o. This is an implicit formulation of the measurement model. Clearly, in this case it is easy to solve for f_o since it is basically a second-degree equation (which is easily

[5]The resonance curve is not exactly parabolic, but in the neighborhood of its maximum it can be approximated by a parabola. The approximation will be better if the frequencies where the response is measured are close to the resonant frequency.

reduced to a first-degree equation). In more complicated cases it might not exist as an explicit solution, or it could be prohibitively complicated.[6]

2.5 Uncertainty

When measuring any measurand, the measurement result will differ not only from the true value, but also from the results of repeated measurements. This characteristic of any measurement process is known as *uncertainty*. Among the causes of uncertainty we can mention:

(1) Causes attributable to the measurand: An incomplete definition of the measurand (for instance lack of specification of some variable that has a significant effect on the measurand); impossibility to warrant that the measurand's realization meets all specifications (for instance, the definition specifies the room temperature but it is known only approximately).
(2) Causes attributable to the measurement procedure: The model underlying the measurement method is incomplete, insufficient, or inaccurate (for example, there is a correction of the effect of temperature but the temperature coefficient is known only approximately); Finite number of repetitions; presence of interferences or disturbances (for instance, background noise); presence of factors that alter the measurand (such as a reflective surface close to the measurement site); excessive effect of environmental conditions.
(3) Causes attributable to the measurement instrument: Finite resolution, lack of precision, manufacturing tolerance, lack of calibration, inadequate, or insufficient frequency response.
(4) Causes attributable to the operator: Alteration of the measurand due to their presence; parallax errors while reading analog instruments; indecision on what value should be recorded from a digital instrument when the least significant digits fluctuate over time.

These sources of uncertainty should not be confused with gross errors, mistakes, or incorrect application of procedures (Bell 2001), for instance, confusing figures when recording readings, confusing measurands, departing from the procedure without a reason (and ignoring the effects that this could cause). In what follows, we assume that all these defects have been prevented and all measurements are conducted by knowledgeable personnel.

As a general rule, uncertainty is the dispersion that can be expected for the results of different measurements of the measurand in specified conditions. Note that in general, it is not necessary to make any explicit mention of the true value.

[6]An example where there are explicit solutions but numerical methods are preferred is the solution of linear equation systems of high order.

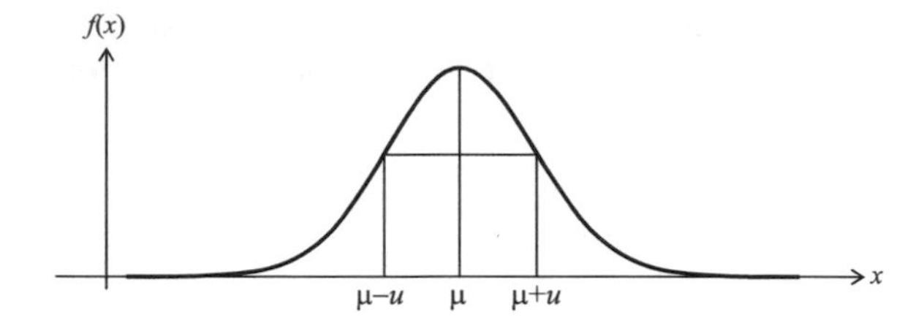

Fig. 2.1 Standard uncertainty example for a normal probability density function

More specifically, we define the *standard uncertainty*, *u*, as the standard deviation of different measurements of the measurand, i.e.[7]

$$u = \sigma = \sqrt{\int_{-\infty}^{\infty} (x - \mu)^2 f(x) dx}, \tag{2.10}$$

where x is a generic result of the measurement conducted in specified conditions, μ is the mean or expected value of x and $f(x)$ is its probability density function. Figure 2.1 shows an example.

Often, the standard deviation of measurement results, understood as the population[8] standard deviation, is not known. However, it is possible to estimate the population parameter from a *sample*, i.e. a finite number n of measurements. In that case, the standard uncertainty is the sample standard deviation:

$$u = s = \sqrt{\frac{1}{n-1} \sum_{k=1}^{n} (x_k - \bar{x})^2}, \tag{2.11}$$

where x_k is the k-th measured value and $\bar{x}$ is the mean of $\{x_k\}$.

Both ways of getting the standard uncertainty are considered correct. Moreover, as we shall see later, the uncertainty may have several components. This gives rise to a general classification of uncertainty components into two categories, called type A uncertainty and type B uncertainty, described in the next section.

2.5.1 Type A Uncertainty

It is any uncertainty component that is determined empirically by applying statistical methods to the results of several measurements. For instance, suppose that we have measured the following ten sound pressure level values:

[7] Appendix 2 introduces basic statistical concepts. In the Glossary of Appendix 1, there is a summary of definitions of common use in metrology.

[8] The term "population" is used in Statistics to refer to all of the homogeneous elements to which some given statistical parameters apply.

73.6	80.2	76.7	75.5	76.7	75.3	75.4	77.8	77.7	77.7

Applying (2.11), we get a mean level of 76.66 dB and a standard deviation of 1.83 dB. A second ten-value sample of measurements corresponding to the same population yields:

76.6	73.8	76.7	78.0	76.3	77.2	76.7	75.1	76.0	74.4

In this case the mean turns out to be 76.08 dB and the standard deviation, 1.29 dB. As can be seen, the estimated values can be rather different between different samples of the same population.

2.5.2 *Type B Uncertainty*

It is any component of uncertainty that is obtained by any method other than the statistical treatment of measurement results. For example,

(a) Previous knowledge or estimation by scientific or technical agreement, for instance the uncertainty in physical constants or arising from procedures;
(b) Information provided in standards;
(c) Instrument specifications;
(d) Information provided in calibration certificates;
(e) Software simulations;
(f) Application of physical laws.

Note: Probably the body of knowledge involved in most components of Type B uncertainty has been gathered originally using statistical methods. The difference is that those methods have been applied to general data or research material, not to the measurement results in a specific occasion, which are what Type A uncertainty refers to.

Examples:

(a) The gas constant R appears in the formula of the velocity of sound c. According to CODATA (Mohr et al. 2012), R = 8.314 462 75 J/(mol·K) with a standard uncertainty u = 0.000 000 91 J/(mol·K). This uncertainty must be considered when calculating c.
(b) International Standard ISO 1996 indicates that in the absence of better data, the standard uncertainty in traffic noise measurements is given by $10/\sqrt{n}$ dB, where n is the number of vehicles that have passed by the observer during the measurement.
(c) A sound level meter rated as Class 1 will have, at 1 kHz, a standard uncertainty of 0.2 dB (IEC 61672-1 2002).
(d) The calibration certificate of a sound calibrator indicates a standard uncertainty of 0.15 dB. Therefore, a sound level meter that has been checked with this calibrator includes an uncertainty component of 0.15 dB due to calibration.

(e) If we measure some noise whose spectrum has a known frequency behavior (for instance, pink noise), it is possible to obtain, by software simulations, the uncertainty in the result of an A-weighted measurement if we know the uncertainty of the instrument's responses at each particular frequency. To this end one could generate many A-weighted frequency responses (a method called Montecarlo) and apply each of them to different realizations of the noise. Finally, we can apply statistical analysis to get the standard deviation. This result could be applied later to measurements of similar noises.

(f) If the quantity to be measured can be computed by means of a known mathematical expression as a function of other quantities to be measured directly, it is possible to propagate the uncertainties to the desired result. We shall see later how it is done (Sect. 2.5.4).

2.5.3 Expanded Uncertainty

Standard uncertainty u, as the standard deviation of measurement results, cannot provide by itself a definite idea of how spread can be the possible measurement results. We could ask, for instance, what is the probability that the result lies in a range $\pm u$ around the mean. The answer depends on the type of statistical distribution. If it is a normal distribution, the probability will be 66.3 %. If it is, instead, a uniform distribution, it will be 57.7 %.

To circumvent this problem, we introduce the *expanded uncertainty*, U, so that the interval $[\mu - U, \mu + U]$ covers a specified percentage of the possible measurement results. In other words, the measurement result will lie in that interval with a specified probability p (which may be expressed in %).

This probability, called *coverage probability*, is not normalized, so one is free to specify it according to the specific need. In any case, the adopted value should be specified and sometimes it is included as a subindex: U_p.

The factor k that converts u into U, which depends on the distribution, is called *coverage factor*:

$$U = k\,u. \tag{2.12}$$

In the frequent case, where the distribution is normal (or approximately normal), it is customary to adopt coverage probabilities of 95.4 % and 99.7 %, which correspond, respectively to $2u$ and $3u$. In this case we can write, for example,

$$U_{95} = 2u, \tag{2.13}$$

where, as is common practice, to alleviate notation, we have written U_{95} instead of $U_{95.4}$. Some examples are presented in Fig. 2.2.

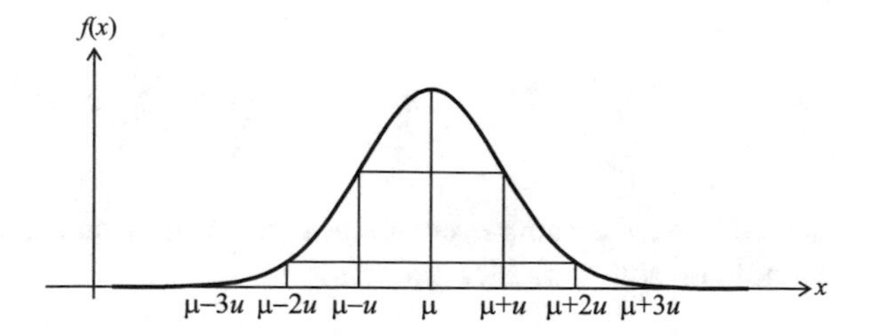

Fig. 2.2 Some examples of intervals corresponding to different coverage probabilities for a normal distribution

2.5.4 Combined Standard Uncertainty

We have already seen that measurements often require to measure several intermediate or subsidiary quantities, and combine them according to a given measurement model. We can ask now to what extent the uncertainty of those intermediate results will contribute to the general uncertainty.

To start with, let us analyze the simplest case, in which the model of the measurand is a linear combination of several independent variables:

$$Y = a_1 X_1 + \cdots + a_N X_N. \tag{2.14}$$

In the first place, note that if μ_k are the means of X_k, then the mean of Y, i.e. μ_Y, is obtained by the same linear combination:

$$\mu_Y = a_1 \mu_1 + \cdots + a_N \mu_N. \tag{2.15}$$

It can also be shown that the standard deviation σ_Y is given by

$$\sigma_Y^2 = \sum_{i=1}^{N} \sum_{j=1}^{N} a_i a_j \operatorname{cov}(X_i, X_j), \tag{2.16}$$

where $\operatorname{cov}(X_i, X_j)$ is the covariance between X_i and X_j, given by

$$\operatorname{cov}(X_i, X_j) = \int_{-\infty}^{\infty} \int_{-\infty}^{\infty} (x_i - \mu_i)(x_j - \mu_j) f(x_i, x_j) \mathrm{d}x \mathrm{d}y, \tag{2.17}$$

where $f(x_i, x_j)$ is the joint probability density function (see Appendix 1).

We define the *combined standard uncertainty*, u_c, as the total standard deviation given in (2.16).

$$u_c = \sigma_Y. \tag{2.18}$$

In the case of a type A uncertainty, where the covariance is estimated from a finite sample, we shall have

$$u_c = s_Y = \sqrt{\sum_{i=1}^{N}\sum_{j=1}^{N} a_i a_j c_{ij}}, \tag{2.19}$$

where c_{ij} is the sample covariance between x_i and x_j calculated from an n-sized sample of both variables according to

$$c_{ij} = \frac{1}{n-1}\sum_{k=1}^{n}(x_{ik} - \bar{x}_i)(y_{jk} - \bar{y}_j). \tag{2.20}$$

In many cases, the variables can be considered statistically independent, thus the cross-covariances (where $i \neq j$) are 0. Since the covariance between a variable and itself is the variance s^2, Eq. (2.19) may be written as

$$s_Y = \sqrt{\sum_{i=1}^{N} a_i^2 s_i^2}. \tag{2.21}$$

The variance of each variable is the square of the standard deviation, so we can rewrite the preceding equation as

$$u_c = \sqrt{\sum_{i=1}^{N} a_i^2 u_i^2}. \tag{2.22}$$

Let us consider now the general case where the measurement model (i.e., the relationship between the measurand and the intermediate quantities that have been directly measured) is not necessarily linear, as in Eq. (2.2),

$$Y = f(X_1, \ldots, X_N). \tag{2.23}$$

This case may be analyzed in a similar fashion. First, we perform a first-order Taylor expansion of f around the measured values. As is shown in Appendix 8, the sample variance of the measurand is given by

$$s_y^2 = \sum_{i=1}^{N}\sum_{j=1}^{N} \frac{\partial f}{\partial x_i}(\bar{x}_i)\frac{\partial f}{\partial x_j}(\bar{x}_j)c_{ij}, \tag{2.24}$$

where c_{ij} is the covariance between variables x_i and x_j which can be estimated by means of the sample covariance expression

$$c_{ij} = \frac{1}{n-1}\sum_{k=1}^{n}(x_{ik} - \bar{x}_i)(x_{jk} - \bar{x}_j). \tag{2.25}$$

In the case in which the random variables $X_1, \ldots, X_N$ are independent, the covariance between pairs of variables vanishes, so the only terms that are preserved are those with $i = j$:

$$s_y^2 = \sum_{i=1}^{N} \left(\frac{\partial f}{\partial x_i}(\bar{x}_i) \right)^2 s_i^2, \tag{2.26}$$

or, in terms of uncertainties,

$$u_c = \sqrt{\sum_{i=1}^{N} \left(\frac{\partial f}{\partial x_i}(\bar{x}_i) \right)^2 u_i^2}, \tag{2.27}$$

The partial derivatives can be computed analytically or numerically. They can be calculated even in the case of implicit functions (see Appendix 9 for an example).

2.6 Examples

In this section, we analyze several situations that appear frequently in acoustical measurements, either at the field or at the laboratory.

2.6.1 Relationship Between Uncertainties in Level and Pressure

Although a microphone generates an electric signal that is proportional to sound pressure, sound level meters calculate and display the sound pressure level L_p. However, sometimes it is necessary to know the sound pressure. It can be computed by means of the inverse formula

$$P_{ef} = P_{ref} 10^{L_p/20}. \tag{2.28}$$

In order to get the uncertainty in P_{ef}, it is better to convert to base e exponential:

$$P_{ef} = P_{ref} e^{\frac{\ln 10}{20} L_p}. \tag{2.29}$$

Derivation is now easier. Once calculated and replaced into (2.27), we arrive at the following expression:

$$u_{P_{ef}} = \frac{\ln 10}{20} P_{ef} u_{L_p}. \tag{2.30}$$

We can see that if the uncertainty in L_p is constant with L_p, then the uncertainty in P_{ef} will be proportional to sound pressure.

2.6.2 Uncertainty in the Correction of Environmental Conditions

Consider the case of example 4 in Sect. 2.4, where the sound pressure level must be corrected to refer it to different environmental conditions from those that prevailed when conducting the measurement. The measurement model was

$$L_{\mathrm{p}}(T_1, h_{\mathrm{r1}}, P_{\mathrm{a1}}) = L_{\mathrm{p}}(T_0, h_{\mathrm{r0}}, P_{\mathrm{a0}}) + K_T \Delta T + K_{h_{\mathrm{r}}} \Delta h_{\mathrm{r}} + K_{P_{\mathrm{a}}} \Delta P_{\mathrm{a}}. \tag{2.31}$$

The problem is that there is some uncertainty in the determination of the environmental conditions. Suppose that the standard uncertainty in the measurement of the respective atmospheric variables is:

$$\begin{aligned} u_{\mathrm{T}} &= 0.8\,^{\circ}\mathrm{C} \\ u_{\mathrm{h_r}} &= 3\ \% \\ u_{\mathrm{P_a}} &= 3\,\mathrm{hPa} \end{aligned}$$

The combined standard uncertainty due to these factors is a type B uncertainty, since it depends on specifications and/or standards fulfilled by the measuring instruments (thermometer, hygrometer and barometer). Its value, calculated by means of Eq. (2.22) under the reasonable assumption that the three variables are independent, is:

$$u_{\mathrm{c,amb}} = \sqrt{(-0.001 \times 0.8)^2 + (0.001 \times 3)^2 + (-0.001 \times 3)^2}\mathrm{dB} = 0.0043\,\mathrm{dB}.$$

We see that the contribution from the measurement errors in the atmospheric variables to total uncertainty is negligible. This is the reason why in general there is no need of sophisticated instruments to measure these variables.

In order to get the total combined uncertainty, we should add the uncertainty due to the instrument, as we shall see later.

2.6.3 Uncertainty in the Calculation of the Equivalent Level

Suppose that we have measured or estimated theequivalent level L_{Aeq,T_i} in N time intervals T_i and we want to get the total equivalent level by means of the formula

$$L_{\mathrm{Aeq},T} = 10 \log\left(\frac{1}{T}\sum_{i=1}^{N} T_i\, 10^{\frac{L_{\mathrm{Aeq},T_i}}{10}}\right). \tag{2.32}$$

This problem arises when we want to determine the equivalent level during a work day from the measurement of a short sample of each individual process into which the daily sequence of activities can be split. These measurements are conducted during relatively short intervals and are extrapolated to the total duration of each process.

We need to calculate the combined standard uncertainty, which has two types of components: one due to levels and the other due to interval durations. Note that the uncertainty in interval durations is attributable not only to the measurement process (including the measurement instrument, in this case a chronometer) but also to some indefinition of the measurand itself, since the duration of a process might depend on factors of difficult control that may adjust according to the intermediate results. We have, therefore,

$$L_{\mathrm{Aeq},T} = g(L_1, \ldots, L_N, T_1, \ldots, T_N) = 10 \log\left(\frac{1}{T}\sum_{i=1}^{N} T_i 10^{\frac{L_{\mathrm{Aeq},T_i}}{10}}\right)$$

In order to calculate the partial derivatives, we express the logarithm as a natural logarithm. The derivatives with respect to L_{Aeq,T_i} are

$$\frac{\partial g}{\partial L_{\mathrm{Aeq},T_i}} = \frac{10}{\ln 10} \frac{\frac{\partial}{\partial L_{\mathrm{Aeq},T_i}}\left(\frac{1}{T}\sum_{i=1}^{N} T_i e^{\ln 10 \frac{L_{\mathrm{Aeq},T_i}}{10}}\right)}{\frac{1}{T}\sum_{i=1}^{N} T_i\, 10^{\frac{L_{\mathrm{Aeq},T_i}}{10}}}$$

$$= \frac{10}{\ln 10} \frac{T_i}{T} \frac{\frac{\ln 10}{10} 10^{\frac{L_{\mathrm{Aeq},T_i}}{10}}}{10^{\frac{L_{\mathrm{Aeq},T}}{10}}},$$

which, after simplification, yields

$$\frac{\partial g}{\partial L_{\mathrm{Aeq},T_i}} = \frac{T_i}{T} 10^{\frac{L_{\mathrm{Aeq},T_i} - L_{\mathrm{Aeq},T}}{10}}. \tag{2.33}$$

The derivatives respect to T_i are, taking into account that $T = \Sigma T_i$,

$$\frac{\partial g}{\partial T_i} = \frac{10}{\ln 10} \frac{\frac{T - T_i}{T^2} 10^{\frac{L_{\mathrm{Aeq},T_i}}{10}}}{\frac{1}{T}\sum_{i=1}^{N} T_i\, 10^{\frac{L_{\mathrm{Aeq},T_i}}{10}}} = \frac{10}{\ln 10} \frac{T - T_i}{T^2} 10^{\frac{L_{\mathrm{Aeq},T_i} - L_{\mathrm{Aeq},T}}{10}}. \tag{2.34}$$

As a numerical example, consider three intervals of durations 2 h, 3 h, and 3 h measured with a standard uncertainty of 12 min. Their equivalent levels were 85.1 dBA, 83.0 dBA and 80.2 dBA, respectively, with a standard uncertainty of 1 dB. Then the total equivalent level turns out to be

$$L_{\mathrm{Aeq},T} = 10 \log\left(\frac{2}{8} 10^{\frac{85.1}{10}} + \frac{3}{8} 10^{\frac{83.0}{10}} + \frac{3}{8} 10^{\frac{80.2}{10}}\right) \mathrm{dBA} = 82.9\, \mathrm{dBA}$$

The derivatives with respect to the levels are

$$\frac{\partial g}{\partial L_{\mathrm{Aeq},T_1}} = \frac{2}{8}10^{\frac{85.1-82.9}{10}} = 0.415,$$
$$\frac{\partial g}{\partial L_{\mathrm{Aeq},T_2}} = \frac{3}{8}10^{\frac{83.0-82.9}{10}} = 0.384,$$
$$\frac{\partial g}{\partial L_{\mathrm{Aeq},T_3}} = \frac{3}{8}10^{\frac{80.2-82.9}{10}} = 0.201.$$

To find the time derivatives we multiply by $10/(T_k \ln 10)$, i.e.,

$$\frac{\partial g}{\partial T_1} = 0.676\frac{\mathrm{dBA}}{\mathrm{h}}.$$
$$\frac{\partial g}{\partial T_2} = 0.347\frac{\mathrm{dBA}}{\mathrm{h}}.$$
$$\frac{\partial g}{\partial T_3} = 0.182\frac{\mathrm{dBA}}{\mathrm{h}}.$$

Therefore,

$$u_c = \sqrt{(0.415^2 + 0.384^2 + 0.201^2) \times 1^2 + (0.676^2 + 0.347^2 + 0.182^2) \times 0.2^2}$$
$$= 0.62\,\mathrm{dBA}$$

Notice that the uncertainty is smaller than that of each single measurement (1 dB). This is due to the effect of dispersion reduction in an average of several variables.

In the absence of temporal errors, if the three levels were equal and had the same duration, the uncertainty would be $1/\sqrt{3}$ dB = 0.58 dB. The duration uncertainty makes it slightly larger.

2.6.4 *Uncertainty in A-Weighting Computed from Octave Spectrum*

Suppose we have measured the octave band levels $\{L_1, \ldots, L_{10}\}$ of a given noise. The A-weighted sound level can be obtained from

$$L_A = 10 \log\left(\sum_{i=1}^{10} 10^{\frac{L_i + A_i}{10}}\right), \tag{2.35}$$

where A_i is the A-weighting correction for the i-th band. The octave band levels are random variables that are not necessarily independent. In order to determine the uncertainty, we need to find the partial derivatives of L_A with respect to the band levels L_i, as well as the covariances of such levels.

As in the previous examples, the partial derivatives can be calculated converting previously the logarithm and exponentials to base e:

$$L_A = \frac{10}{\ln 10} \ln\left(\sum_{i=1}^{10} e^{\ln 10 \frac{L_i + A_i}{10}}\right).$$

Applying the chain rule,

$$\frac{\partial L_A}{\partial L_i} = \frac{10}{\ln 10} \frac{1}{\left(\sum_{i=1}^{10} e^{\ln 10 \frac{L_i + A_i}{10}}\right)} e^{\ln 10 \frac{L_i + A_i}{10}} \frac{\ln 10}{10} = \frac{10^{\frac{L_i + A_i}{10}}}{10^{\frac{L_A}{10}}}$$

from where we finally get

$$\frac{\partial L_A}{\partial L_i} = 10^{\frac{L_i + A_i - L_A}{10}}. \tag{2.36}$$

The covariance matrix for calculating a type B uncertainty can only be obtained if some properties of the phenomenon originating the noise to be measured are known. Otherwise, it will be necessary to conduct a series of measurements and estimate the covariance matrix from them. If $L_{i,k}$ and $L_{j,k}$ are the octave band levels of the i-th and j-th bands corresponding to the k-th measurement, then the covariances can be estimated by

$$u(L_i, L_j) = \frac{1}{n-1} \sum_{k=1}^{n} (L_{ik} - \overline{L}_i)(L_{jk} - \overline{L}_j), \tag{2.37}$$

where $\overline{L}_i$ and $\overline{L}_j$ are the mean levels of the i-th and j-th bands respectively, and n, the number of measured spectra. Finally, we can obtain the combined standard uncertainty as

$$u_c = \sqrt{\sum_{i=1}^{10} \sum_{j=1}^{10} \frac{\partial L_A}{\partial L_i} \frac{\partial L_A}{\partial L_j} u(L_i, L_j)}. \tag{2.38}$$

This uncertainty corresponds to the estimation of the A-weighted level from a single measurement. However, having taken n measurements, it is natural to estimate the A-weighted as an average of all measurement results. In such case the covariance are divided by n, increasing the confidence of L_A.

Note that for specific noises (such as traffic noise) the different spectral bands are not independent, so the cross-correlations ($i \neq j$) do not vanish.

2.6.5 Uncertainty in the Measurement of Sound Transmission Loss

The sound transmission loss of a sound insulating material is measured in the laboratory through the measurement of the octave or one-third octave band levels L_{1k} and L_{2k} at the source and receiver chambers, respectively, and the reverberation time T_k of the latter. Also needed are the area S_{12} of the sample and the volume V_2 of the receiver chamber. For the k-th band we have

$$R_k = L_{1k} - L_{2k} + 10 \log_{10} \frac{T_k S_{12}}{0.161\, V_2}, \tag{2.39}$$

The derivatives with respect to L_{1k} and L_{2k} are, respectively, 1 and −1. The derivatives with respect to T_k, S_{12} and V_2 are, respectively,

$$\frac{\partial R_k}{\partial T_k} = \frac{10}{\ln 10} \frac{1}{T_k}. \tag{2.40}$$

$$\frac{\partial R_k}{\partial S_{12}} = \frac{10}{\ln 10} \frac{1}{S_{12}}. \tag{2.41}$$

$$\frac{\partial R_k}{\partial T_k} = -\frac{10}{\ln 10} \frac{1}{V_2}. \tag{2.42}$$

Example: Suppose that the k-th band has been measured, with the following results: T_k = 2.3 s, L_{1k} = 87.3 dB, L_{2k} = 45.6 dB, S_{12} = 10.4 m^2, V_2 = 51.84 m^3. As regards the uncertainties, it is known that u_L = 1 dB, u_T = 0.2 s, u_S = 0.099 m^2, u_V = 0.38 m^3. Replacing into the previous equations, we have

$$\frac{\partial R_k}{\partial T_k} = 1.888\, \frac{\text{dB}}{\text{s}}$$
$$\frac{\partial R_k}{\partial S_{12}} = 0.418 \frac{\text{dB}}{\text{m}^2}$$
$$\frac{\partial R_k}{\partial V_2} = -0.0838 \frac{\text{dB}}{\text{m}^3}$$

Hence, the combined standard uncertainty in R_k will be

$$u_{c,R_k} = \sqrt{\frac{\partial R_k}{\partial L_{1k}}^2 u_{L_{1k}}^2 + \frac{\partial R_k}{\partial L_{2k}}^2 u_{L_{2k}}^2 + \frac{\partial R_k}{\partial T_k}^2 u_{T_k}^2 + \frac{\partial R_k}{\partial S_{12}}^2 u_{S_{12}}^2 + \frac{\partial R_k}{\partial V_2}^2 u_{V_2}^2}$$
$$= \sqrt{1^2 \times 1^2 + 1^2 \times 1^2 + 1.888^2 \times 0.2^2 + 0.418^2 \times 0.099^2 + 0.0838^2 \times 0.38^2}$$
$$= \sqrt{1 + 1 + 0.143 + 0.0017 + 0.00101} = 1.46\,\mathrm{dB}$$

The intermediate results are detailed in order to show that the greatest contribution to uncertainty corresponds to the acoustic components, i.e., the measurement of levels at both sides of the partition under test.

2.7 Uncertainty and Resolution

The finite resolution of a measuring instrument implies an uncertainty, since every true value of the measurand that falls between two consecutive divisions or between two least significant digits is assigned the same value. We can assume that the statistical distribution of the error is uniform, extended to an interval whose amplitude is equal to the resolution. As detailed in Appendix 2, if we call the resolution Δx_{min}, the standard uncertainty will be

$$u_{\mathrm{resol}} = \frac{\Delta x_{min}}{\sqrt{12}}. \tag{2.43}$$

Since we are dealing with a uniform distribution, the expanded uncertainty is smaller than for a normal distribution. For instance, U_{95} is attained with a coverage factor k such that

$$k\frac{\Delta x_{\mathrm{min}}}{\sqrt{12}}\frac{1}{\Delta x_{\mathrm{min}}} = \frac{0.95}{2},$$

Hence,

$$k = \frac{0.95\sqrt{12}}{2} = 1.65.$$

However, this component of uncertainty is not isolated but is part of several components. The *central limit theorem* implies that when adding several random variables together the resulting distribution tends to be normal. Besides, the resolution component of sound level meters uncertainty is much smaller than the other components.

Example: If a sound level meter has a resolution of 0.1 dB, then the uncertainty due to resolution will be

$$u_{\mathrm{resol}} = \frac{0.1\,\mathrm{dB}}{\sqrt{12}} = 0.029\,\mathrm{dB}.$$

It is vanishingly small compared to the uncertainty of the instrument itself, which is of the order of 1 dB.

2.8 Uncertainty and Systematic Error

Many measurements present the problem of the presence of a *systematic error*, i.e. some constant or predictable bias or offset that affect equally all measurements conducted under given conditions. It can be due to the operator, the instrument, the procedure, or a combination of them.

In the case of the operator, the cause is usually a parallax error when reading an analog (needle) instrument as well as a biased decision while reading a fluctuating digital display. In the case of the instrument itself, the typical reason is the lack of calibration. In the case of the measurement procedure, the reason may be some uncorrected effect (known or not), for instance, some temperature drift.

The systematic error may be additive (zero error), multiplicative (scale error) or nonlinear (linearity error).

2.8.1 *Additive Systematic Error*

We can define the additive systematic error as the difference between the mean of the measured values, μ_o and the true value x of the measurand,[9] i.e.

$$\varepsilon_{\mathrm{sys}} = \mu_o - x. \tag{2.44}$$

In order to apply (2.44), we would need to know the population mean of all possible measurements of the measurand, μ_o, and its true value, x, which is not possible. Instead, we can apply (2.44) in particular to a *standard reference*, for instance, the length of a standard bar, the mass of a standard weight or the level of a calibration tone. In any case, it is a quantity with a known conventional value and with an also known standard uncertainty (in general, substantially smaller than a

[9]Equivalently, it can be defined as the mean of the difference between the measured value and the true value.

typical instrument's uncertainty). If we call that value x_{cal}, we can estimate the mean of the observed values $\mu_{xcal,o}$ with the mean $\bar{x}_{cal,o}$ of a sample of m measurements, and adopt the conventional value as the true value. Our estimate of the systematic error will be, hence,

$$\hat{\varepsilon}_{sys} = \bar{x}_{cal,o} - x_{cal}. \tag{2.45}$$

where $\bar{x}_{cal,o}$ is the mean of m measurements of the reference. The uncertainty of this estimate will be

$$u_{\hat{\varepsilon}_{sys}} = \sqrt{u_{\bar{x}_{cal,o}}^2 + u_{x_{cal}}^2}. \tag{2.46}$$

If u_o is the uncertainty of a single measurement,

$$u_{\hat{\varepsilon}_{sys}} = \sqrt{\frac{u_o^2}{m} + u_{x_{cal}}^2}, \tag{2.47}$$

Once the systematic error has been estimated, we can correct the value of the measurand to remove the systematic effects:

$$x_c = \bar{x}_o - \hat{\varepsilon}_{sys}. \tag{2.48}$$

where $\bar{x}_o$ is the mean of n observations of the measurand (n is not necessarily the same as the previous m).

The uncertainty in this measurement will be

$$u_{x_c} = \sqrt{u_{\bar{x}_o}^2 + u_{\hat{\varepsilon}_{sys}}^2}, \tag{2.49}$$

or, in terms of the general uncertainty of a single measurement,

$$u_{x_c} = \sqrt{\frac{u_o^2}{n} + u_{\hat{\varepsilon}_{sys}}^2}. \tag{2.50}$$

It is interesting to note that although it is possible to correct the systematic error (and it *must* be corrected), it is not possible to remove its uncertainty.

Example: Suppose that we take four measurements of a noise in identical conditions to get 83.3 dB, 83.6 dB, 84.1 dB and 83.0 dB with an instrument whose standard uncertainty is 1 dB. Previously, the calibration is checked with an acoustic calibrator whose nominal value is 93.85 dB and whose recent calibration certificate indicates a standard uncertainty of 0.15 dB and an error of −0.1 dB. It is checked three times, getting in all cases the same value 94.1 dB.

Let us first calculate the correction and its uncertainty. The nominal value of the reference level is 93.85 dB, but as the systematic error is −0.1 dB, a perfect instrument would read

$$L_{\text{cal}} + \varepsilon_{L_{cal},\text{sys}} = 93.75\,\text{dB}.$$

Our instrument reads, instead,

$$\bar{x}_{\text{cal,o}} = \frac{94.1 + 94.1 + 94.1}{3}\,\text{dB} = 94.1\,\text{dB}$$

The estimated systematic error due to lack of calibration will be, hence

$$\hat{\varepsilon}_{\text{sys}} = 94.1\,\text{dB} - 93.75\,\text{dB} = 0.35\,\text{dB}.$$

The uncertainty of this estimate will be

$$\begin{aligned} u_{\hat{\varepsilon}_{\text{sys}}} &= \sqrt{\frac{1^2 + (0.1/\sqrt{12})^2}{3} + 0.15^2} \\ &= \sqrt{0.3333 + 0.0003 + 00.0225} = 0.597\,\text{dB} \end{aligned}$$

The second term inside the root symbol corresponds to the finite resolution uncertainty. We can see that in this case this effect is negligible.

As regards the measured and corrected value of the measurand, we have:

$$x_{\text{c}} = \frac{83.3 + 83.6 + 84.1 + 83.0}{4} - 0.35 = 83.15\,\text{dB}$$

The standard uncertainty is

$$\begin{aligned} \varepsilon_{x_c} &= \sqrt{\frac{1^2 + (0.1/\sqrt{12})^2}{4} + 0.597^2} \\ &= \sqrt{0.25 + 0.0002 + 0.3564} = 0.78\,\text{dB} \end{aligned}$$

Once more, the uncertainty due to finite resolution is not important compared to the instrument's and calibrator's uncertainty. We can also see that the latter prevails when we average several measurements. In this example, by averaging only four measurements the instrument's uncertainty drops below the uncertainty due to the systematic error of the calibrator.

2.8.2 *Multiplicative Systematic Error*

Also known as *scale error* or *relative systematic error*, it is present when the physical principle of the measurement has the form,

$$y = Kx, \tag{2.51}$$

where x is a measurand that cannot be directly measured (for instance, a sound pressure), y, an associated quantity that can be directly measured (for instance, a voltage) and K, the transducer sensitivity that converts the measurand into a measurable quantity.

What is actually measured is, thus, y, after which it is possible to get the desired value as

$$x = \frac{1}{K} y, \tag{2.52}$$

The error arises when the real constant K in (2.51) differs from the one used in (2.52). In general, in (2.51) there is an unknown constant K_{real}, but when we compute x in (2.52) we assume a nominal or ideal constant K_{ideal}. In consequence, the relationship between the observed value x_o and the true value will be

$$x_o = \frac{K_{real}}{K_{ideal}} x. \tag{2.53}$$

A new difficulty arises: the measured value of y has also some uncertainty since the instrument is not an ideal one. Hence, (2.51) must be rewritten as

$$y = K_{real} x + \varepsilon_y, \tag{2.54}$$

where ε_y is a random variable centered at 0 that represents the dispersion in the measurement of y. Hence, the observed value of x will be

$$x_o = \frac{1}{K_{ideal}} y = \frac{K_{real}}{K_{ideal}} x - \frac{1}{K_{ideal}} \varepsilon_y. \tag{2.55}$$

Since ε_y has mean 0, we could remove the random measurement error taking the population mean μ_o of all observed values x_o of x:

$$\mu_o = \frac{K_{real}}{K_{ideal}} x. \tag{2.56}$$

Finally, we define the *relative systematic error* ε_{rel} as

$$\varepsilon_{rel} = \frac{\mu_o}{x} - 1 = \frac{K_{real}}{K_{ideal}} - 1. \tag{2.57}$$

As can be noted, it represents the relative error between the real and ideal constants. The relative error is dimensionless and is 0 in the ideal case in which $K_{real} = K_{ideal}$. For acceptable measurements it should be much less than 1.

Since μ_o is the population mean of x_o, it cannot be known, so we shall attempt to estimate ε_{rel}. To that end, instead of an arbitrary value of x, we chose a reference calibration value x_{cal}, and take the average of m observations of x_{cal}. We assume that x_{cal} and its uncertainty are known. We get

$$\hat{\varepsilon}_{rel} = \frac{\bar{x}_{cal,o}}{x_{cal}} - 1. \tag{2.58}$$

The combined standard uncertainty is computed by means of Eq. (2.27), yielding

$$\begin{aligned} u_{\hat{\varepsilon}_{rel}} &= \sqrt{u_{\bar{x}_{cal,o}}^2 \left(\frac{1}{x_{cal}}\right)^2 + u_{x_{cal}}^2 \left(\frac{-\bar{x}_{cal,o}}{x_{cal}^2}\right)^2} \\ &= \frac{1}{x_{cal}} \sqrt{u_{\bar{x}_{cal,o}}^2 + u_{x_{cal}}^2 (1 + \hat{\varepsilon}_{rel})^2} \end{aligned} \tag{2.59}$$

Now that we have estimated the value of the relative systematic error, it is possible to compute the corrected value of a measurand x in general:

$$x_c = \bar{x}_o (1 + \hat{\varepsilon}_{rel}). \tag{2.60}$$

The combined standard uncertainty turns out to be

$$u_{x_c} = \sqrt{u_{\bar{x}_o}^2 (1 + \hat{\varepsilon}_{rel})^2 + \bar{x}_o^2 u_{x_{cal}}^2} \tag{2.61}$$

Or, making explicit the reduction of the mean with respect to that of a single measurement,

$$u_{x_c} = \sqrt{\frac{u_{x_o}^2}{n} (1 + \hat{\varepsilon}_{rel})^2 + \bar{x}_o^2 u_{x_{cal}}^2}. \tag{2.62}$$

Note that since

$$\bar{x}_o = \frac{x_c}{1 + \hat{\varepsilon}_{rel}}, \tag{2.63}$$

the uncertainty in x_c increases with x_c, unlike the relative systematic error, which is independent of x_c.

Example: Consider the measurement of a sound pressure with a microphone whose sensitivity S differs from the nominal one. In that case, if v is the voltage generated in the microphone when excited with a pressure p, we will have [10]

[10]It is interesting to note that in this case, we are interested in measurements of sound pressure instead of sound pressure level.

$$p = \frac{1}{S} v. \tag{2.64}$$

Suppose a Rion UC-53A microphone whose nominal (ideal) sensitivity is −28.0 dB referred to 1 V/Pa, and whose calibration certificate indicates a real sensitivity (at 1 kHz) of −26.2 dB with a standard uncertainty of 0.2 dB. We are interested in obtaining the relative systematic error. We need the ideal and real sensitivities and uncertainty in V/Pa. The ideal sensitivity is

$$S_{\text{ideal}} = 10^{\frac{S_{\text{ideal,dB}}}{20}} 10^{\frac{-28.0}{20}} \text{V/Pa} = 0.03981 \text{ V/Pa}$$

The real sensitivity,

$$S_{\text{real}} = 10^{\frac{S_{\text{real,dB}}}{20}} = 10^{\frac{-26.2}{20}} \text{V/Pa} = 0.04898 \text{ V/Pa}$$

The standard uncertainty can be obtained by applying Eq. (2.27), in this case with a single variable:

$$u_{S_{\text{real}}} = \frac{\partial S_{\text{real}}}{\partial S_{\text{real,dB}}} u_{S_{\text{real,dB}}} = \frac{\ln 10}{20} S_{\text{real}} u_{S_{\text{real,dB}}} = 0.0011 \text{ V/Pa}$$

If we suppose that the microphone generates a voltage v, the observed pressure will be

$$p_{\text{o}} = \frac{v}{S_{\text{ideal}}},$$

while the real pressure is

$$p = \frac{v}{S_{\text{real}}}.$$

The estimated relative error turns out to be

$$\hat{\varepsilon}_{\text{rel}} = \frac{p_{\text{o}}}{p} - 1 = \frac{S_{\text{real}}}{S_{\text{ideal}}} - 1 = 0.2303.$$

The standard uncertainty of this error is

$$u_{\hat{\varepsilon}_{\text{rel}}} = \frac{u_{S_{\text{real}}}}{S_{\text{ideal}}} = 0.0276.$$

If, for example, we measured a sound pressure $p_{\text{o}} = 1$ Pa, the corrected value would be

$$p_{\text{c}} = 1 \text{ Pa} \times (1 + 0.2303) = 1.2303 \text{ Pa}$$

The uncertainty component due to the relative error will be

$$u_{p_c,\mathrm{rel}} = 1\,\mathrm{Pa} \times 0.0276 = 0.0276\,\mathrm{Pa}$$

The total uncertainty must include, besides, the uncertainty due to nonsystematic random effects, inherited from the uncertainty in the measurement of v.

Note: The multiplicative systematic error is converted into additive when the quantities are presented in logarithmic version, since the error in the scale factor is transformed into an additive term.

2.8.3 Nonlinear Systematic Error

The *nonlinear systematic error* or *linearity error* is present in those situations in which a linear model is assumed:

$$y = Kx, \tag{2.65}$$

but it is actually nonlinear and presents dispersion:

$$y = g(x) + \varepsilon_y. \tag{2.66}$$

In the preceding equations y is a quantity that can be directly measured, and x, the measurand to be computed from y. To determine x it is generally assumed that (2.65) holds with a nominal or ideal constant K_o, so that the observed value, x_o, will be

$$x_o = \frac{y}{K_o} = \frac{g(x)}{K_o} + \frac{\varepsilon_y}{K_o}. \tag{2.67}$$

Suppose, for instance, that the function $g(x)$ can be approximated by a third-degree polynomial without independent term,

$$y = Kx + K_1x^2 + K_2x^3. \tag{2.68}$$

For a cubic equation such as this one, it is possible to solve for x, but in the most interesting case where K_1 and K_2 are small (a quasi-linear relationship), we can approximate

$$x \cong Ay + By^2 + Cy^3. \tag{2.69}$$

Although the actually measured quantity is y, the instrument in general provides an observed value x_o of the measurand, obtained internally from y with the proportional relationship of Eq. (2.67). Thus, the true value of x can be approximated by

$$x \cong a x_o + b x_o^2 + c x_o^3. \tag{2.70}$$

If we knew K, K_1, and K_2, we could obtain a, b, and c as follows. First we replace x_o by y/K_o according to (2.67) and y as a function of x using (2.68). The second member is a ninth degree polynomial. Keeping up to the cubic term and equating coefficients, we arrive at the following correction for the instrument's reading x_o:

$$x_c = \frac{K_o}{K} x_o - \frac{K_1 K_o^2}{K^3} x_o^2 + \frac{2K_1^2 - K_2 K}{K^5} K_o^3 x_o^3. \tag{2.71}$$

Note that we have replaced the sign ≅ by an equality. This is because we do not claim that x_c is equal to x, but only a corrected value that, by taking into account the nonlinear systematic effect, is a better approximation of the true value.

In this case, we do not attempt to introduce a specific quantification of the error since it would require several parameters. However, it is still possible to use the concept of relative systematic error with respect to the linear term.

In general, K, K_1, and K_2 are unknown, so the model must be obtained empirically. Instead of the model (2.68) we will directly use (2.70), since it provides a corrected value directly from the observed value (the instrument's reading). In order to find a, b, and c, we need to measure three reference values whose conventional values x_1, x_2, x_3 and uncertainties u_{x1}, u_{x2}, u_{x3} are known.[11] Let x_{1o}, x_{2o}, x_{3o} be the measured (observed) values (or the respective means of m measurements). We can pose the equation system

$$\begin{cases} a x_{1o} + b x_{1o}^2 + c x_{1o}^3 = x_1 \\ a x_{2o} + b x_{2o}^2 + c x_{2o}^3 = x_2 \\ a x_{3o} + b x_{3o}^2 + c x_{3o}^3 = x_3 \end{cases} \tag{2.72}$$

This system of three equations and three unknowns, a, b, and c can be solved numerically. We first rewrite it in matrix form,

$$\begin{bmatrix} x_{1o} & x_{1o}^2 & x_{1o}^3 \\ x_{2o} & x_{2o}^2 & x_{2o}^3 \\ x_{3o} & x_{3o}^2 & x_{3o}^3 \end{bmatrix} \begin{bmatrix} a \\ b \\ c \end{bmatrix} = \begin{bmatrix} x_1 \\ x_2 \\ x_3 \end{bmatrix}, \tag{2.73}$$

from where

$$\begin{bmatrix} a \\ b \\ c \end{bmatrix} = \begin{bmatrix} x_{1o} & x_{1o}^2 & x_{1o}^3 \\ x_{2o} & x_{2o}^2 & x_{2o}^3 \\ x_{3o} & x_{3o}^2 & x_{3o}^3 \end{bmatrix}^{-1} \begin{bmatrix} x_1 \\ x_2 \\ x_3 \end{bmatrix}. \tag{2.74}$$

[11] Note that only the conventional values are known, not the real ones. However, the conventional values have less dispersion and do not have any known systematic effect.

This completes the data that are necessary to correct for the nonlinear systematic effects:

$$x_c = ax_o + bx_o^2 + cx_o^3. \tag{2.75}$$

Now let us calculate the uncertainty. Note, first, that once the constants a, b, and c are obtained we have a model that can be applied to many observed data x_o, but that model is not the only one that we could have arrived to using the same procedure, because the measurements of the reference values are subject to uncertainty. So what we really have is

$$x_c = g(x_o, x_{1o}, x_{2o}, x_{3o}). \tag{2.76}$$

The functional dependency of x_{1o}, x_{2o} and x_{3o} is through the coefficients a, b, and c:

$$x_c = a(x_{1o}, x_{2o}, x_{3o})x_o + b(x_{1o}, x_{2o}, x_{3o})x_o^2 + c(x_{1o}, x_{2o}, x_{3o})x_o^3. \tag{2.77}$$

The uncertainty is, then,

$$u_{x_c}^2 = \left(\frac{\partial g}{\partial x_o}\right)^2 u_{x_o}^2 + \left(\frac{\partial g}{\partial x_{1o}}\right)^2 u_{x_{1o}}^2 + \left(\frac{\partial g}{\partial x_{2o}}\right)^2 u_{x_{2o}}^2 + \left(\frac{\partial g}{\partial x_{3o}}\right)^2 u_{x_{3o}}^2. \tag{2.78}$$

The basic uncertainties u_{xo}, u_{x1o}, u_{x2o}, u_{x3o}, are those corresponding to the measurement (may be a direct reading or an average of several readings). They might depend on the measurand value in each case. The derivatives are:

$$\frac{\partial g}{\partial x_o} = a + 2bx_o + 3cx_o^2 \tag{2.79}$$

$$\frac{\partial g}{\partial x_{io}} = \frac{\partial a}{\partial x_{io}} x_o + \frac{\partial b}{\partial x_{io}} x_o^2 + \frac{\partial c}{\partial x_{io}} x_o^3, \tag{2.80}$$

where $i = 1, 2, 3$. The functions a, b, and c are defined implicitly through system (2.72). Even if it is possible to derive an analytic expression, it is preferable to apply implicit derivation. Let us rewrite (2.72) as:

$$\begin{cases} ax_{1o} + bx_{1o}^2 + cx_{1o}^3 - x_1 = 0 \\ ax_{2o} + bx_{2o}^2 + cx_{2o}^3 - x_2 = 0 \\ ax_{3o} + bx_{3o}^2 + cx_{3o}^3 - x_3 = 0 \end{cases} \tag{2.81}$$

Deriving each expression with respect to x_{1o} we get

$$\begin{cases} \frac{\partial a}{\partial x_{1o}} x_{1o} + \frac{\partial b}{\partial x_{1o}} x_{1o}^2 + \frac{\partial c}{\partial x_{1o}} x_{1o}^3 = -(a + 2bx_{1o} + 3cx_{1o}^2) \\ \frac{\partial a}{\partial x_{1o}} x_{2o} + \frac{\partial b}{\partial x_{1o}} x_{2o}^2 + \frac{\partial c}{\partial x_{1o}} x_{2o}^3 = 0 \\ \frac{\partial a}{\partial x_{1o}} x_{3o} + \frac{\partial b}{\partial x_{1o}} x_{3o}^2 + \frac{\partial c}{\partial x_{1o}} x_{3o}^3 = 0 \end{cases} \tag{2.82}$$

This is a system of three equations and three unknowns that allow to solve for the derivatives of a, b, and c with respect to x_{1o}. In matrix version:

$$\begin{bmatrix} x_{1o} & x_{1o}^2 & x_{1o}^3 \\ x_{2o} & x_{2o}^2 & x_{2o}^3 \\ x_{3o} & x_{3o}^2 & x_{3o}^3 \end{bmatrix} \begin{bmatrix} \partial a/\partial x_{1o} \\ \partial b/\partial x_{1o} \\ \partial c/\partial x_{1o} \end{bmatrix} = - \begin{bmatrix} a + 2bx_{1o} + 3cx_{1o}^2 \\ 0 \\ 0 \end{bmatrix} \tag{2.83}$$

Interestingly, we see that the system's matrix is the same as for system (2.73), originally used to solve for a, b, and c. This is advantageous, since it suffices to compute the inverse only once.

Proceeding similarly, we get and solve systems for the derivatives respect to x_{2o} and x_{3o}. After replacing in (2.80) and similar equations the values obtained from (2.83) and similar systems, we substitute (2.79) and (2.80) in (2.78), arriving finally at the desired uncertainty. The calculations are rather lengthy so they are implemented by software. In Appendix 9 there is an example.

2.8.4 Uncorrected Systematic Errors

Sometimes there is a systematic error that cannot be fully corrected. A typical example is when an instrument's calibration is checked with a calibrator whose actual level does not coincide with the nominal one. This situation is dealt with taking into account the calibrator's uncertainty.

A similar situation happens when the calibrator suffers long-term drifts. For instance, the specification chart of calibrator Brüel and Kjær Type 4231 (Brüel and Kjær 2014) indicates that the long-term stability (one year) is better than 0.05 dB with a confidence level of 96 %. We can assume that this implies a coverage factor $k = 2$. Hence, within a year, the standard uncertainty will increase by 0.025 dB with respect to the standard uncertainty attributable to the reference level of the calibrator, so the uncertainty due to the uncorrected systematic error will be

$$u_{\mathrm{cal}} = \sqrt{u_{\mathrm{initial}}^2 + u_{1\,\mathrm{yr}}^2} \tag{2.84}$$

With the data of the example given in 2.8.1, it turns out to be $u_{\mathrm{cal}} = 0.152$ dB. This represents a negligible increment compared to the uncertainty of 0.15 dB indicated in the most recent calibration certificate.

2.9 Chain Calculation of Uncertainty

In complex uncertainty calculations, it is often possible to split the problem into a series of chained calculations through the use of nested composite functions. Suppose, for instance, that we have a measurand that depends on several variables which in turn depend on other variables and so on, for instance:

$$\begin{aligned} w &= f(z_1, \ldots, y_L), \\ z_k &= z_k(y_1, \ldots, y_M), \\ y_j &= y_j(x_1, \ldots, x_N). \end{aligned}$$

Thus,

$$\begin{aligned} u_w^2 &= \sum_{i=1}^{L} \left(\frac{\partial w}{\partial x_i} \right)^2 u_{x_i}^2 = \sum_{i=1}^{L} \left(\sum_{j=1}^{M} \frac{\partial w}{\partial y_j} \frac{\partial y_j}{\partial x_i} \right)^2 u_{x_i}^2 \\ &= \sum_{i=1}^{L} \left(\sum_{j=1}^{M} \left(\sum_{k=1}^{N} \frac{\partial f}{\partial z_k} \frac{\partial z_k}{\partial y_j} \right) \frac{\partial y_j}{\partial x_i} \right)^2 u_{x_i}^2 \end{aligned}$$

i.e.,

$$u_w^2 = \sum_{i=1}^{L} \left(\sum_{j=1}^{M} \sum_{k=1}^{N} \frac{\partial f}{\partial z_k} \frac{\partial z_k}{\partial y_j} \frac{\partial y_j}{\partial x_i} \right)^2 u_{x_i}^2$$

However, even if x_i are statistically independent, y_j and z_k not necessarily are, since they depend functionally on the same variables. In cases as this one, it is important that the intermediate variables depend on different independent variables, which sometimes can be achieved grouping together in a single function variables that depend on the same variable.

Problems

(1) Imagine you are measuring traffic noise in a street across a park. Identify all possible factors of uncertainty for your measurement and try to figure out how you would minimize their effects.
(2) In the same scenario of problem (1), try to define as thoroughly as possible your measurand. Remember that the measurement conditions are a part of the definition.
(3) Describe in as much detail as possible the measurement method and procedure in the case of problem (1).
(4) Repeat problems (1) to (3) for a different type of measurement of your choice, such as one you perform frequently.

(5) Explain why a high precision measurement instrument could yield very inaccurate results.
(6) Indicate which of the following situations correspond to direct measurement methods: (a) Measuring the equivalent level of a room's ambient noise using an integrating sound level meter. (b) Measuring the A-weighted equivalent level extended to 8 h in a factory which has three well-identified processes by measuring each process during a short interval and then combining the results according to the specified duration of each process. (c) Measuring the octave band spectrum of the immission noise in an office using a spectrum analyzer. (d) Measuring the statistical level L_{90} by taking 100 measurements at 5 s intervals and selecting the largest of the 10 smallest results. (e) Measuring the equivalent noise level of a laboratory chamber applying a correction for temperature. (f) Measuring the statistical level L_{10} in a street using an instrument with statistical measurement capabilities. (g) Measuring the peak level of conversational speech at 1 m with an instrument capable of measuring peak levels.
(7) Indicate the measurement model used to measure the D-weighted sound pressure level from the measurement of the one-third octave spectrum of the signal.
(8) The noise reduction coefficient (*NRC*) is defined as the average of the absorption coefficients at 250, 500, 1000, and 2000 Hz. Describe a possible measurement model for the *NRC*.
(9) Find the resolution of a method of measurement that consists in averaging 10 direct measurements with a sound level meter whose resolution is 0.1 dB
(10) Find the type A uncertainty of the following series of 15 s measurements of traffic noise: 70.5, 79.3, 74.7, 73.1, 74.6, 72.8, 73.0, 76.2, 76.0, 76.0.
(11) Find the type B uncertainty of a Class 2 sound level meter associated to the weighting network for different frequencies.
(12) Consider a class of noise sources that have a 1/3 octave noise spectrum increasing at a rate of 6 dB/oct until a peak is reached at 250 Hz, after which it starts decaying at −3 dB/oct. Assume that the standard deviation for all bands is 1 dB and that the deviation from the preceding model is independent for all bands. (a) Find the type B uncertainty in the indirect measurement of the A-weighted sound level from the 1/3 octave spectrum using the partial derivatives approach. (b) Simulate the outcome of a large number of spectrum measurements by adding to each band a different random number with a normal distribution with standard deviation 1 dB, then compute the A-weighted sound level for each measurement and find the uncertainty as the standard deviation of all the results. Compare with the result of (a).
(13) In order to measure the sound power of an acoustic source in a semi-anechoic room, a hemispherical surface equidistant to the projection on the floor of the center of the source is first divided into regions of area S_k. Then the spectrum is measured with a microphone located at the geometric center of each region

yielding levels L_{ik} where i corresponds to the different bands and k to the different regions. Then the following formula is applied.

$$\overline{L_{pi}} = 10 \log\left(\frac{1}{S}\sum_{k=1}^{N} S_k 10^{L_{ik}/10}\right),$$

where $\overline{L_{pi}}$ is the mean sound pressure level of the i-th band on the whole surface S. Finally the sound power level is computed with the formula

$$L_{w} = \overline{L_{pi}} + 10 \log\frac{2\pi r^2}{S_o},$$

where r is the hemisphere radius and $S_o = 1\ \mathrm{m}^2$. Find the uncertainty in L_w given the uncertainty u_{Li} of each band level and the uncertainty on the radius r, u_r. Note: The uncertainty may differ between the different bands.

(14) A sound level meter has temperature, relative humidity, and ambient pressure coefficients $K_T = -0.001$ dB/ °C, $K_{hr} = 0.001$ dB/%, $K_{Pa} = -0.001$ dB/hPa. Suppose the environmental conditions cannot be measured but it is known from past climate statistics that the mean conditions at the location and current season are $T = 22$ °C, $h_r = 60$ %, $P_a = 1010$ hPa with standard deviations $\sigma_T = 5$ °C, $\sigma_{hr} = 15$ % and $\sigma_{Pa} = 10$ hPa. Find the measurement uncertainty due to lack of knowledge of the environmental conditions.

(15) A sound level meter has a resolution of 0.1 dB. A measurement method is implemented to measure random noise by averaging 20 instantaneous measurements of L_p using fast time constant taken at regular intervals. (a) Indicate which is the method resolution. (b) Find the uncertainty due to the resolution. Hint: The distribution is not uniform but approximately normal (Can you tell why?).

(16) Suppose we have a sound calibrator whose nominal sound pressure level within its acoustic coupler is 94.0 dB. The calibration certificate states that the calibrator has an error of 0.2 dB and an expanded uncertainty U_{95} of 0.3 dB. We perform 10 independent measurements of the calibrator tone with a free-field sound level meter (removing each time the microphone from the coupler) getting eight times a value of 94.2 dB and twice a value of 94.1 dB. Afterwards we measure 10 times the noise of an electric pump at a distance of 1 m, getting the following values: 87.3, 86.8, 86.7, 86.9, 87.1, 87.5, 87.3, 87.4, 87.2, 87.2 dBA. Assuming the sound level meter has an expanded uncertainty of 1.4 dB and a resolution of 0.1 dB, find: (a) The systematic error of the calibrator tone and its uncertainty; (b) The systematic error of the sound level meter and its uncertainty; (c) The average sound level of the pump noise corrected for systematic effects and its uncertainty.

References

Bell S (2001) Measurement good practice guide No. 11 (Issue 2). A Beginner's guide to uncertainty of measurement. National Physical Laboratory. Teddington, Middlesex, United Kingdom, 2001. http://resource.npl.co.uk/cgi-bin/download.pl?area=npl_publications&path_name=/npl_web/pdf/mgpg11.pdf. Accessed 26 Feb 2016

Brüel & Kjær (1995) Microphone handbook. Brüel & Kjær Falcon™ range of microphone products. BA 5105–12. Revision February 1995. http://www.bksv.com/doc/ba5105.pdf. Accessed 26 Feb 2016

Brüel & Kjær (1996) Mcrophone handbook—vol 1: theory. Technical documentation Brüel & Kjær BE 1447–11. Nærum, Denmark, 1996. http://www.bksv.es/doc/be1447.pdf. Accessed 26 Feb 2016

Brüel & Kjær (2014) Sound calibrator Type 4231. Brüel & Kjær BP 1311–17 Nærum, Denmark. http://www.bksv.jp/doc/bp1311.pdf. Accessed 26 Feb 2016

IEC 61672-1 (2002) Electroacoustics—sound level meters—Part 1: Specifications. International Electrotehcnical Commission

ISO-GUM (1993) Guide to the expression of uncertainty in measurement, prepared by ISO Technical Advisory Group 4 (TAG 4), Working Group 3 (WG 3), October 1993

JCGM 100:2008. Evaluation of measurement data—guide to the expression of uncertainty in measurement. Working Group 1 of the Joint Committee for Guides in Metrology. http://www.bipm.org/utils/common/documents/jcgm/JCGM_100_2008_E.pdf. Accessed 26 Feb 2016

Mohr PJ, Taylor BN, Newell DB (2012) CODATA recommended values of the fundamental physical constants: 2010. Reviews of Modern Physics, vol. 84, October–December 2012. http://physics.nist.gov/cuu/pdf/RevModPhysCODATA2010.pdf. Accessed 26 Feb 2016

Popescu A (2007) Some considerations on measurement uncertainty in LSP microphones calibration by reciprocity method. SISOM 2007 and Homagial Session of the Commission of Acoustics, Bucharest 29–31 May. http://www.imsar.ro/SISOM_Papers_2007/SISOM_2007_A_22.pdf. Accessed 10 Oct 2016

Taylor B, Kuyayt C (1994) NIST technical note 1297. Guidelines for evaluating and expressing the uncertainty of NIST measurement results. Physics Laboratory, National Institute of Standards and Technology, Gaithersburg, MD. http://www.nist.gov/pml/pubs/tn1297/. Accessed 17 Mar 2016

Chapter 3
Digital Recording

3.1 Introduction

Software-based acoustical measurements require that the acoustic signal be transformed into the electric domain and then converted to digital audio using an analog/digital converter and recorded in some physical medium. Only then it is possible to apply algorithms to complete the measuring procedure. In this chapter, we shall introduce the fundamentals of digital audio and digital recording.

3.2 Digital Audio

A digital signal is, in the first place, a signal represented with numbers instead of a continuous variable such as sound pressure or voltage. While an analog signal goes through infinite arbitrarily close values, a digital signal can only adopt a finite number of values, according to the available number of digits. Moreover, the analog signal has usually a continuous-time domain, i.e., it is defined for all time instants. If it is to be represented with numbers to be stored in a memory device, we need to sample it, i.e., to select a finite number of instants and discard the rest, getting a *discrete signal*. Thus, a digital signal is a signal that has *discrete magnitude values* in *discrete time*.

Digital signals have several advantages with respect to analog signals that make them very interesting in many applications (in particular, metrological ones):

(1) If a proper storage medium is used, the signal is virtually inalterable. This contrasts with analog recording media such as phonograph records or magnetic tapes.
(2) It is possible to make as many backup copies as desired, even successive copies of copies, all of them identical to the original one. In the case of analog recordings, each new copy implies a degradation due to noise and distortion.

F. Miyara, *Software-Based Acoustical Measurements*, Modern Acoustics and Signal Processing, DOI 10.1007/978-3-319-55871-4_3

(3) Digital signals can be transmitted by cable or wireless without being altered by the presence of noise or distortion in the transmission channel, in contrast with analog signals.
(4) In the case of noisy channels or fragile storage media (such as an optical disc), it is possible to implement sophisticated error correction algorithms based on redundancy. In analog media, redundancy can reduce but not remove signal degradation.
(5) It is possible to implement a huge variety of digital processes such as filters, noise reduction or removal, spectrum analysis, statistical analysis, integration, convolution, postprocessing, labeling, graphic display, automatic application of criteria, etc.

In contrast, the hardware required for any digital processing is generally more sophisticated, including filters, converters and digital signal processors, microprocessors, microcontrollers, or other computing devices. The time discretization implicit in the sampling process imposes, also, a severe limitation in the high frequency spectrum. Yet, the advantages largely outweigh the drawbacks.

We shall start analyzing the digitization process in its two basic stages: sampling and analog/digital conversion.

3.3 Sampling

The sampling process consists in retaining the signal value after the instant of the sample. This is achieved by means of a *sample and hold* circuit. This step is necessary because the analog/digital converters require some finite time to complete the conversion, and during that time the signal must remain constant.

Signal hold can be attained by connecting a condenser during a very brief time to the signal source (in our case, the microphone's output, previously amplified) and then disconnecting it. The condenser retains its electric charge during some time, allowing the conversion process. This process repeats itself once and again after a time T_s called *sampling period*. Figure 3.1 depicts the sampling result.

A very important issue in sampling theory is how many samples must be taken per unit time. This is called *sample rate*, F_s (also referred to as *sampling frequency*), and is given by

$$F_s = \frac{1}{T_s}. \tag{3.1}$$

From the point of view of both the memory size to store the signal and the computing power of the system that later will process the signal (number of operations per unit time), it would be preferable that F_s be kept small. Yet, a very small F_s would cause the loss of important details of the signal. Figure 3.2 shows the same signal from the example of Fig. 3.1 sampled at smaller sample rates. It can be seen that as the sample rate becames smaller, less signal details are preserved.

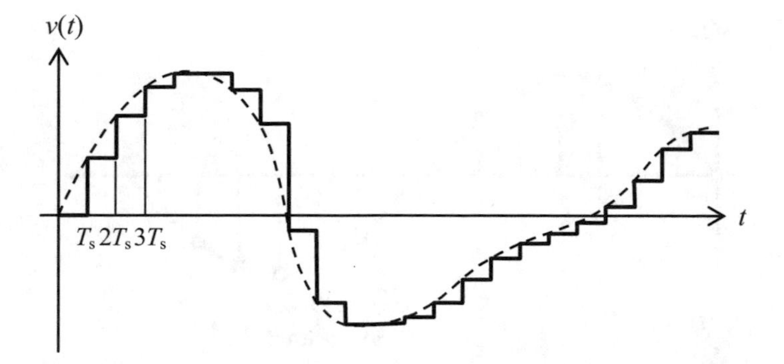

Fig. 3.1 Example of the sample and hold process. At each instant kT the sample value is updated and the new value is retained until $(k + 1)T$

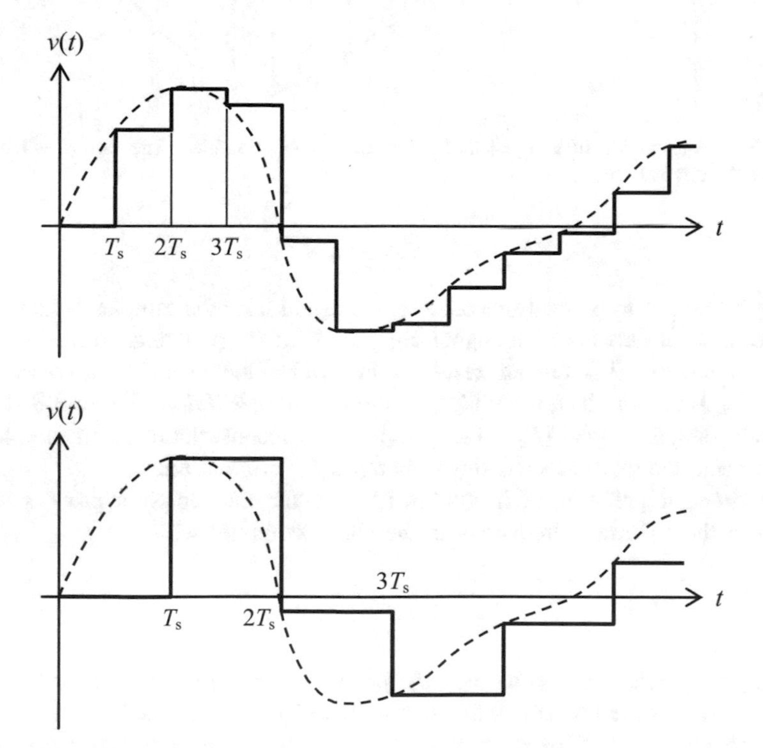

Fig. 3.2 The same signal as in Fig. 3.1 sampled at smaller sample rates. The result exhibits decreasing fidelity to the original signal

It can be proven that if the sample rate is greater than or equal to twice the maximum frequency contained in the continuous-time signal spectrum, i.e.,

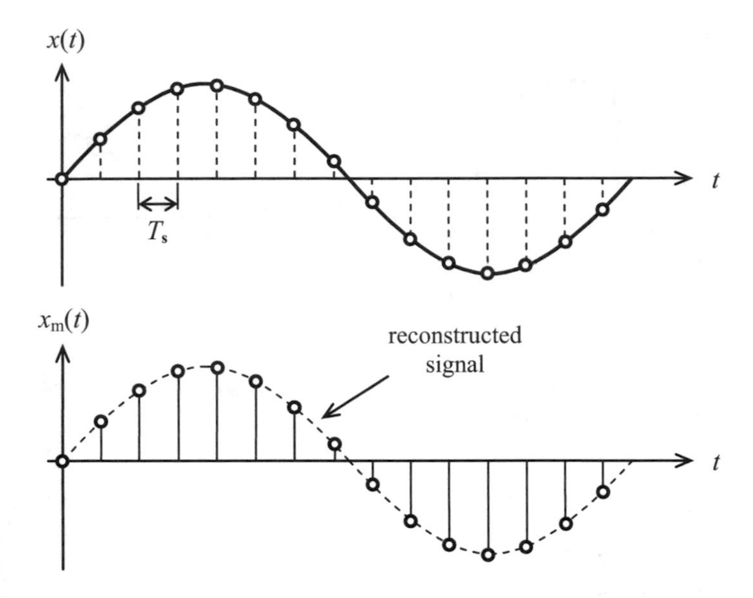

Fig. 3.3 Reconstruction of a signal that has been correctly sampled. The reconstructed signal matches the original one

$$F_s \geq 2f_{max} \tag{3.2}$$

then it is possible to perfectly reconstruct the signal from the samples, which in turn implies that all details of the signal are preserved (in particular, those of metrological interest). This crucial result is known as the *sampling theorem* (see Appendix J) and the inequality (3.2) as the *Nyquist condition*. Figure 3.3 shows a correctly sampled signal (F_s = 14.7 $f_{máx}$) and its reconstruction from samples. As can be seen, the reconstructed signal matches the original one.

Another interpretation of inequality (3.2) is that the sampling process will be correct if the maximum frequency in the signal complies with

$$f_{max} \leq \frac{F_s}{2}. \tag{3.3}$$

Frequency $F_s/2$ is called the *Nyquist frequency* and also the *Nyquist limit*.

If the signal does not satisfy the inequality (3.3), the signal will be affected by a phenomenon called *aliasing*, consisting in the appearance, in the reconstructed signal, of frequencies that were not present in the original signal. These frequencies are called *alias frequencies* and are equivalent to reflections over $F_s/2$ of every frequency exceeding $F_s/2$. Given a frequency f_o such that

$$\frac{F_s}{2} < f_o < F_s, \tag{3.4}$$

the reconstructed signal will be plagued with an alias frequency

$$f_{\text{alias}} = \frac{F_s}{2} - \left(f_o - \frac{F_s}{2}\right) = F_s - f_o \tag{3.5}$$

This phenomenon is illustrated in Fig. 3.4, where a 35 kHz signal has been sampled at a sample rate F_s = 40 kHz. The reconstruction contains a 5 kHz tone.

Frequencies above F_s are reflected over $F_s/2$, then over 0, then over $F_s/2$, and so on, so that the alias frequency will fall always inside the interval from 0 to $F_s/2$. If

$$\begin{aligned} F_o &= f_o \bmod (F_s/2) \\ k &= \frac{F_o}{F_s/2} \end{aligned} \tag{3.6}$$

where mod is the modulus operation (the rest of the integer division). Then we can write,

$$f_{\text{alias}} = \begin{cases} \frac{F_s}{2} - F_o & k\,\text{odd} \\ F_o & k\,\text{even} \end{cases} \tag{3.7}$$

It is important to bear in mind that because of this effect the maximum frequency in inequality (3.3) does not refer only to the maximum frequency *of interest*, but rather to the maximum frequency *effectively* present in the signal to be sampled, even if that frequency comes from some inaudible noise. If the Nyquist criterion is not

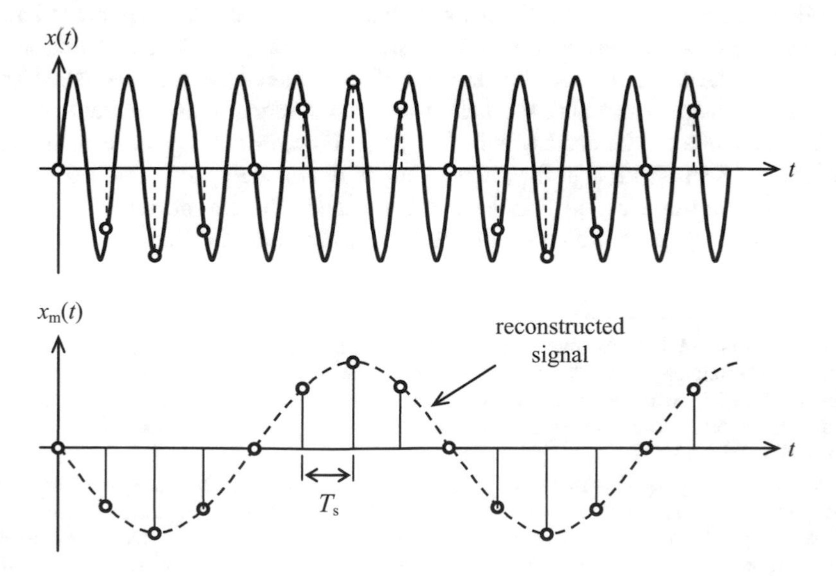

Fig. 3.4 Effect of sampling a signal with a sample rate smaller than twice the maximum frequency in the signal (undersampling). A 35 kHz signal is sampled at a sample rate of 40 kHz; When attempting to reconstruct it, we get an *alias* frequency of 5 kHz

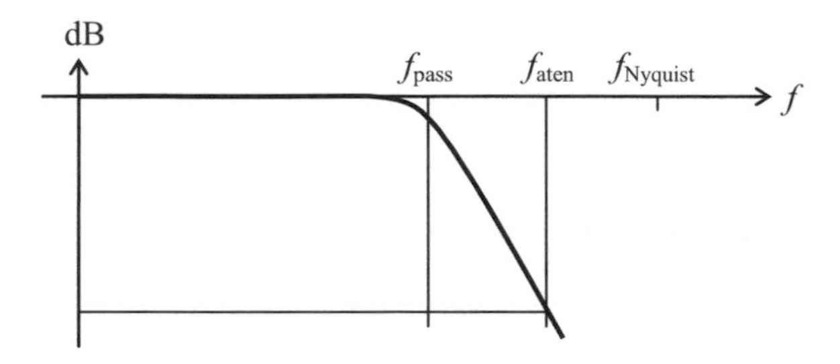

Fig. 3.5 Frequency response of a low pass anti-aliasing filter. The attenuation frequency in the stop band must be under the Nyquist frequency

satisfied, the reconstructed signal will contain spurious frequencies in the audible range. A typical case is when the audio signal is contaminated by wide-band electric noise such as thermal noise generated in the microphone or in the electronics.

In order to prevent aliasing, a necessary stage of the signal conditioning block previous to sampling is a low pass filter that effectively cuts off any spectral content above the Nyquist limit, $F_s/2$. This filter, called *anti-aliasing filter* (see Fig. 3.5), must attenuate signal components above $F_s/2$ enough to turn them irrelevant as compared to other uncertainty sources.

Although the anti-aliasing filter might seem to be a relatively simple part of the whole system, in practice is the weakest link, causing the greatest distortion in a software-based measurement system. Indeed, in order to save storage space, the sample rate must be chosen only slightly above the Nyquist condition. For instance, for audio signals whose relevant limit is close to 20 kHz, a frequency of 44.1 kHz is often used.[1] This implies a Nyquist frequency of 22.05 kHz. This allows a very narrow margin of 2.05 kHz for the filter to attenuate from unity gain to a typical stop band gain of −90 dB. The filter turns to be of a very high order; for instance, an elliptic filter of order 13 or a Chebychev filter of order 29. These filters have a quite strong transient response, causing distortion in random signals. Additionally, some of the second-order cells that are part of such filters have a large quality factor Q, which translates into high resonant peaks and dynamic range limitations.

This problem can be dealt with in two ways. First, increasing the sample rate to 48 kHz or even 96 kHz, at the expense of more storage space. The second is to work with oversampling[2] techniques that use much higher sample rates, pushing the Nyquist limit upwards, which simplifies the specification of the anti-aliasing filters.

[1]The exact value of 44.1 kHz obeys to the fact that the first digital audio tape recorders used the video tape technology. They located three 12 bit samples within each horizontal scan line by means of a PCM (pulse code modulation) adaptor. There were 294 active lines at a vertical scan frequency of 50 Hz (an interleaved scan equivalent to 588 lines at 25 kHz). Then we have $3 \times 294 \times 50 = 44\,100$.

[2]It consists in sampling at a much higher sample rate (for instance, 128 times the target sample rate), but using a very low bit count (typically, 1 bit). Instead of representing the complete value of the sample, what is represented is the difference between successive samples. Due to the extremely high sample rate, the increase between successive samples is very small, that is why it can be represented with only 1 bit. The complete value is computed by digital integration, and finally the signal is subsampled by the same oversampling factor.

3.4 Digitization

Once sampled, the discrete signal is converted to the digital domain using an analog/digital converter whose mission is to assign to each analog sample a number that is proportional to the analog value through a conversion constant. Although we could use a number system to any base, all known converters use the binary system, since it allows to use binary logic circuitry with two voltage levels: 0 and V_{cc}. This is equivalent to using the concept of open and closed switches, which correspond, respectively, to the binary digits 0 and 1. This allows a very high noise immunity since it would be necessary a noise amplitude exceeding $V_{cc}/2$ in order that a level could be confounded with the other.[3] Each binary digit is called a *bit* (contraction of binary digit). A group of bits is called generically *digital word*. Most often digital words have a multiple of 8 bits.

In order to carry out the conversion, we first choose an analog reference voltage V_{ref}. Then the reference is divided into 2^n steps of amplitude $V_{ref}/2^n$, where n is the number of bits or *resolution* of the digital data.

To convert nonnegative signals, the digital zero (i.e., word of n zeros) is put in correspondence with the analog zero, and the digital maximum, $2^n - 1$ (i.e., a word of n ones) with the analog maximum $V_{ref}\,(1 - 1/2^n)$. For bidirectional signals, the digital 0 corresponds to $-V_{ref}/2$ and the digital maximum corresponds to $V_{ref}\,(1/2 - 1/2^n)$. In this case, the analog zero corresponds to 2^{n-1}, i.e., a one followed by $n - 1$ zeros.

Figure 3.6 depicts the analog/digital conversion of a sine wave of maximum amplitude ($V_{ref}/2$) for resolutions of 1 bit, 2 bit, 3 bit, and 4 bit. In this case, to each analog value corresponds the highest step not exceeding it:

$$D = \begin{cases} 0 & x < -V_{ref}/2 \\ \left[\frac{x - (-V_{ref}/2)}{V_{ref}} 2^n\right] & -V_{ref}/2 \le x < V_{ref}/2\,, \\ 2^n - 1 & V_{ref}/2 \le x \end{cases} \tag{3.8}$$

where D is the digital data and $[u]$ is the integer part of u. The definitions for $x < V_{ref}/2$ and $V_{ref}/2 \le x$ correspond to saturation effects. We see that as the number of bits increases, the signal will be represented with greater fidelity.

A more symmetric approach is to locate the analog zero half way between the equivalent to digital values $2^{n-1} - 1$ and 2^{n-1}. However, this solution is seldom used because it is more complex to implement and it cannot represent correctly a null signal.

The lack of symmetry, which is very noticeable for low resolution digitization, becomes irrelevant for higher resolutions such as 16 bit.

---

[3] From the point of view of Information Theory, it can be shown that most efficient code, i.e., the one that maximizes the amount of information per digit is a hypothetical code to the base $e = 2.71828\ \ldots$, so the most efficient code of integer base is the ternary code (base 3).

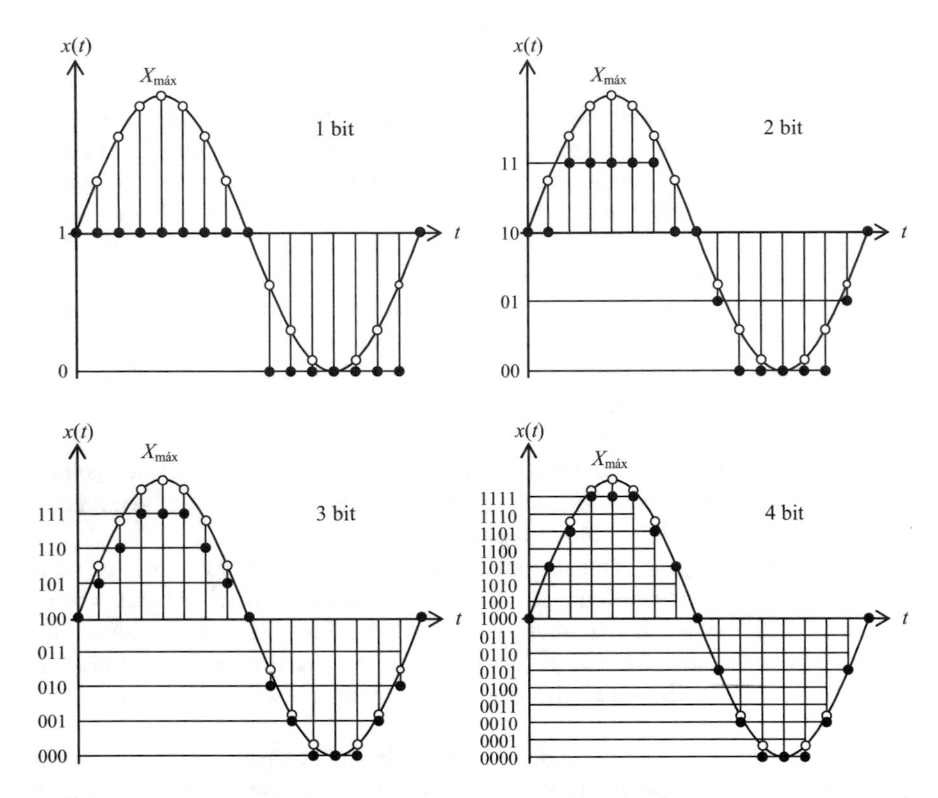

Fig. 3.6 Examples of analog/digital conversion with resolutions of 1 bit, 2 bit, 3 bit, and 4 bit. The *empty circles* represent the exact (analog) value of each sample, and the *filled circles*, the corresponding digital value. In this example, each analog sample has been approximated by the closest step that does not exceed the analog value

3.5 Signal-to-Noise Ratio

When the amplitude of the signal is relatively high with respect to the least significant step (the analog step that corresponds to the least significant bit, 1 LSB), the difference between the original signal $x(t)$ ad the digitized signal $x_d(t)$ is a quasi random variable that behaves as a noise:

$$\varepsilon(t) = x(t) - x_d(t). \tag{3.9}$$

In Fig. 3.7 there is an example in which a sine wave of maximum amplitude has been digitized with a resolution of 6 bit.

The portions close to the peak present a distinct pattern, but as the resolution rises it reduces to a smaller fraction of signal duration. Hence, the residue will approach random noise with uniform distribution between 0 and $X_{max}/2^{n-1}$. Its effective value, equal to its standard deviation (see Appendix B), is

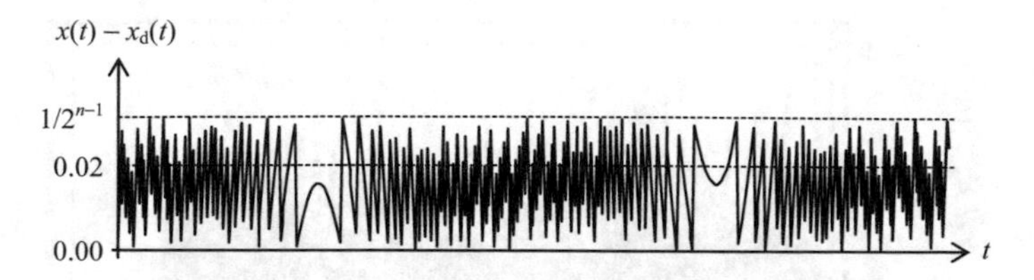

Fig. 3.7 Difference or error between a sine wave signal with amplitude 1 and its digitized version with resolution $n = 6$ bit. NOTE: To prevent confounding time discretization effects the sampling rate was much higher than the signal frequency

$$E_{ef} = \frac{X_{max}/2^{n-1}}{\sqrt{12}}. \tag{3.10}$$

The effective value of a sine wave of maximum amplitude is

$$X_{ef} = \frac{X_{max}}{\sqrt{2}}, \tag{3.11}$$

from which the signal-to-noise ratio may be calculated as

$$S/R|_{dB} = 20 \log \frac{V_{ef}}{E_{ef}} = 20 \log \sqrt{\frac{3}{2}} + 20 \log 2^n$$

i.e.,

$$S/R|_{dB} = 1,76 + 6.02\, n. \tag{3.12}$$

A frequently used rule of thumb to get an approximation of the digitization signal-to-noise ratio is

$$S/R|_{dB} \cong 6\, n. \tag{3.13}$$

For instance, for the standard resolution of 16 bit, the digitization signal-to-noise ratio is close to 96 dB, high enough to correctly represent all the quantitative aspects of the acoustic phenomena that are relevant to human perception and metrological needs.

3.6 Low-Level Distortion

When signal amplitude is small, of the order of a few LSB's, the digitized signal presents severe distortion, as can be seen in both upper spectra of Fig. 3.8.

This distortion is clearly audible and in the case of more than one tone, it can cause distinct distortion products (intermodulation distortion). It is a quite unnatural

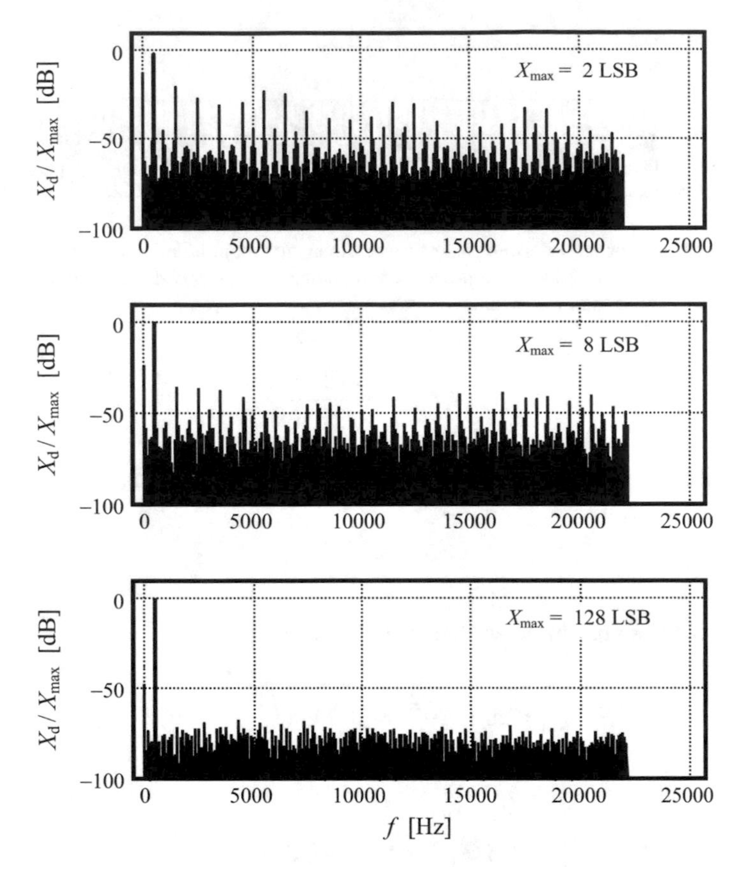

Fig. 3.8 Spectrum of a 500 Hz sine wave digitized for several amplitudes expressed in LSB. Shown values are equivalent to digitizing a maximum amplitude sine wave with 2 bit, 4 bit and 8 bit. In the first two cases, the harmonics of the original signal are clearly observed. In the case of 8 bit, the spectrum presents no discernible harmonics

type of distortion since instead of happening near saturation, it arises at very low levels. In general, starting from a resolution of 8 bit the distortion effect for a tone of maximum amplitude is no longer a problem, behaving approximately as white noise (see Fig. 3.8, bottom). However, low amplitude tones will still present distortion. Nowadays all digitization systems support at least 16 bit and many of them even 24 bit, so this distortion only takes place at extremely low levels.

3.7 Dither

A way to attack the problem of low-level distortion is to add a small amount of random noise $x(t)$ before digitizing. This noise is called *dither* and the process of adding it, *dithering*. The effective value of that noise is, in general, a fraction of an

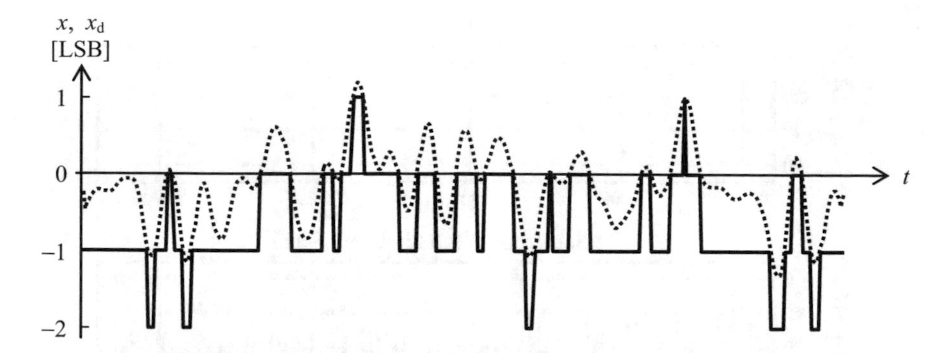

Fig. 3.9 An example of dithering signal (a filtered Gaussian noise with effective value of 0.6 LSB) and its digitized version

LSB, typically between 0.2 LSB and 0.7 LSB, and its statistical distribution $f(x)$ may be normal, triangular or uniform. If x_{d} is the digital version of the noise, we can compute its effective value from

$$\begin{aligned} X_{\mathrm{d,ef}}^2 &= \int_{-\infty}^{\infty} x_{\mathrm{d}}^2(x) f(x) \mathrm{d}x = \sum_{k=-\infty}^{\infty} \int_{k}^{k+1} x_{\mathrm{d}}^2(x) f(x) \mathrm{d}x \\ &= \sum_{k=-\infty}^{\infty} \int_{k}^{k+1} k^2 f(x) \mathrm{d}x = \sum_{k=-\infty}^{\infty} k^2 \int_{k}^{k+1} f(x) \mathrm{d}x \\ &= \sum_{k=-\infty}^{\infty} k^2 (F(k+1) - F(k)) \end{aligned} \tag{3.14}$$

This series converges very quickly. In Fig. 3.9, an example of dither and its digitization is shown. In this case, for the sake of clarity the noise has been previously filtered.

Adding dither tends to decorrelate the strong nonlinear effect that takes place when the signal amplitude is a few LSB's. The net result is that it no longer behaves as distortion but as noise. The advantage is that while distortion concentrates its energy in isolated harmonics or partials, noise spreads it over the whole spectrum, which is much more tolerable.

In Fig. 3.10, there is an example that illustrates this effect for a signal consisting in two tones of frequencies f_{o} and 1.25 f_{o} and amplitudes 1.1 LSB and 0.7 LSB. The second panel shows that after digitization there are a large number of distortion products present, superposed to an approximately white noise. When a Gaussian dither of 0.4 LSB is added, the spectrum before digitization presents a white noise that after digitization is practically the same. All distortion products have disappeared.

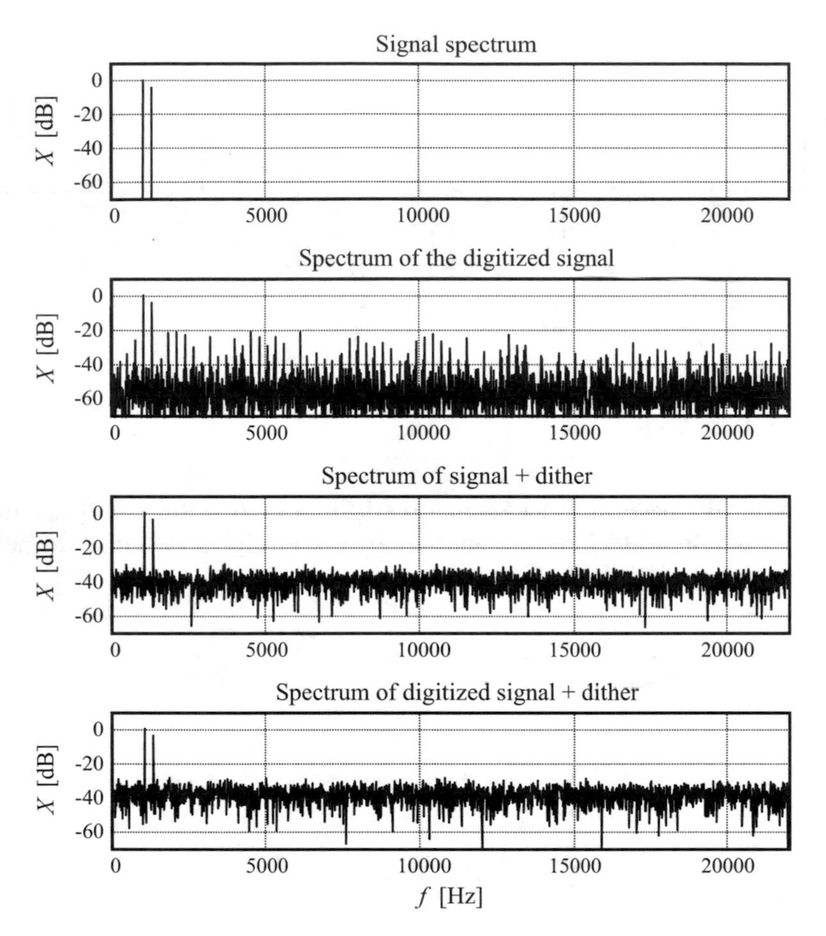

Fig. 3.10 Linearizing effect of dithering applied to a signal that contains two tones of 1077 and 1346 Hz and amplitudes 1.1 LSB and 0.7 LSB respectively. From *top to bottom* Line spectra of signal before and after digitization, and then the signal with the addition of Gaussian dither of 0.4 LSB, before and after digitization

Table 3.1 presents the effect of several effective values of a Gaussian dither on the signal-to-noise ratio for a sine wave of maximum amplitude. The S/N falls in a larger proportion than before digitizing. It is the price to pay for changing distortion for wide-band noise.

From the metrological point of view, this slight reduction in the signal-to-noise ratio is in general irrelevant, provided that the gain level is adjusted to make use of the full dynamic range available. In the case of signals with a wide dynamic range (such as symphonic music) it can happen that in order to prevent saturation in high intensity portions of the signal, the gain is kept relatively low. In this case the lowest levels could be partially masked by noise. A possible solution to this problem is to work with a higher resolution. Modern digital recorders and computer

Table 3.1 Effective value in LSB of a Gaussian dither and of its digitized version. Also shown is the signal-to-noise ratio that results for a sine wave of maximum amplitude

X_{ef}	$X_{d,ef}$	$S/N\|_{dB}$ (16 bit)
0.20	0.70711	96.330
0.25	0.70720	96.328
0.30	0.70832	96.315
0.35	0.71313	96.256
0.40	0.72446	96.119
0.45	0.74335	95.895
0.50	0.76893	95.602
0.55	0.79949	95.263
0.60	0.83343	94.902
0.65	0.86963	94.533
0.70	0.90745	94.163
0.75	0.94650	93.797
0.80	0.98658	93.437
0.85	1.0275	93.083
0.90	1.0693	92.738
0.95	1.1117	92.400
1.00	1.1547	92.070

sound cards offer resolutions up to 24 bit. However, while such a large resolution ideally allows a signal-to-noise ratio of 24 × 6 dB = 144 dB, the accompanying analog electronics does not currently allow such an outstanding noise performance.[4] Thus, at best the least significant bits are representing the noise of the electronics and the transducer itself, which finally plays the role of a sort of natural dither.

3.8 Jitter

Ideally, a sampling system would take the samples at perfectly regular intervals. In any real system, this is not possible because of clock frequency instability and the presence of noise at several analog and digital stages. The result is a random

[4]Indeed, assuming a maximum signal of 20 V, a signal-to-noise ratio of 144 dB implies that the output noise must be of 1.26 μV. For a gain of 100, it is equivalent to an input noise voltage of 12.6 nV. Assuming thermal noise, the effective noise voltage of a resistance R is $V_{ef} = \sqrt{4kTRB}$, where $k = 1.38 \times 10^{-23}$ J/K is Boltzmann's constant; T, the absolute temperature; and B, the bandwidth. It turns out that at ordinary temperature the input resistance should be 0.5 Ω, hence the microphone impedance should be at least one or two orders of magnitude smaller. It is quite a challenge for state-of-the-art technology (as of 2017).

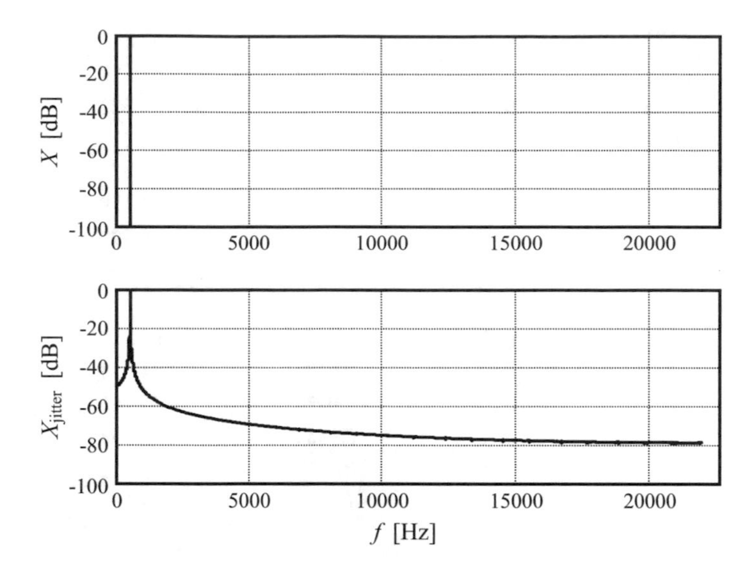

Fig. 3.11 Equivalent noise caused by the presence of jitter. In this example, a tone of frequency $f_o = 44{,}100 \times 100/8192$ Hz = 538.33 Hz has been modulated by a Gaussian noise $n(t)$ so that the instantaneous frequency is $f(t) = f_o(1 + n(t))$. The effective value of $n(t)$ is 0.005. Line spectra with $N = 8192$ of both the signal an its version with jitter are shown

frequency or phase modulation of the sample rate known as *jitter*. This effect is equivalent to the presence of noise superposed to the signal.

Figure 3.11 depicts the spectrum of a tone

$$x(t) = \text{sen}\, 2\pi f_o t$$

and the long-term average spectrum of the same tone with frequency modulation by a Gaussian noise:

$$f(t) = f_o(1 + n(t)).$$

As can be seen, the spectrum of the original signal has been contaminated by a nonconstant wide-band noise. The energy content of the noise is concentrated around the original frequency f_o.

The jitter due to noise in the clock transmission lines or to instability in the clock frequency at the receiver or player can be almost completely removed by means of buffering and re-synchronization. However, the jitter added during the sampling process cannot be corrected since there is no way to know the actual values of the signal at the ideal sampling instants.

3.9 D/A and A/D Conversion

3.9.1 Digital/Analog Conversion

Although it could seem that from the point of view of software-based metrology we are interested only in analog/digital conversion, it is frequently necessary to generate digital signals (like a tone sweep) that have to be later converted to the analog domain for their use as test signals.

Digital/analog conversion is implemented in general by summing weighted electric currents (with factors $1/2^k$) that go through switches that are on if the corresponding bit is 1, and off if it is 0. Figure 3.12 shows a widely used circuit, where n is the bit resolution of the digital data to convert. It is a clever structure called *R-2R ladder network.* By alternating series resistors R with parallel resistors $2R$, it achieves successive current dividers by factors of 2. All currents are proportional to a reference voltage V_{ref}.

Ideally, we get

$$v_{\text{o}} = -\frac{V_{\text{ref}}}{R}\sum_{k=1}^{n}\frac{d_k}{2^{n-k+1}}. \tag{3.15}$$

where d_k is the k-th bit, starting from the least significant one (LSB).

A problem that plagues this kind of circuit is the need of a very tight tolerance in the resistors. If, for example, the tolerances were 1 %, a maximum positive error in the first $2R$ resistor and a maximum negative error in the rest would imply that a

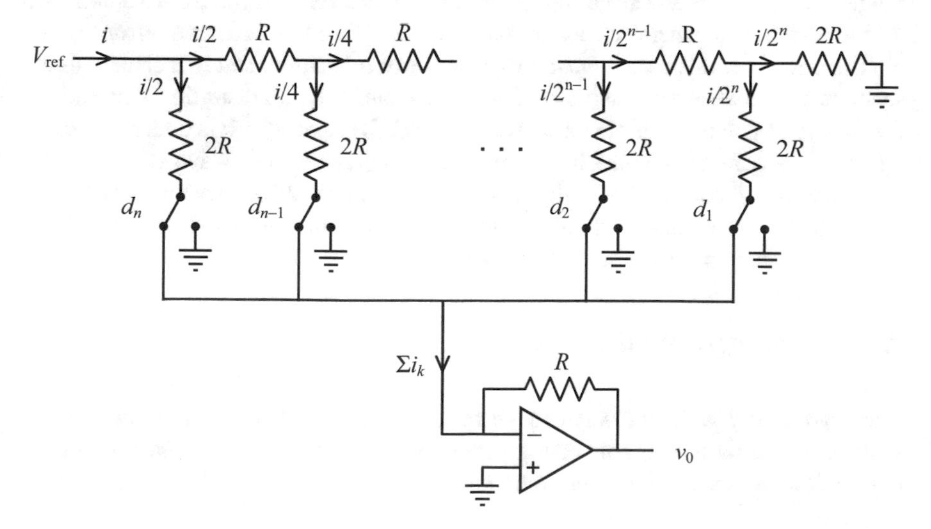

Fig. 3.12 Current-mode R-2R analog/digital converter. The virtual ground at the inverting input of the operational amplifier warrants that the halving property of the R-2R network holds whatever the switches states

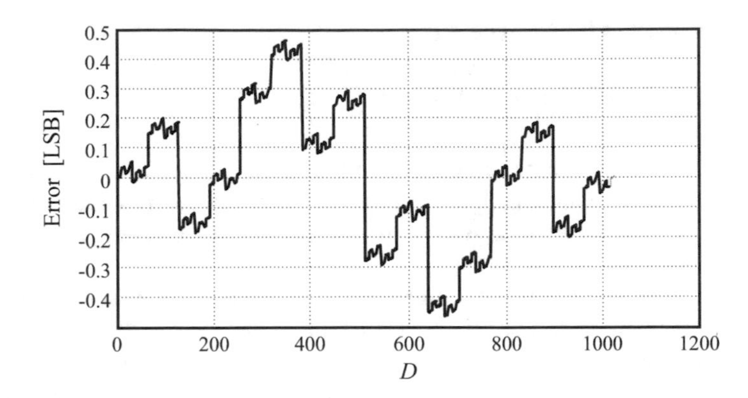

Fig. 3.13 Example of the error (in LSB) in a 10 bit *R*-2*R* converter when the resistor tolerance is 0.2 % (this tolerance corresponds to a relative uncertainty with $\sigma = 0.001$ for a 95 % confidence)

change from 0111 1111 1111 1111 to 1000 0000 0000 0000, instead of causing an output increment of 1 LSB would produce a 0.5 % decrement, i.e., −328 LSB for a 16 bit converter!

In order to achieve much tighter tolerances, the most critical resistances are adjusted automatically during the manufacturing process using a technique known as *laser-trimming*. It consists in splitting each resistor into several segments that are short-circuited by means of laser burning while the resistance value is being monitored until specifications are guaranteed.

Figure 3.13 shows an example of the error caused by resistor tolerances, obtained by Montecarlo simulation.[5] The error is plotted versus the digital input D, in the range from 0 to $2^n - 1$. The curious pattern is completely changed from a simulation to the next, but in general we can arrive at a bound for the maximum error.

The error is in this case a *linearity error*, since it is equivalent to a change in the scale factor between successive samples. This causes some distortion that adds to the distortion due to the digitization process itself. However, if it is under 1 LSB its effect is, in general, negligible. In this example, for a sine wave of maximum amplitude the distortion due to digitization plus a dither of 0.3 LSB is 0.116 %. If we include the nonlinearity error due to resistor tolerance it increases to 0.127 %. The difference is not significant (Fig. 3.13).

3.9.2 Analog/Digital Conversion

There are many different ways to perform an analog/digital conversion, but in systems requiring a high conversion rate they reduce to only two: Flash (parallel) conversion and successive approximation.

[5]The Montecarlo method consists in selecting the parameter values (in this case, the resistances) by means of a random number generator with the same statistical distribution as the parameters.

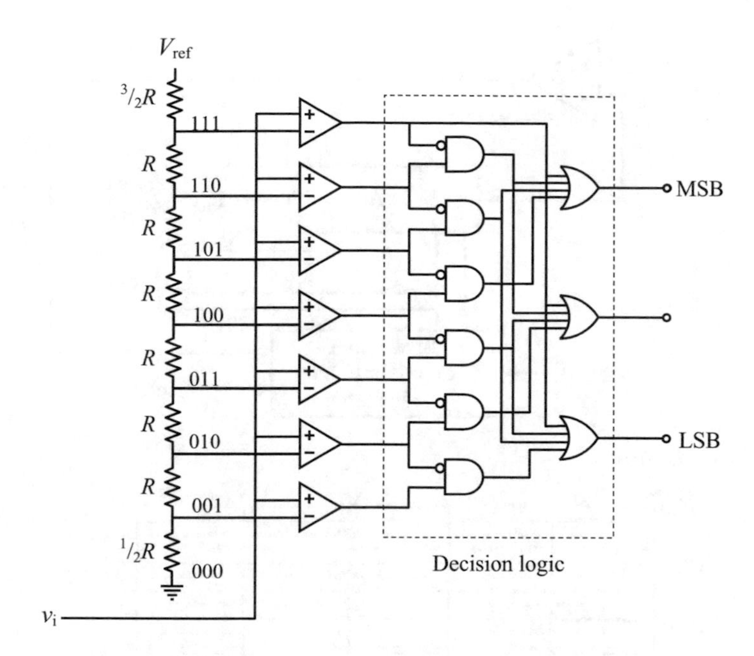

Fig. 3.14 Structure of a 3 bit flash (parallel) analog/digital converter

Flash converters consist in a series of voltage comparators that compare the input voltage simultaneously with each one of the 2^n possible levels with n bit. These levels are obtained using a multiple resistive divider. The comparator outputs are 1 for levels smaller than the input and 0 for the greater ones. Then a combinational logic circuit transforms the 2^n intermediate values in an n bit binary representation. Figure 3.14 illustrates, as an example, a 3 bit converter.

The main advantage of the flash converters is their high conversion speed, since the only latency (input-output delay) to complete the conversion is the commutation time of the comparators and the delay of the logic circuit.

The main drawback is that for high bit resolutions the number of comparators and the complexity of the logic circuit grow exponentially. There are flash converters up to 10 bit. These implementations often divide the problem by performing first a range preselection, and then an assignment of upper and lower reference values of the resistive divider according to the range.

Successive approximation converters (Fig. 3.15) compare the input analog signal with the output of a digital/analog converter that is fed with digital values that approximate in a dichotomous way the final value. In the first step, a 1 is applied to the most significant bit (MSB) of the subsidiary D/A converter and 0 to the other bits. If the input is greater than the D/A converter output that bit is confirmed. Otherwise, it is assigned a definitive value of 0. In the next step, a 1 is applied to the following bit. Again, if the input is greater than the D/A converter output, the second bit is confirmed, otherwise it returns to 0. This process is iterated until the LSB is reached. As can be seen, the complete conversion requires n steps. The device that runs this algorithm is called *successive approximation register*.

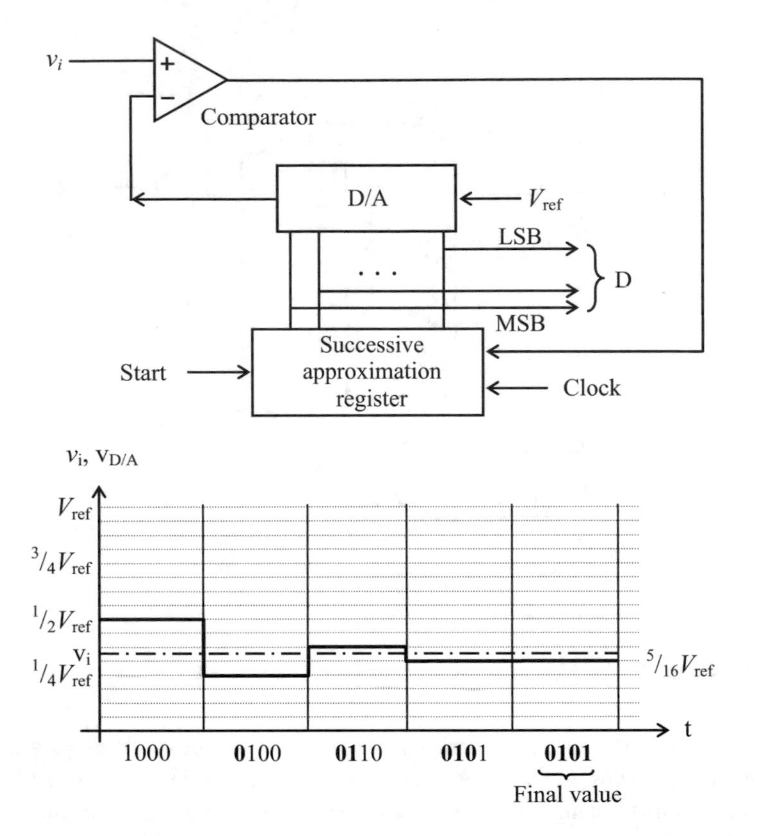

Fig. 3.15 *Top* Block diagram of a successive approximation analog/digital converter. *Bottom* Example of the approximation process in the case of a 4 bit converter. Digits in **bold** font represent those that at each stage are definitive

It is to be noted that unlike the flash converter, in this case it is required that the input stays rigorously constant, otherwise we could have gross errors. Indeed, once the most significant bits have been set, they cannot be changed until the whole conversion cycle ends, so the algorithm keeps attempting to find the best possible approximation with the rest of the bits, which may easily conduct to a meaningless result. This is the reason why a *sample and hold* is required at the input.

3.10 Pulse Code Modulation (PCM)

When analog/digital conversion is realized repeatedly over time, we get a synchronized sequence of digital data. That sequence may be represented in parallel, using a wire for each bit (with a total of n wires), where the voltage can be 0 or V according to the value of the bit, or can be represented in series, using a single wire which transmits the n bits in succession (the bit rate is equal to n times the

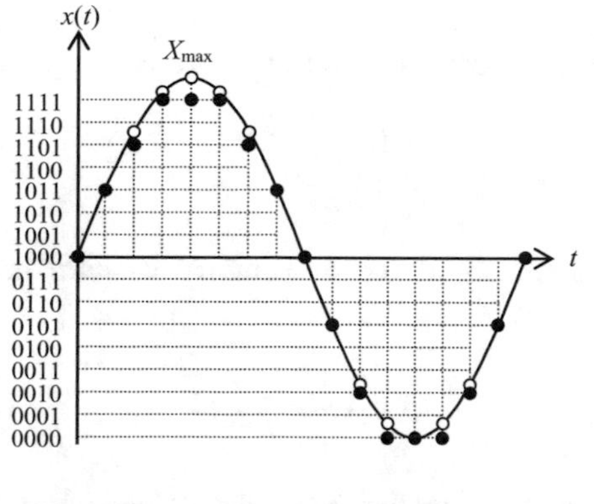

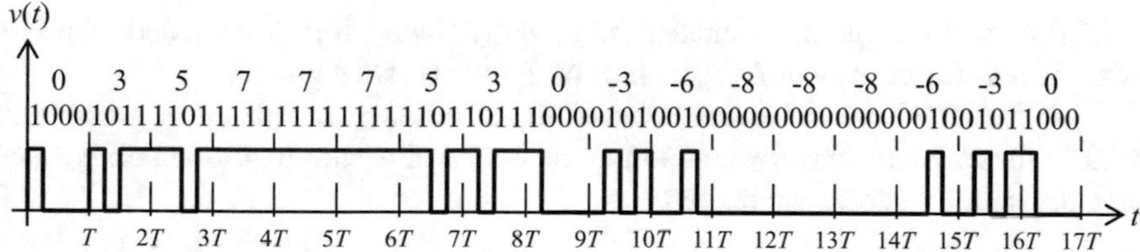

Fig. 3.16 An example of pulse code modulation (PCM). A sine wave of maximum amplitude has been coded with 4 bit and a sampling period T. The bits are represented as low (0) and high (1) voltage levels. Each bit requires 1/4 of the sampling period. Above the binary data is the corresponding decimal value

sample rate). This process, called *pulse code modulation*, *PCM*, is shown in Fig. 3.16.

In this code, it is necessary to guarantee the synchronism of the successive bits, since the displacement of a single bit turns the whole signal meaningless. Frequently, there are additional bits, such as a parity bit to detect isolated errors, as well as bits intended for keeping the DC component half way between both signal levels. Sometimes there is a thorough recodification of the signal according to an equivalence table where the bits are not interpreted as belonging to a positional number system any longer. The advantage is that it is possible to add redundancy as well as error-correcting capabilities. It is also possible to add a specific synchronization code. In these cases some processing block will decode the information for later use.

3.11 Differential Pulse Code Modulation (DPCM)

Instead of digitizing the sampled value of the signal at a given instant, sometimes it is convenient to digitize the difference between the present sample and the previous one. The result is known as *differential pulse code modulation*, *DPCM*. In the case of signals with bounded variation rate, this may imply an interesting reduction of the resolution. Suppose, for instance, a signal $v(t)$ of maximum amplitude V_{max} that has been filtered by a first-order low pass filter with time constant τ. Then its maximum slope (corresponding to a transition from $-V_{max}$ to V_{max}) is

$$\left.\frac{dv}{dt}\right|_{max} = \frac{2V_{max}}{\tau}$$

If the sampling period is T, the maximum transition between to successive samples will be

$$\Delta v|_{max} = 2V_{max}\frac{T}{\tau}.$$

If $T \ll \tau$, $\Delta v|_{max}$ is much smaller than $2V_{max}$. Hence, it may be coded with less bits. Indeed, for each twofold reduction of T we can save one bit.

Note that from the difference between the successive samples it is possible to retrieve the original signal by computing the cumulative sum of all such differences plus the initial value of the signal:

$$v(t_n) = v(t_1) + \sum_{k=1}^{n-1} (v(t_{k+1}) - v(t_k)). \tag{3.16}$$

The conversion of the difference between present and previous samples implies that one has to have both analog values available at the same time. A way to do that would be to use an analog delay line with two cascaded sample and hold circuits. An analog subtractor would get the difference to convert. This is, however, impractical.

A better way to do so is as shown in Fig. 3.17, where the previous analog value is reconstructed from the digitized difference. First, it is converted to the analog domain and then it is integrated. The latch block retains the difference during one sampling period.

Figure 3.18 shows the waveforms at different points of the preceding block diagram during the conversion of a sine wave. At the initial instant, the value $x(0) - \int y(t)\mathrm{d}t$ is sampled and converted to the digital domain (it is 0 as $x(0) = 0$ and $\int y(t)\mathrm{d}t = 0$, since the integrator has not received any signal yet). This result is

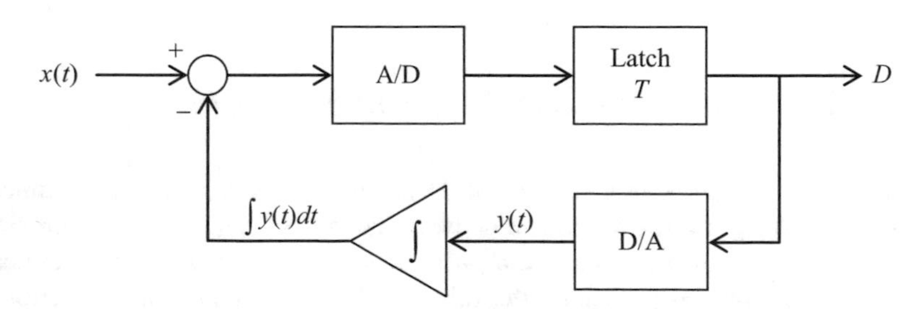

Fig. 3.17 Block diagram of a DPCM converter. The A/D converts the difference between the present sample and the reconstruction of the previous one

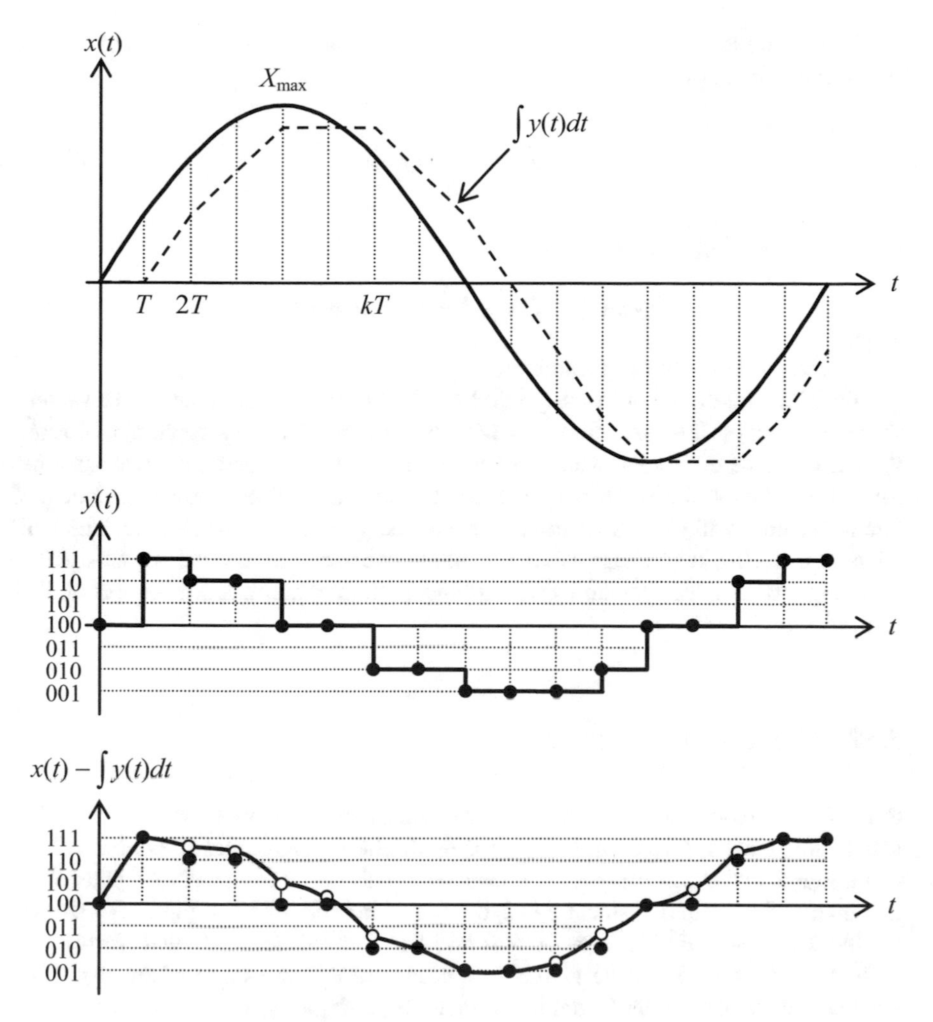

Fig. 3.18 Waveforms of the DPCM converter for a sine wave of maximum amplitude ($X_{max} = X_{ref}$). The waveform in *dashed line* is the integrator's output

retained by the latch during a sampling period T, converted to the analog domain and integrated (the result is so far 0 because the integrand is 0). At the end of the first sampling period, the new difference $x(T) - \int y(t)\mathrm{d}t$ is sampled, converted, and retained. This new value is converted to the analog domain and integrated. Because the integrand is constant, the integral is a ramp. Note that at the end of each period the integral approximates, the input at $t - T$, so $x(T) - \int y(t)\mathrm{d}t \approx x(T) - \mathrm{x}(t - T)$. At the end of this second sampling period, the third sample of $x(t) - \int y(t)\mathrm{d}t$, is taken, converted to the digital domain and retained until $3T$. This is integrated, creating a new ramp which is subtracted from the signal and, at $3T$, sampled and integrated. This process is repeated, creating the successive values of the DPCM code.

From the digital DPCM signal, it is possible to retrieve the PCM signal by means of a discrete integral or cumulative sum:

$$D_{\mathrm{PCM}}(k) = \sum_{h=1}^{k} D_{\mathrm{DPCM}}(h), \tag{3.17}$$

or by the following recursive expression:

$$D_{\mathrm{PCM}}(k) = D_{\mathrm{PCM}}(k-1) + D_{\mathrm{DPCM}}(k). \tag{3.18}$$

that is computationally more efficient.

DPCM modulation requires in general less bits than PCM modulation. However, in order to exploit this possibility it is necessary to increase the sample rate or limit the derivative of the signal, which in turn implies that the spectral content must be limited way below the Nyquist frequency. The bit rate will not improve, since the bits per sample will be smaller but at the expense of a larger sample rate. The real advantage is that if the sample rate is much greater than needed to satisfy the sampling theorem, then the anti-aliasing filter is simpler and has less impact on the in-band signal.

3.12 Delta Modulation (DM)

Since DPCM modulation requires less bits than PCM modulation, we can push the idea to the extreme were a single bit is required. The analog/digital converter will be just a sign detector that assigns 1 to positive signals and 0 to negative signals. The modulation thus obtained shall be called *delta modulation*, *DM*. Figure 3.19 shows the block diagram of a delta modulator and Fig. 3.20 displays the waveforms.

In order for the idea to be viable, it is necessary that the slope of the signal be low enough so that the maximum increase during a single sampling period does not exceed 1 LSB, i.e.,

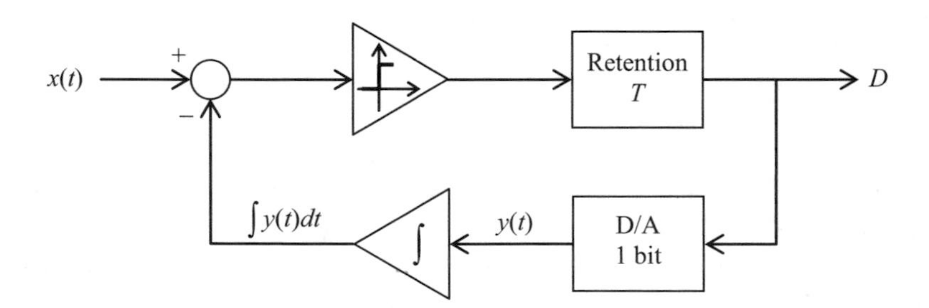

Fig. 3.19 Block diagram of a DM converter. The D/A converter simply shifts downward ½ LSB the binary output so that a 0 corresponds to –½ LSB and a 1 to ½ LSB

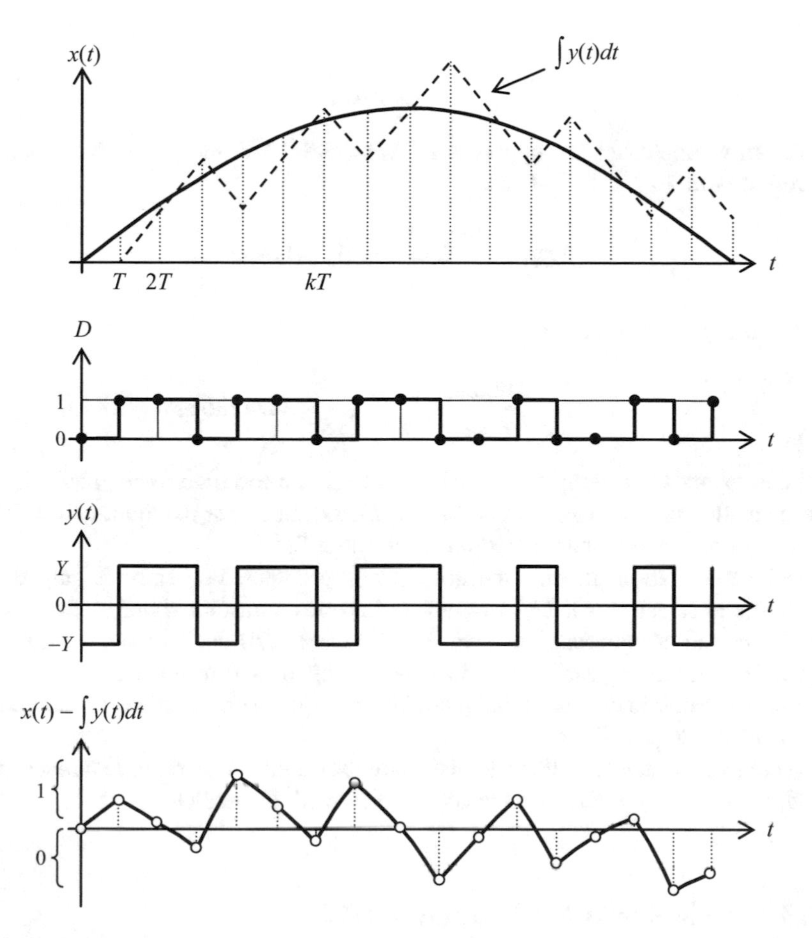

Fig. 3.20 Waveforms in the DM converter for a sine wave. The signal in *dashed line* is the integrator's output

$$\left|\frac{\mathrm{d}x}{\mathrm{d}t}\right| < \frac{1\,\mathrm{LSB}}{T}. \tag{3.19}$$

This will require a very high sample rate.

It is interesting to note that unlike other types of coding, in delta modulation there is no intrinsic amplitude limit, since a signal could grow indefinitely without exceeding the maximum slope of 1 LSB per sampling period. However, for a given frequency, for instance the maximum frequency of interest in audio (20 kHz), the maximum slope is proportional to the amplitude and to the frequency:

$$\left|\frac{\mathrm{d}}{\mathrm{d}t} V \sin \omega t\right| \leq V\omega. \tag{3.20}$$

From Eqs. (3.19) and (3.20), we get the following upper bound for the sampling period:

$$T < \frac{1\,\text{LSB}}{V_{\text{máx}}\omega_{\text{máx}}} \tag{3.21}$$

As an example, let us suppose a 20 kHz and 10 V signal which we want to represent with 16 bit. In that case

$$1\,\text{LSB} = \frac{10\,\text{V}}{2^{15}} = 0.00030518509\,\text{V}$$

The sampling period must be

$$T < \frac{0.00030518509}{10\,\text{V} \times 2\pi \times 20000\,\text{Hz}} = 0.24286\,\text{ns} \tag{3.22}$$

that corresponds to a sample rate of 4.1176 GHz, far too high to be practical. If we use a smaller sample rate, we will have a distortion in the maximum slope of the same type as the slew-rate distortion of an amplifier.

Delta modulation presents the *granularity* problem, i.e., since it only detects positive and negative values, it cannot represent a constant signal such as a null signal. Instead, it represents such waves with a ± 1 LSB oscillation around the real value. Granularity appears as an added noise superposed to the signal.

Decoding of this type of digital modulation is quite simple, since it is attained by means of a low pass filter.

As regards to noise, it allows to distribute the quantization noise in a much wider band, so the noise in the audio band, once filtered, is smaller.

3.13 Sigma-Delta Modulation (SDM)

Sigma-delta modulation, SDM, is based, again, on the concept of feeding back the converted signal (de la Rosa 2011; Boser and Wooley 1988). However, in this case the integration is located after the comparator, instead of before, as shown in Fig. 3.21.

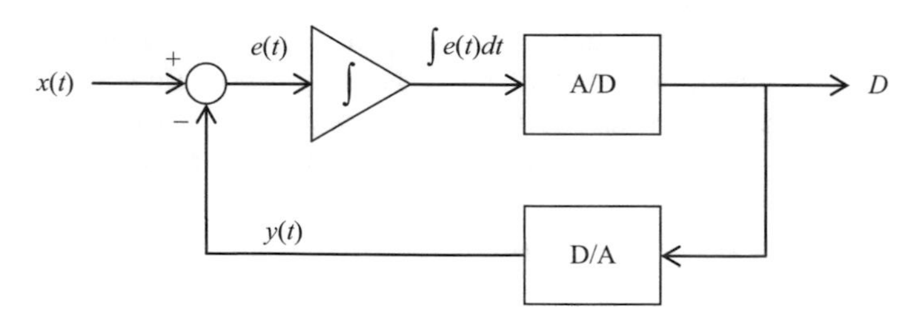

Fig. 3.21 Block diagram of a sigma-delta modulator

The original signal is compared with the analog version of the digitized signal, getting an error signal that is integrated prior to the conversion to the digital domain. The sigma-delta converter uses the concept of *oversampling*, i.e., sampling at a much higher sample rate F_{os} than the desired (target) sample rate F_s (Gray 1987). The ratio between both of them is the *oversampling factor*, N_{os},

$$F_{os} = N_{os} F_s. \tag{3.23}$$

The oversampling factor is usually a relatively high power of 2, such as 64 or 128, and the resolution of the A/D and D/A converters is low, typically between 1 bit and 5 bit. After the modulation, it is necessary to reduce the sampling rate to F_s (a process called *downsampling* or *decimation*) and, at the same time, increase the resolution as required by the final format. This is done in three stages. First, a digital anti-aliasing filter is applied to the oversampled signal in order to remove spectral content above $F_s/2$. Second, this signal is digitally downsampled taking every N_{os}-th sample. Finally, the signal is digitally requantized to have the desired bit resolution. Note that the requantization is a process equivalent to analog/digital conversion. Although in principle one could just truncate the digitally filtered signal[6] to the desired bit count, in practice it is convenient to apply digital dithering to reduce low-level distortion.

Figure 3.22 shows the results of a software simulation of the sigma-delta modulator response to a 1 kHz sine wave. The A/D and D/A converters have a resolution of 3 bit, the oversampling factor is $N_{os} = 4$ and the desired sampling rate, $F_s = 44\ 100$ Hz. The feedback signal, $y(t)$, exhibit high frequency oscillations between the two digital steps that are closer to the input signal. These oscillations will be removed by the digital anti-aliasing filter at the output.

The advantage of this structure is that it reduces the low frequency error due to the presence of an integrator in the direct path. Indeed, given a low frequency signal, whenever the error tends to rise, it is rapidly integrated so that the feedback signal gets closer to the input, reducing the error signal.[7] This can be checked analytically with the linearized model of Fig. 3.23. The joint action of the A/D and D/A converters has been replaced by the addition of a quantization noise source $r(t)$.

Since the integrator's Laplace transfer function is $1/\tau s$ and feedback is unitary, we can write

$$Y(s) = \frac{1}{1+\tau s} X(s) + \frac{\tau s}{1+\tau s} R(s). \tag{3.24}$$

The transfer function for the input $x(t)$ is a low pass filter with cutoff angular frequency $\omega_c = 1/\tau$, while the transfer function for the noise $r(t)$ is a high pass filter

[6]Bear in mind that the digital filter works with a much higher resolution than the low bit output signal D. Such resolution is dictated by the arithmetic processor.

[7]This idea is inspired in control theory, where an integrator is used in the direct path in order to cancel the static error and reduce the error at low frequency.

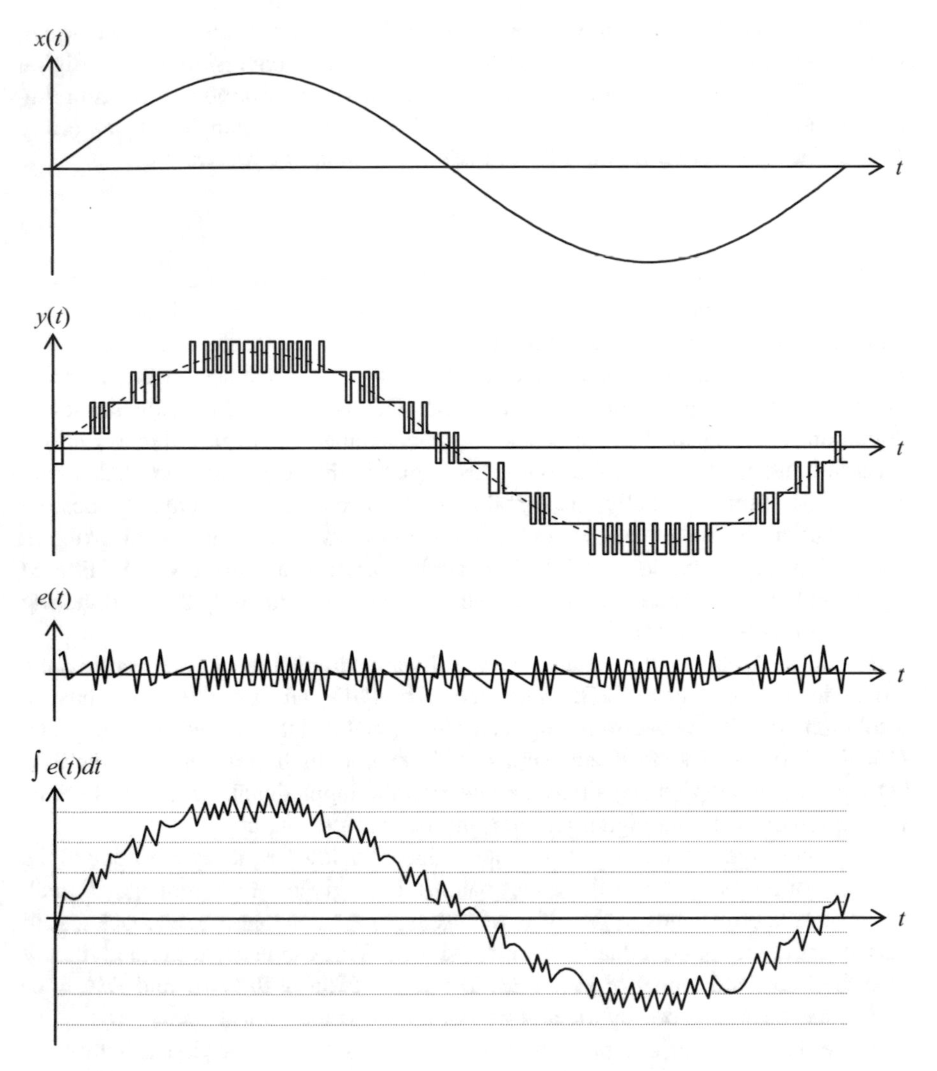

Fig. 3.22 Waveforms of a 3 bit sigma-delta converter with $F_s = 44\ 100$ Hz and $N_{os} = 4$. From *top to bottom* Input 1 kHz sine wave $x(t)$; feedback signal $y(t)$, analog equivalence of the digital output; error signal $e(t)$; integral of the error and quantization levels of the A/D converter

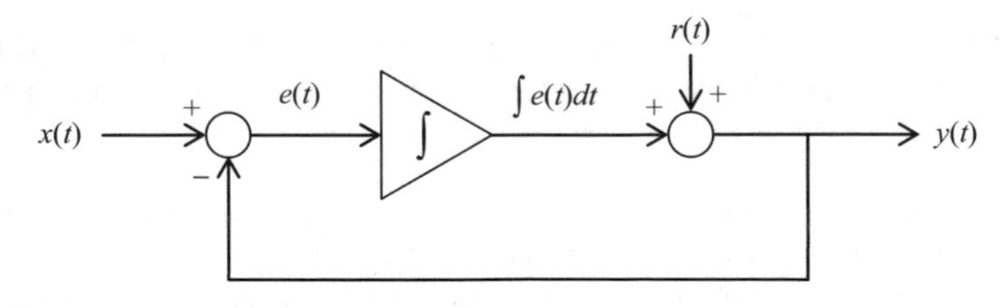

Fig. 3.23 Linear model of the sigma-delta converter

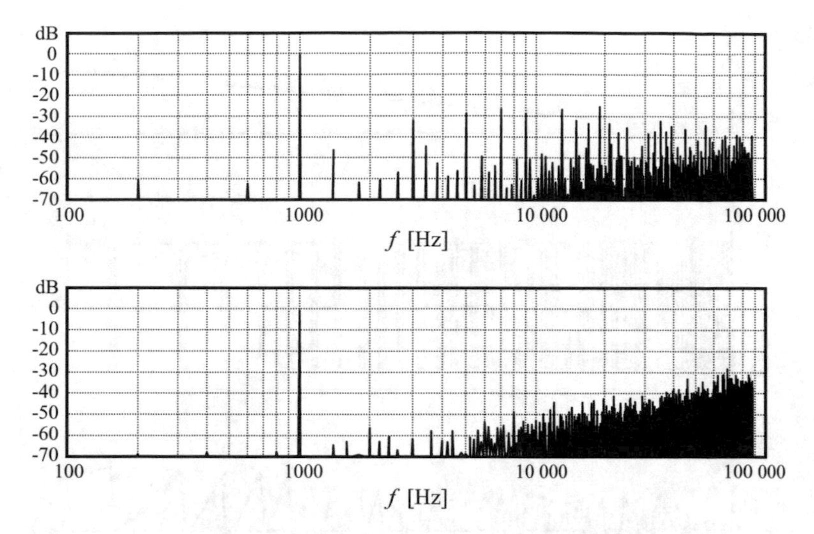

Fig. 3.24 *Top* Spectrum of a 1 kHz tone after 3 bit PCM quantization. *Bottom* Spectrum of the same tone after 3 bit sigma-delta quantization. In both cases the target sample rate is $F_s = 44\ 100$ Hz and the oversampling factor is $N_{os} = 4$

with the same cutoff frequency. This implies that only the high frequency components of the digitization noise are preserved at the output This is very convenient, since as commented earlier, the digital version of $y(t)$ is applied to a digital anti-aliasing filter that removes high frequency components.

Figure 3.24 compares the spectrum of a sine wave that has been digitized with two different converters. The first one is a conventional 3 bit PCM converter that works by level classification. The second one is a 3 bit sigma-delta modulator as described above. In the case of the PCM converter, the spectrum exhibits the typical low-level distortion, with many distinct partials (harmonic and inharmonic) in the baseband ($f < F_s/2$). In the case of sigma-delta converter, the baseband partials are much smaller, at the expense of an increase in high frequency noise. However, this is not a problem since high frequency spectral content will be removed by the anti-aliasing filter.

This idea can be pushed further substituting the integrator (a kind of first-order filter) by a high-order filter. The idea of this filter is that finally the spectral content of the quantization error is shifted toward the high frequency end. In this way, the baseband quantization noise will be substantially attenuated. This is called *noise shaping*.

A frequent case in sigma-delta modulators is when the number of bits is 1. In this case, the A/D converter is a simple comparator that outputs a voltage V_{max} (corresponding to 1) if its input is positive and $-V_{max}$ (corresponding to 0) if it is negative. Figure 3.25 shows the waveforms when the input signal is a ramp. As can be seen, the output signal presents more consecutive 1's when the input is larger.

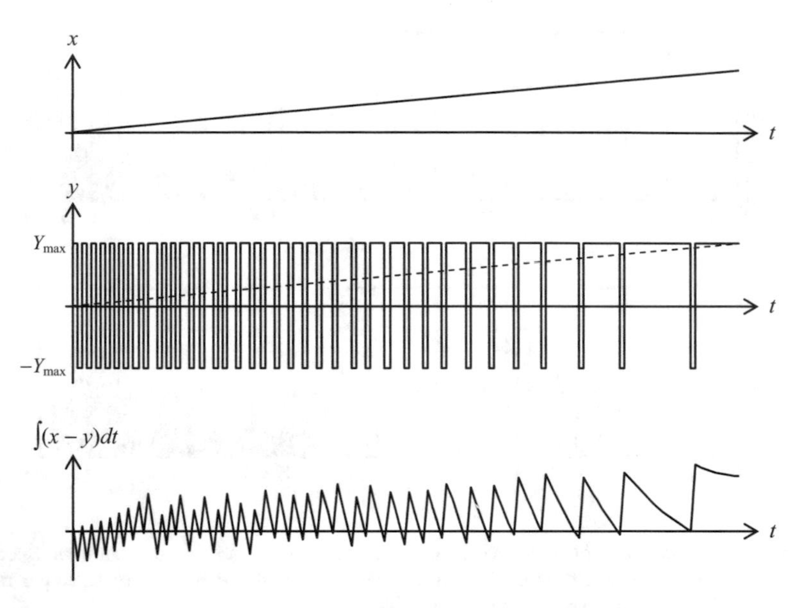

Fig. 3.25 Waveforms of a 1 bit sigma-delta modulator. *Top* A ramp input. *Center* Feedback signal, equivalent to the output. *Bottom* Integral of the error. The 1 bit A/D converter outputs 0 if the input is negative and 1 if it is positive. The D/A converter transforms 0 and 1 in two opposite levels, $-Y_{max}$ and Y_{max}

3.14 Digital Recording Devices

Traditionally, the acquisition of audio signals for subsequent processing and analysis has been possible through the use of instrumentation-grade tape recorders, for instance, the portable tape recorder Nagra IV-SJ, an analog piece of equipment of the 70s. With the introduction of digital technology in the 90s, other option was the *digital audio tape recorder* (DAT).

Several portable recorders commercially available such as the DAT recorder, the Minidisc recorder (MD) and hard disc based (HD) recorders have been studied as regards to their suitability for metrological purposes (Miyara 2001a, b). The DAT and the HD have proved adequate for almost all applications, whereas the MD exhibits a somewhat irregular frequency response at high and low frequencies as well as a reduced effective dynamic range due to the fact that it uses data lossy compression. This is an issue when trying to retrieve a tone inside a narrow spectral well by means of filtering. Another problem of portable MDs is that their output is analog and must be digitized a second time with a computer sound card (which implies a new signal degradation). Besides, many MD recorders include anti-copy resources that prevent a second digital copy of a recording.[8]

[8]The reason for this is to prevent duplication of copyrighted material.

In recent years, a new technology has become increasingly popular. It uses a removable solid-state *flash memory* as a recording medium. Some examples are SD (*Secure Digital*), SDHC (*High Capacity Secure Digital*), CF (*Compact Flash*) or SDXC (*Secure Digital Extended Capacity*). The low-power consumption and the high storage capacity (up to 64 GB for SDHC and 2 TB for SDXC) allow continuous recording[9] during several days or even months. This technology is, thus, adequate for long-term field recordings. As these recorders have no moving parts, they are completely silent, making them ideal for recordings in very low-noise environments.

While there are in the market sound level meters that allow to record the calibrated signal in a flash memory, this feature increases substantially the instrument cost, even if the option requires only a firmware update or just enabling preinstalled software. For instance, the license to enable the recording option of the sound level meter Brüel and Kjær 2250 (Brüel and Kjær 2016) has a cost of $2500, while digital recorders such as the two-channel Zoom H4n can be purchased for about $500 (prices in Argentina as of 2013) and used to record the calibrated AC output of a sound level meter.

There are also some very low-cost experimental recorders completely developed using free software and free hardware[10] that record in lossless compressed formats such as FLAC, with WiFi wireless transfer capabilities, including uploading the digital audio recording to a specified Internet site (Raynaudo and Treffiló 2013).

Since low-price digital audio recorders are oriented to the consumer market, particularly for personal recording of live performances or music demos, there is legitimate concern about the real quality of the recordings for measurement purposes. Particularly, the specification chart often omits some parameters that are of vital importance for our needs, such as signal-to-noise ratio, distortion or frequency response.[11]

In addition, international standards such as ISO 1996-2:2007 state that in order for a piece of equipment to be suitable for measurement purposes, the complete system including that part must fully comply with International Standard IEC 61672 on sound level meters. In particular, this means that the tolerance limits for the frequency response are very strict. There are also other conditions such as self-generated noise, linearity and transient response (see Chap. 9).

[9]The file system FAT 16/32 in use in flash memories has a maximum limit of 4 Gb for file size, hence when file reaches that size it is necessary to stop recording. Some recorders circumvent this difficulty starting a new file once the limit has been reached so that afterwards both files can be joined.

[10]Free hardware is a recent form of cooperative hardware that provides processing and connectivity solutions with increasing access to wide documentation. Some examples are the Arduino, BeagleBone, and Raspberry Pi boards. Many of them allow the installation of free operating systems such as GNU Linux and hence much software, including mathematical processing software such as Scilab, Octave, and Euler, which allow to perform many interesting signal processing and analysis tasks such as those described in this book.

[11]Probably the reason is that high fidelity audio equipment is now very cheap, so it is at everybody's reach, even those without much technical knowledge to understand a specification sheet.

It is important to bear in mind that the digitization involves some processes that could impair the signal or add artifacts. One of them is the anti-aliasing filtering. Intended to prevent severe aliasing distortion, this filter could alter the frequency and phase response, particularly in the upper range of the audible spectrum. Other process is the analog/digital conversion itself, since it limits the available dynamic range because of quantization noise. This noise tends to correlate with the signal at low levels, causing distortion. This effect is reduced with dithering, which in turn reduces even more the dynamic range.

In order to check that these limitations do not significantly affect the quality of the recorded signal, a complete test (see Chap. 9) of three different units of the digital audio recorder Zoom H4 has been performed (Miyara et al. 2010). The study has shown that such model is suitable for its use as part of a measuring system. Its frequency response, for instance, turns out to be flat with a tolerance that is at least a factor of 10 lower than the admissible one (see Fig. 120 in Chap. 9). The effect of temperature was also negligible, since the difference between the responses at 24 and 40 °C was smaller than 0.05 dB in the 20 Hz–10 kHz range, and only above 15 kHz it reached 0.2 dB, a value way below the acceptable tolerance in that spectral range.

The unweighted signal-to-noise ratio turns out to be in excess of 90 dB, and 95 dB with A-weighting. Linearity, expressed in terms of the real minus the ideal response, as specified in International Standard IEC 61672, is less than 0.2 dB for signal levels (referred to the maximum) between −10 dB and −70 dB. Expressed as total harmonic distortion (THD); it is of the order of 0.02 dB between −10 dB and −50 dB and rises to 5 % for signals of −73 dB. This distortion increase for low levels is an expected result since the continuous waveform is replaced by a stepped one. The only serious problem that was found is some distortion at the high frequency end (above 10 kHz) that cause some aliasing effects. This problem disappears completely if the recording level is kept under −20 dB, at the expense of reducing the dynamic range to about 70 dB. This dynamic range is acceptable in most situations, particularly in urban noise measurement, where the background noise is considerable in comparison with the recorder's self noise.

3.15 Sound File Formats

Whatever the storage medium, the digital information of the successive samples must adhere to certain format that is understandable by the user or the software used for its analysis or postprocessing. Generally speaking, the audio file formats may be classified into *compressed* and *uncompressed* formats.

Uncompressed formats present the successive PCM data exactly as they are created by the analog/digital converter, sometimes without any additional information (.pcm and .raw extensions); in other cases including a header that contains data such as bit resolution, number of channels, sample rate, and number of audio samples (.wav extension). The first case (.pcm and .raw) is not convenient for files to be shared with others or intended for long-term storage, because without additional information they may become meaningless. In the second case (.WAV), the file itself contains the additional information necessary for its interpretation by any other user or software.

The advantage of uncompressed formats is that no decoding is necessary to retrieve the original information. The main drawback is that the file size is large. For instance, if we use a sample rate of 44 100 Hz and a resolution of 16 bit (2 B) per sample, it turns out that we need 5.292 MB per recorded minute per channel, i.e., 315.52 MB per hour per channel. Such a high information rate is problematic when we need to store large amounts of audio, particularly for long-term storage, or when it has to be transmitted through data networks.

Compressed formats are used to reduce the amount of information needed to represent some amount of audio, and consequently, to save storage space or transmission bandwidth. All compression methods exploit in some way signal redundancy or predictability. Typically, they use some modeling strategy that allows predicting the following sample value from a number of preceding samples. The model uses few parameters so if the parameters of the model are stored instead of the samples, it is possible to save storage space. Of course, it cannot be expected that the prediction be perfect, so the method is completed with a representation of the error, which is normally small, requiring fewer bits.

Compression may be, in turn, *lossy* or *lossless*. Lossy formats, such as the popular .mp3 (MPEG-1 Layer 3), the free format .ogg (OGG Vorbis), or ATRAC (the compression used in the Minidisc) use psychoacoustic criteria to discard information that will normally be inaudible because of masking (Zwicker and Fastl 1999). They are ideal for storage in small portable players and for download from Internet repositories of audio files for casual listening. Reductions in a factor of 5 or more are achieved without audible effects, and 10 or more with reasonable fidelity. Their use for metrological purposes, however, is not recommended in general, especially in the case of specialized measurements such as line spectrum, since the compression may completely suppress regions of the spectrum that could contain useful information. For instance, a sound transmission path could be apparent from a line spectrum analysis, if one finds traces of some tone that is generated remotely, although it is inaudible.

Lossless compression formats are based in statistical considerations and information theory. They allow perfect signal reconstruction. They usually achieve data rate reductions by a factor of 2. One of the most popular ones are .flac (Free Lossless Audio Coding), which unlike other proprietary formats, are an open and free format that presents a fast coding and decoding. Lossless formats are used for long-term storage, for high fidelity music repositories on the Internet, for sound databases and for network transmission, since they reduce to about 50 % the resource usage. The only drawback is that it requires a specific algorithm for coding and decoding, but the computational load is low enough to support even real-time streaming applications (Marengo et al. 2011).

3.15.1 WAV Format

The WAV format (from *wave*) is a particular type of resource interchange file format (RIFF)[12] (Microsoft Corporation 1994). Even though it is widely recognized by most digital audio editors and programming languages, in some cases it is not a native format and must be exported and imported.

A .wav file is divided into *chunks*, each one of which contains a specific type of information, for instance, format, audio data, *cues*, *labels*, *notes*. Each chunk has an identifier, formed by a character string,[13] and a series of numerical data. These data are expressed in the hexadecimal number system with 1 byte to 4 bytes, where each byte (8 bit) includes two hexadecimal digits from the set {0, 1, 2, …, 9, A, …, F}. For instance, the byte 01101101 represents the hexadecimal 0x6D (the 0x is used to indicate that what follows is a hexadecimal), i.e.,

$$\underbrace{0110}_{0\text{x}6}\underbrace{1101}_{0\text{xD}} = 6 \times 16 + 13 = 109.$$

The bytes within each multibyte hexadecimal number use the *little endian*[14] order of the Intel 80x86 microprocessors, i.e., the bytes are ordered from the least significant to the most significant (inverse order). Hence, the hexadecimal number 0xA34E will be stored or transmitted as 4EA3. This is important when it is necessary to retrieve the information from a .wav file using nonspecific software (such as mathematical software).

WAV files always begin with a header with the identifier "RIFF" (represented in ASCII code as 0x52494646),[15] followed by 4 bytes that represent the total number of bytes that follow (i.e., the file size minus 8 bytes) and the identifier of the type of file, i.e., "WAVE" (in ASCII, 0x57415645). See an example in Table 3.2.

Then comes the *format chunk* which begins with the identifier "fmt " (in ASCII, 0x666D7420) and the number of bytes that follow in the format chunk (i.e., the total size of the format chunk minus 8 bytes; it is always equal to 16, i.e., 0x10000000). Then we have two bytes that code the format type. The code for uncompressed PCM (the only one of interest for measurement purposes) is 1 (0x0100, note the little endianness). Following, we have the number of channels (2 bytes), the sample rate (4 bytes), the number of bytes per second, equal to the number of channels times the sample rate (4 bytes), the number of bytes per sample (2 bytes), and the number of bits per sample (2 bytes).

---

[12]RIFF files are used for audio, video, images, MIDI and multimedia information.

[13]Each character is represented by a byte using the ASCII code.

[14]The phrases *little endian* and *big endian* are inspired in the novel "Gulliver's Travels" by Jonathan Swift. The Big endians were the ones that opposed to the Emperor's decree according to which the eggs should be broken from the little end and kept breaking them stealthily on the big end.

[15]Note that little endianness applies to numbers, but not to ASCII code strings.

Table 3.2 Example of .wav of 1 channel, sample rate 48 000 Hz, 24 bit resolution. Only the first two and the last audio data are shown

Chunk	Type	Content	Hex value	Decimal
Header	Identifier	“RIFF”	0x52494646	–
	Length	4 bytes	0XC2601A00	1 728 706
	Identifier	“WAVE”	0x57415645	–
Format	Identifier	“fmt”	0x666D7420	–
	Length	4 bytes	0x10000000	16
	Format type	2 bytes	0x0100	1
	Channels	2 bytes	0x0100	1
	Sample rate	4 bytes	0x80BB0000	48 000
	Bytes per second	4 bytes	0x80320200	144 000
	Bytes per sample	2 bytes	0x0300	3
	Bits per sample	2 bytes	0x1800	24
Audio data	Identifier	“Data”	0x64617461	–
	Audio data length	4 bytes	0x005E1A00	1 728 000
	Sample value	3 bytes	0x5B8E00	36 443
	Sample value	3 bytes	0xC97701	96 201
	…	…	…	…
	Sample value	3 bytes	0x0685f6	16 155 910

After the format chunk, we have the *data chunk*. It begins with the identifier “data” (in ASCII, 0x64617461) followed by the audio data length (4 btyes), and the audio data themselves. In the case of mono recordings (a single channel), there are as many bytes per sample as indicated in the format chunk. In the case of stereo recordings, the number is doubled. Channels are interleaved: first comes the left channel’s first sample, then the right channel’s first sample; then the left channel’s second sample, then the right channel’s second sample, and so on. The same criterion applies if there were more channels.

In all cases, the data length of each chunk is expressed in bytes and excludes the chunk identifier and the 4 bytes indicating the length. In the example of Table 3.2, the length of the format chunk is 16 bytes, which correspond to the format type, channels, sample rate, bytes per second, bytes per sample, and bits per sample. The total length of the chunk is 8 bytes larger, i.e., 24 bytes.

The audio data length, in this case 1 728 000, corresponds to the total bytes, not to the number of samples. To find the number of samples, it is necessary to divide by the bytes per sample (in this case, 3) and by the number of channels (in this case, 1). Therefore, we have 1 728 000/3/1 = 576 000 samples. At a sample rate of 48 000 Hz, it is equivalent to 12 s of audio.

The interpretation of the audio data depends on the number of bits per sample per channel. In the case of 8 bit per sample, they are unsigned integers, which means that it is necessary to subtract 128 to get the actual signed value. This case is irrelevant for metrological purposes because of the poor signal-to-noise ratio. In the case of more bits per sample, for instance, 16 bit, they are signed integers in the numeric code known as *two's complement*. In this code, the values between 0 and $2^{n-1}-1$ (those having the MSB equal to 0) are interpreted as ordinary positive binary numbers. The values between 2^{n-1} and 2^n-1 (those ones having the MSB equal to 1) are interpreted by subtracting 2^n, getting, thus, negative numbers. Let us consider, for example, the case of $n = 16$. In normal notation (big-endian):

$$\begin{aligned} 0x0000 &\equiv 0 \\ 0x0001 &\equiv 1 \\ 0x7FFF &= 2^{15}-1 \equiv 32\,767 \\ 0x8000 &\equiv 2^{15}-2^{16} = -32\,768 \\ 0xFFFF &\equiv 2^{16}-1-2^{16} = -1 \end{aligned}$$

The values thus obtained are integer numbers from -2^{n-1} to $2^{n-1}-1$, with 0 being approximately halfway between the minimum and the maximum. While they are proportional to the instantaneous value of the corresponding sample, they differ from the real value (in pascals) in a scale factor that needs to be applied to get physically meaningful results. Such scale factor involves the transducer sensitivity, S, in V/Pa, the gain of the signal conditioning circuit (preamplifier), G, and the voltage reference of the analog/digital converter, V_{ref}. If the sound pressure corresponding to sample k is p_k and its digital representation with n bits is $p_{\text{D}k}$, then

$$p_k = \frac{V_{\text{ref}}}{2^{n-1}S\,G}p_{Dk}, \tag{3.25}$$

i.e.,

$$p_k = K\,p_{Dk}, \tag{3.26}$$

Equation (3.25) may be useful when designing the system's hardware, but in practice when recordings are made with closed-architecture commercial equipment, the constants S, G, and V_{ref} are not known. Even if they are specified or can be derived from specifications, they are subject to manufacturing tolerances and drifts over time, increasing the uncertainty. It is better to record a calibration signal of known (nominal) level, and use it to get the constant K. Calling p_{cal}, the calibration pressure and $p_{\text{cal, D}}$ its digital value, then

$$K = \frac{p_{\text{cal}}}{p_{\text{cal, D}}}. \tag{3.27}$$

Instead of the instantaneous pressure of a particular sample, it is better to use the effective value extended to an interval of several seconds (since the calibration signal in general is a tone whose sound pressure level is known). This can be done with the formula

$$P_{\mathrm{cal,D,ef}} = \sqrt{\frac{1}{N}\sum_{k=1}^{N} p_{\mathrm{cal,D}}^{2}(k)}. \tag{3.28}$$

Then,

$$K = \frac{P_{\mathrm{cal,ef}}}{P_{\mathrm{cal,D,ef}}}. \tag{3.29}$$

It is convenient to choose N as close as possible to an integer number of cycles, to reduce the residual oscillating error as shown in Fig. 1.3. If N is also large, the uncertainty is reduced almost to the uncertainty of the calibration signal, since in practice it is equivalent to averaging many measurements.

The information included in a basic WAV file as described (including all the audio data) is enough to have a valid digital audio file. However, it is frequently desirable to be able to include some supplementary nonaudio information, for instance, comments associated with specific time instants or intervals. This can be done by means of *cues*, *labels*, and *notes*.

They can be used to identify sources, provide a time stamp and GPS information of the site where the recording has been made, synchronization signals, localization of calibration signals, etc. This information is generally referred to as *metadata* (data relative to other data, such as descriptions, classification, etc.).[16]

While the original file obtained from the recording process is created automatically, the metadata information has to be provided interactively by the operator at a later time, using specific sound editor software components. These components provide dialog windows that allow adding cues and editing their labels and notes.

The way to generate the metadata depends on the measurement protocol. One possibility is to orally record the events, either on the same channel during irrelevant intervals (for instance, before or after a calibration or measurement) or in an independent channel synchronized with the main one. Another way is to keep a careful written record with timing and description of the events of interest. In both cases, this information must be copied later to the WAV file metadata section using the dialogs mentioned above. Some recorders allow introducing cues in real time during recording, but they do not always inform the correlative number of the last cue in order to associate it with a manually written note. This possibility is not recommended because of confusion risk.

A good measurement protocol will allow to extract automatically the metadata stored in the file. To this end, it is necessary to adopt a standard format, including well-defined descriptors and data fields. For instance, if the calibration tones are identified with a cue with label "CAL" with which one can associate the level in the note field (for instance, "93.9"), then it will be possible to develop an algorithm that

[16]The WAV format allows other metadata such as recording device, author, performer, copyright, etc., intended for the record production industry. In our case, they are irrelevant.

Table 3.3 Example of cues distribution. This information appears immediately after the last audio data shown in Table 3.2

Chunk	Type	Content	Hex value	Decimal
Cue	Identifier	"cue"	0x63756520	–
	Length	4 bytes	0xC4000000	196
	Number of cues	4 bytes	0x41000000	8
	Cue number	4 bytes	0x01000000	1
	Associated sample	4 bytes	0x61C00000	52 321
	Identifier	"Data"	0x64617461	–
	Chunk start	4 bytes	0x00000000	0
	Block start	4 bytes	0x00000000	0
	Offset	4 bytes	0x61C00000	52 321
	Cue number	4 bytes	0x02000000	2
	Associated sample	4bytes	0x43C10100	115 011
	Identifier	"Data"	0x64617461	–
	Chunk start	4 bytes	0x00000000	0
	Block start	4 bytes	0x00000000	0
	Offset	4 bytes	0x43C10100	115 011
	…	…	…	…
	Cue number	4 bytes	0x08000000	8
	Associated sample	4 bytes	0xA73C0500	343 207
	Identifier	"Data"	0x64617461	–
	Chunk start	4 bytes	0x00000000	0
	Block start	4 bytes	0x00000000	0
	Offset	4 bytes	0xA73C0500	343 207

reads the level and applies to any other sample the adequate correction so that we can directly compute equivalent levels, spectra, etc.

Since most of the audio editing software is intended for music production, sound tracks, voice overs, etc., cues are thought of mainly as edit points and are usually involved in interactive processes (insertion, substitution, application of effects). In our case, it is necessary to be able to access or even generate or modify automatically the metadata. This is why it is important to know how they are organized within the file. Tables 3.3, 3.4 and 3.5 give some examples of the the metadata chunks in a WAV file.

The first element (see Table 3.3) are the *cues*. They appear in a new chunk, the *cue chunk*, located immediately after the audio data chunk (near the end of file). It is possible to know where it begins looking for the identifier "cue" after the audio data.[17] In the example of Table 3.2, after the byte

[17]Note that the string "cue " might appear in the data chunk since its ASCII code, 0x63756520, could be part of the numeric audio data.

Table 3.4 Example of ltxt chunks that provide the number of samples associated with each cue. This information begins after the last data in Table 3.3. Only cue 3 has a nonzero interval

Chunk	Type	Content	Hex value	Decimal
Associated data list	Identifier	"LIST"	0x4C495354	–
	Length	4 bytes	0xCA010000	458
	Identifier	"adtl"	0x6164746C	–
Cue length	Identifier	"ltxt"	0x6C747874	–
	Length	4 bytes	0x14000000	20
	Cue number	4 bytes	0x01000000	1
	Number of samples	4 bytes	0x00000000	0
	Text purpose	"rgn"	0x72676E20	–
	Country	2 bytes	0x0000	0
	Language	2 bytes	0x0000	0
	Dialect	2 bytes	0x0000	0
	Code page	2 bytes	0x0000	0
	…	…	…	…
	Identifier	"ltxt"	0x6C747874	–
	Length	4 bytes	0x14000000	20
	Cue number	4 bytes	0x03000000	3
	Number of samples	4 bytes	0xA6BD0000	48 850
	Text purpose	"rgn"	0x72676E20	–
	Country	2 bytes	0000	0
	Language	2 bytes	0000	0
	Dialect	2 bytes	0x0000	0
	Code page	2 bytes	0x0000	0
	…	…	…	…
	Identifier	"ltxt"	0x6C747874	–
	Length	4 bytes	0x14000000	20
	Cue number	4 bytes	0x08000000	8
	Number of samples	4 bytes	0x00000000	0
	Text purpose	"rgn"	0x72676E20	–
	Country	2 bytes	0x0000	0
	Language	2 bytes	0x0000	0
	Dialect	2 bytes	0x0000	0
	Code page	2 bytes	0x0000	0

$$\underbrace{8+4}_{\text{header}} + \underbrace{8+16}_{\text{format}} + \underbrace{8+1\ 728\ 000}_{\text{audio data}} = 1\ 728\ 044\ .$$

The next 4 bytes after the "cue" contain the length of the cue chunk thereafter. The next 4 bytes indicate the total number of cues. Then we have groups of 24 bytes, each one containing information of a cue. The first 4 bytes indicate the

Table 3.5 Example of subchunks "labl" of labels associated with each cue. This information begins after the last byte in Table 3.4. Only are present the nonempty labels

Chunk	Type	Content	Hex value	Decimal
Labels	Identifier	"labl"	0x6C61626C	–
	Length	4 bytes	0x0C000000	12
	Cue number	4 bytes	0x01000000	1
	Content	"Marca 1"	0x4D617263612031	–
	Null termination	1 byte	0x00	–
	Identifier	"labl"	0x6C61626C	–
	Length	4 bytes	0x0D000000	13
	Cue number	4 bytes	0x02000000	2
	Content	"Marca 1a"	0x4D61726361203161	–
	Pad byte	1 byte	0x00	–
	Null termination	1 byte	0x00	–
	Identifier	"labl"	0x6C61626C	–
	Length	4 bytes	0x0D000000	12
	Cue number	4 bytes	0x03000000	3
	Content	"Marca 4"	0x4D617263612034	–
	Null termination	1 byte	0x00	–
	…	…	…	…
	Identifier	"labl"	0x6C61626C	–
	Length	4 bytes	0x0D000000	12
	Cue number	4 bytes	0x08000000	8
	Content	"Marca 6"	0x4D617263612036	–
	Null termination	1 byte	0x00	–

number of cue, in the order of creation (which not necessarily is chronological, since it is possible to insert a new cue anywhere). The next 4 bytes indicate the number of associated sample, i.e., its position in the time axis expressed in samples. Then, preceded by the identifier "data" (not to be confused with audio data), there are three 4 byte data that are only relevant for compressed formats or for files that have several independent audio blocks.

Next, there is an *associated data list* (ADTL), containing the labels, notes, and intervals associated with each cue (see Table 3.4). The list begins with the identifier "LIST," followed by the length of the data thereafter and the identifier "adtl."

Then there are three types of subchunks. The first type, which begins with the identifier "ltxt" (*labeled text chunk*) followed by the number of bytes of the subchunk, contains the cue number and the number of samples of the interval associated with the cue. This interval may have any length from 0 to the rest of the remaining duration. This is useful to select regions of interest, for example, a calibration tone or a portion of signal where the equivalent level is to be computed. After this information, there are 10 bytes with information that is irrelevant for metrological purposes (they provide compatibility with other formats).[18] There are as many ltxt subchunks as cues.

The second type of subchunk in the associated data list is the one that corresponds to *labels* (see Table 3.5). It begins with the identifier "labl" (*label chunk*), followed by the number of bytes thereafter. Then comes the cue number (4 bytes) followed by the label content (the ASCII representation of the corresponding string). The label content is arbitrary, but in general is a brief description of the main information that will be contained in the note subchunk (described later). In all cases, the subchunk ends with a null string (0x00). However, since the total number of bytes in the subchunk has to be even, it may be necessary to add a null *pad byte*. The pad byte is not considered in the declared number of bytes. For instance, in the second subchunk of Table 3.5 (corresponding to cue 2) the length is 13. If the pad byte were included, the total byte count would be 14.

Note that the label list includes only those that are nonempty. If some cue does not have a label, the corresponding label subchunk is absent.

The third type of subchunk in the associated data list is the one that corresponds to *notes* (see Table 3.6). It is very similar to the label subchunks. It begins with the identifier "note" (*note chunk*), followed by the number of bytes thereafter. Then follows the cue number (4 bytes) and the note content, i.e., an alphanumeric string in ASCII code. Usually notes are longer or more specific than labels, providing detailed information as related to the cue or to the associated interval. It may contain the name of a place or its address, GPS coordinates, time and date, the description of a noise source, etc. Like the label subchunk, it must terminate with a null string and it may include a pad byte to ensure that the total byte count is even. As the label list, the note list includes only nonempty notes.

3.15.2 FLAC Format

FLAC (Free Lossless Audio Codec)[19] is a lossless compressed format that is convenient for permanent or documentary storage as well as for the transmission of massive amounts of audio, since it saves nearly a 50 % of memory or communication channel bandwidth (for instance, a wireless local area network or the Internet) (Salomon 2007; Marengo et al. 2011; Roveri 2011; Raynaudo and Treffiló 2013). Unlike the WAV format, FLAC needs a codec (coder-decoder) to compress the original PCM signal and to retrieve it at a later time.

The compression method consists in breaking the signal into consecutive blocks, approximating each of them by means of a predictor with a few parameters, and

[18]These data are an identifier of the purpose of the text, in this case "rgn" (region), and codes for country, language and dialect, page code, and the text, which in the case of audio files is not present.

[19]This format adheres to the philosophy of *free software*, i.e., free-license, open-code, and free distribution software.

Table 3.6 Example of subchunks "note" of notes associated with each cue. This information begins immediately after the last byte in Table 3.5. Only are present the nonempty notes

Chunk	Type	Content	Hex value	Decimal
Notes	Identifier	"Note"	0x6E6F7465	–
	Length	4 bytes	0x0D000000	13
	Cue number	4 bytes	0x01000000	1
	Content	"12:22:35"	0x31323A32323A3335	–
	Pad byte	1 byte	0x00	–
	Nul termination	1 byte	0x00	–
	Identifier	"Note"	0x6E6F7465	–
	Length	4 bytes	0x13000000	19
	Cue number	4 bytes	0x03000000	3
	Content	"48,550 muestras"	0x3438353530206D75657374726173	–
	Pad byte	1 byte	0x00	–
	Nul termination	1 byte	0x00	–
	Identifier	"Note"	0x6E6F7465	–
	Length	4 bytes	0x0B000000	11
	Cue number	4 bytes	0x08000000	8
	Content	"Máximo"	0x4DE178696D6F	–
	Pad byte	1 byte	0x00	–
	Nul termination	1 byte	0x00	–

coding the *residuals* between the original signal and the approximation with a low bit count, since residuals are, hopefully, much smaller than the signal.

This description above is, however, oversimplified, since what really happens is that there are *statistically* much more small residuals than large residuals. In a PCM code, the required number of bits is determined by the maximum possible value to represent, so a single residual of maximum amplitude will force the use of the maximum bit resolution for *all* residuals.

This problem can be solved using a *variable length code* that uses less bits for the most frequent values, i.e., the smaller ones in this case. The selection of such a code is determined by the statistical distribution of the values to code. It has been found empirically that, for the typical predictors, the errors or residuals have an approximate Laplace distribution[20] (see Fig. 3.26), whose probability density function is

[20]Strictly speaking, it is not a Laplace distribution, since we are working with discrete variables, so the residuals are discrete. However, the general shape of the histogram is similar to the Laplace probability density function.

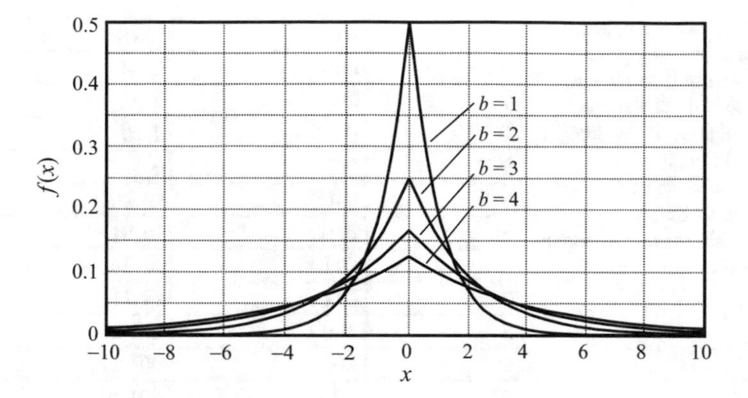

Fig. 3.26 Probability density function of the Laplace distribution of mean 0 for four values of the parameter b

$$f(x) = \frac{1}{2b} \mathrm{e}^{-\frac{|x-\mu|}{b}}, \tag{3.30}$$

where μ is the mean and b is a scale parameter proportional to the standard deviation:

$$\sigma = \sqrt{2}b. \tag{3.31}$$

3.15.2.1 Golomb–Rice Code

For a variable with Laplace distribution, one of the best codes available is the Golomb-Rice code. It is a hybrid code in which the number to be coded is first divided by a properly selected divisor D (in general, a power of 2). The integer quotient is represented in *unary code* and the remainder in binary code. In principle, the unary code represents numbers with a single symbol that is repeated as many times as the number to represent. However, since its length is variable and unknown, a different symbol is necessary to indicate the end of the code. In a typeset version, such symbol could be a space, but in a binary-based system, the main and end symbols must be chosen from {0,1}. If we choose 0 as the main symbol and 1 as the end symbol, then 3 will be represented as 0001.

Table 3.7 illustrates the Golomb–Rice code with divisor $D = 4$ for numbers between 0 and 15. We see that while the largest values require 6 bit, when they could be represented with only 4, the smaller ones require only 3 bit. The idea is that if small numbers are much more frequent than large ones, then the code will help to reduce the average bit count. This is exactly what happens in a Laplace distribution.

Table 3.7 An example of Golomb–Rice coding for numbers from 0 to 15 with divisor $D = 4$. The space separation between the integer quotient in unary representation and the remainder in binary representation is only to make the visual analysis easier

N	Binary	Golomb–Rice
0	0000	1 00
1	0001	1 01
2	0010	1 10
3	0011	1 11
4	0100	01 00
5	0101	01 01
6	0110	01 10
7	0111	01 11
8	1000	001 00
9	1001	001 01
10	1010	001 10
11	1011	001 11
12	1100	0001 00
13	1101	0001 01
14	1110	0001 10
15	1111	0001 11

Note that if the divisor is known, a string of 0's and 1's representing a series of numbers can be decoded uniquely. Given, for instance, the string

$$0010000101010010110101111 0000$$

we can rewrite it, for the sake of clarity, as

$$00100\,|\,00101\,|\,0100\,|\,101\,|\,101\,|\,0100\,|\,00111$$

Indeed, we know that the first number begins with some 0's (might be zero 0's) followed by a 1 (end symbol) and two arbitrary bits. In this case, we have two 0's and a 1, followed by the binary 00. This can be repeated as long as the string is consistent with a Golomb–Rice code. According to the code meaning, the numbers are

$$2 \times 4 + 0,\ 2 \times 4 + 1,\ 1 \times 4 + 0,\ 1 \times 4 + 1,\ 1 \times 4 + 1,\ 1 \times 4 + 0,\ 2 \times 4 + 3,$$

i.e.,

$$8,\quad 9,\quad 4,\quad 1,\quad 1,\quad 4,\quad 11$$

The selection of divisor D depends on the parameter b of the Laplace distribution. The smaller b, the larger the proportion of very small values, hence it will be convenient to choose a small divisor. Since D is selected as a power of 2,

$$D = 2^d, \tag{3.32}$$

it suffices to specify d, known as *Rice parameter*. FLAC uses the following estimation formula as the optimum d (Salomon 2007; Robinson 1994):

$$d = \log_2(\ln 2\, E(|e(k)|)), \tag{3.33}$$

where $E(|e(k)|)$ is the mean of the absolute values of the prediction errors. For a Laplace distribution, we have

$$E(|e(k)|) = b = \frac{\sigma}{\sqrt{2}}. \tag{3.34}$$

If, for example, the standard deviation of the residuals is 8, then $d \cong 2$, from where $D = 2^2 = 4$, which corresponds to the example of Table 3.7.

3.15.2.2 Predictors

FLAC implements four different predictors to approximate each signal block: (1) the signal itself (verbatim); (2) a constant value; (3) a polynomial approximation; and (4) a linear prediction model, LPM, also called linear predictor coding, LPC. The encoder tries them all and selects the one that provides the largest compression.

3.15.2.3 Verbatim

This predictor is appropriate for completely random signals, such as white noise. In this case, where the signal has maximum entropy,[21] it makes no sense attempting to compress it using any other predictor, since the attainable compression would be minimal and would probably be spoiled after the inclusion of the predictor's parameters. Besides, it would increase unnecessarily the decoding computational load.

[21]A signal's entropy is an indicator of its unpredictability, i.e., its lack of redundancy. The larger the entropy, the smaller the redundancy and the lesser the effectivity of any compression attempt. For discrete signals, it is defined as $H = -\Sigma p_k \log_2(p_k)$, where p_k is the probability of the k-th discrete value of the signal. It can be interpreted as the theoretical minimum of the average bits per sample that can be accomplished by a code. Actual codes often require more bits per sample (MacKay 2005).

3.15.2.4 Constant Predictor

This case applies particularly to silence intervals, which can be compressed just counting the 0 samples. This type of compression is known as *run-length* and is similar to the commonly used one in general purpose ZIP files or GIF images files.

3.15.2.5 Polynomial Predictor

This predictor consists in an n-th degree polynomial that goes through the $n + 1$ previous samples ($n = 0, \ldots, 3$), extrapolating to the present sample (Fig. 3.27). Since the samples are separated by uniform intervals, it can be shown that the polynomials adopt a simple form and the residuals can be computed very easily.

Calling k the discrete time, we need to predict sample $y(k)$ as a function of the $n + 1$ previous samples, i.e., $y(k - 1), \ldots, y(k - n - 1)$ by means of a n-degree polynomial.

For instance, for $n = 2$, we want to find the coefficients of a quadratic polynomial

$$P(t) = a_0 + a_1(t - k) + a_2(t - k)^2 \tag{3.35}$$

such that

$$\begin{aligned} P(k-1) &= y(k-1) \\ P(k-2) &= y(k-2) \\ P(k-3) &= y(k-3) \end{aligned} \tag{3.36}$$

Substituting (3.36) in (3.35), we have the following equation system:

$$\begin{aligned} a_0 - a_1 + a_2 &= y(k-1) \\ a_0 - 2a_1 + 4a_2 &= y(k-2) \\ a_0 - 3a_1 + 9a_2 &= y(k-3) \end{aligned} \tag{3.37}$$

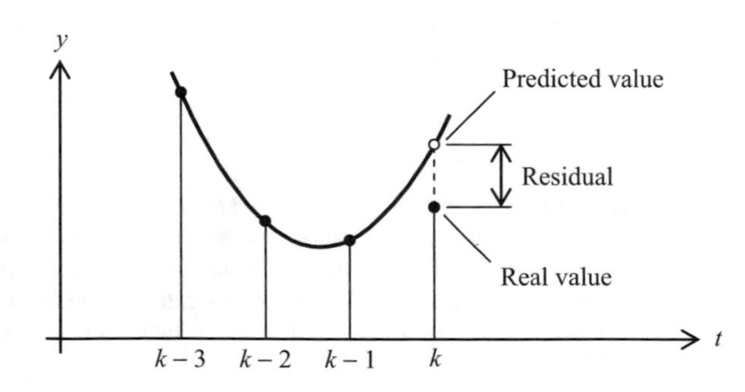

Fig. 3.27 Second-degree polynomial predictor

whose solution is

$$
\begin{aligned}
a_0 &= 3y(k-1) - 3y(k-2) + y(k-3) \\
a_1 &= \frac{5}{2}y(k-1) - 4y(k-2) + \frac{3}{2}y(k-3) \\
a_2 &= \frac{1}{2}y(k-1) - y(k-2) + \frac{3}{2}y(k-3)
\end{aligned}
\tag{3.38}
$$

The predicted value in $t = k$ is

$$\widehat{y}(k) = P(k) = a_0 = 3y(k-1) - 3y(k-2) + y(k-3) \tag{3.39}$$

This approximation is called of order 3 because it uses three previous samples. We can compute the polynomial approximations of other degrees in a similar fashion, with results shown in Table 3.8. Note that even though the approximation is polynomial, the result turns to be a *linear combination* of past samples.

Since the concept of "past samples" depends on the present sample, the model must be updated on a sample-by-sample basis. At the encoding stage, it is intended solely for the prediction of a sample in order to compute the residual.

It is natural to ask what order is better for a given case. The answer is that the encoder tries them all and selects the one that achieves the smaller cumulative absolute residual, extended to the samples in the present block. Although this would seem to be a brute-force method (and it certainly is!), an interesting twist is that the residuals for degree n can be easily computed from the residuals for degree $n - 1$. Calling $e_n(k)$, the residual in the prediction of sample $y(k)$ with a polynomial of degree n, i.e.,

$$e_n(k) = y(k) - \widehat{y}_n(k), \tag{3.40}$$

then it can be shown that

$$
\begin{aligned}
e_0(k) &= y(k) - y(k-1) \\
e_1(k) &= e_0(k) - e_0(k-1) \\
e_2(k) &= e_1(k) - e_1(k-1) \\
e_3(k) &= e_2(k) - e_2(k-1)
\end{aligned}
\tag{3.41}
$$

Table 3.8 Expressions for the polynomial approximations up to degree 3

Order	Degree	$\widehat{y}(k)$
0	–	0
1	0	$y(k-1)$
2	1	$2\,y(k-1) - y(k-2)$
3	2	$3\,y(k-1) - 3\,y(k-2) + y(k-3)$
4	3	$4\,y(k-1) - 6\,y(k-2) + 4\,y(k-3) - y(k-4)$

Therefore, this iterative process allows to find the residuals for all orders in a computationally efficient way.

Once the degree n of the polynomial that achieves the minimum cumulative residual has been determined, the only parameter that is needed is n.

3.15.2.6 LPC Predictor

The linear predictor is a more general variant of the polynomial predictor. It consists in expressing the sample to predict as a linear combination of p previous samples:

$$\widehat{y}(k) = \sum_{h=1}^{p} a_h y(k-h) \tag{3.42}$$

The difference with the polynomial model is that the coefficients a_h are not fixed any longer. Instead, they are determined from all the samples in the block in such a way that the quadratic prediction error

$$e^2 = \sum_{k=1}^{N} \left(y(k) - \sum_{h=1}^{p} a_h y(k-h) \right)^2, \tag{3.43}$$

is minimized, where N is the number of block samples. The coefficients are computed using an algorithm due to Levinson and Durbin (Proakis and Manolakis 1996).

The user can choose a maximum p between 1 and 32. The encoder selects the optimum p from the dynamic range and the block size (Salomon 2007). The larger is p, the larger is the computing time. It is generally observed that for $p \geq 10$ the improvement is only marginal and is not worth the extra computing time.

In the case of a Laplace distribution, the minimization of the quadratic prediction error also implies the minimization of the average absolute error (Eq. (3.34)), taking full profit of the Golomb–Rice code. The price to pay is the need for more information. While polynomial predictors require a single parameter (the degree), LPC predictors require p coefficients. Hopefully, this is not significant since there are about 10 LPC coefficients versus hundreds or thousands of samples in a block.

Note that in order to be able to work with integer coefficients (to prevent arithmetic implementation issues in the processor), the coefficients must be scaled and quantized, so the theoretical optimality is not really achieved.

3.15.2.7 Interchannel Decorrelation

So far we have analyzed the compression technique for a single audio channel. In the case of stereo channels, there is generally much redundancy between the left and right channels, y_L and y_R, from which one can profit, getting further information rate reduction. To that end, we can use the transformation from left-right to mid-side:

$$\begin{aligned} y_{\mathrm{M}} &= \frac{y_{\mathrm{L}} + y_{\mathrm{R}}}{2} \\ y_{\mathrm{S}} &= y_{\mathrm{L}} - y_{\mathrm{R}} \end{aligned} \tag{3.44}$$

If both channels are highly correlated (the usual case), then y_{S} is significantly smaller, allowing larger compression. However, this is not always the case, so the FLAC encoder analyzes both versions, (y_{L}, y_{R}) and (y_{M}, y_{S}), encoding finally the one with better compression. Left and right channels can be later retrieved using the equations

$$\begin{aligned} y_{\mathrm{L}} &= \frac{2y_{\mathrm{M}} + y_{\mathrm{S}}}{2} \\ y_{\mathrm{R}} &= \frac{2y_{\mathrm{M}} - y_{\mathrm{S}}}{2} \end{aligned} \tag{3.45}$$

Since FLAC works with integers, if $y_{\mathrm{L}} + y_{\mathrm{R}}$ is odd, the division by 2 in Eqs. (3.44) is fractional and must be truncated, which would lead to an imperfect reconstruction. It can be prevented by subtracting 1 to y_{L} during the conversion to (y_{M}, y_{S}) and then, if y_{S} is odd, adding 1.

3.15.2.8 Structure of a FLAC File

The compression method just described needs a series of parameters that indicate the exact meaning of the compressed data. Among them, there are the sampling specification (sample rate, bit resolution, number of channels), the type of predictor adopted for each block, its variant and its parameters. There is also a series of metadata including comments, application used to generate the file (for instance, the reference or official codec), tables that simplifies the search process (necessary because the residual coding is of variable length, making the location of specific samples impossible without decoding first the audio data). It is also possible to include images embedded in the file, for instance, the cover of the CD box. In the metrological case, they could be photographs of the measurement site, the location on a map, etc. The details of the format that allow to decode a FLAC file can be found in Appendix K.

3.16 Project Files

Some digital audio editors, such as Audacity or Cubase, work with the concept of *project* instead of a single sound file. A project is a file structure that contains the digital audio itself as well as other associated information. An advantage of the project approach for audio editing is that it is possible to store some *edit events*, allowing recording and documenting the applied processes. The main drawback is that the formats are proprietary or specific for each software, so they are not recognized by other programs. A solution to this problem is the import and export features, which allow creating or reading files of widely accepted formats such as WAV or FLAC. However, the editing features are lost.

We will here refer specifically to the case of Audacity, since it is a free, open-source, and cross-platform software (it has builds for Linux, Windows, and Mac OS X, as well as other operating systems) and with an active development agenda (2017).

Audacity projects consist of a file with extension .aup (Audacity project) and a directory whose name is the same as the project's name ended with "_data." For instance, the project test.aup will be accompanied by the directory test_data. That directory contains a subdirectory structure where the audio data is stored in a number of files with extension .au or else some files with extension .auf associated with external audio files.

3.16.1 AUP Format

A .aup file is a text file in XML[22] format. As any XML document, it is formed by syntactic units called *elements*, each of which is affected by some modifiers called *attributes*. In turn, each element may contain other *nested elements* (called *child elements*). There is a main element that contains all the other ones, called *project*, among whose attributes are the state of the visualization windows of the different project tracks (horizontal position, selection limits, and zoom percentage) the last time it was saved, as well as the project's sample rate. Note that this could differ from the sample rates of different tracks, which in turn could differ from each other, particularly when they are imported from various original sources. This possibility is not too neat nor convenient in metrological applications, since the export of the project as a unique sound file involves resampling operations that might alter the original signal. The specification of AUP files is given in detail in Appendix L.

Within the project, we can find *tags*, *timetrack*, *wavetrack*, and *labeltrack* elements. The tags element contains typical metadata of the record industry, such as artist, genre, etc. In its present status, there is not support for other metadata that could be interesting in metrology, such as location, timestamp, microphone position, and orientation, etc., but since this is free software, it is possible that the community collaborates to include them in future versions.

The timetrack element refers to a *time control* track. It allows the user to input a time envelope that modifies the reproduction speed. While this is a very useful

[22]The XML format (eXtensible Markup Language) is a metalanguage suitable for defining specific languages (such as the present one or those for text documents, musical scores, or databases). A markup language is a code that includes text as well as metadata that refer to structure, semantics, or format. Each XML file has a specific reference to a *document type definition*, DTD, which sets its syntax, i.e., the description of each of its elements and their logical and hierarchical sequence. See more details in Appendix L.

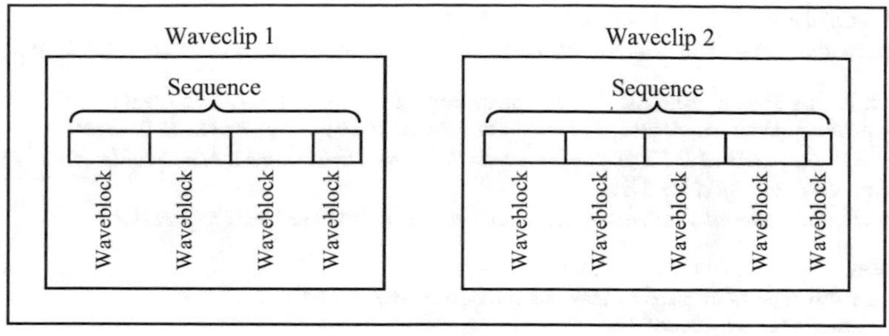

Fig. 3.28 *Wavetrack* structure in an Audacity project

feature for musical applications (particularly for electroacoustic music), it has no metrological application, since it introduces serious alterations in time and spectrum.

Each wavetrack element corresponds to an audio track. It contains the track name, the audio channel with which it is in principle associated,[23] as well as information as whether the track is configured as *solo*[24] or if it has been *muted.*[25] It also specifies the height in pixels of the track window, and whether it is minimized or selected. Finally, it contains the sample rate of the track, the gain and the panning status as a number between -1 and 1, where -1 means totally to the left channel, 1, totally to the right channel, and 0, both channels with equal signal proportion.

Wavetrack elements (Fig. 3.28) do not point to a single complete sound file. Instead, Audacity projects split each track into a series of *waveclip* elements. These are audio fragments, in general not contiguous. This allows saving storage space when between two fragments there is a silence. Representing that silence in an uncompressed format would require an inordinate storage of a long run of zeros.

In turn, each *waveclip element* is split into small audio segments of a maximum size of 2^{20} bytes, each one of which has an associated sound file in the project's data directory, or refers to part of an external sound field that has been imported without letting the software copy it in the data directory.

Next, an example of an AUP project file is provided. Interspersed with the content, there are notes and explanations. The original XML document's entries are copied in bold font, while the explanations are in italics and preceded by a dash. For the sake of clarity, most attributes of each element have been split into different lines and indented. The elements and sub elements have been indented as well. For a more formal specification, see Appendix L.

---

[23]Note that a channel corresponds to a physical input or output (for instance left and right), while a track contains a signal that may be assigned partially to each one of the channels by means of the *pan* control.

[24]When a channel is configured as *solo*, all other channels not configured as *solo* are automatically silenced.

[25]When a channel is *muted*, it is silenced.

```
<?xml version="1.0" standalone="no" ?>
```

— *Uses version 1.0 of XML and this file is not standalone (requires an external DTD definition)*

```
<!DOCTYPE project PUBLIC "-//audacityproject-1.3.0//DTD//EN"
"http://audacity.sourceforge.net/xml/audacityproject-1.3.0.dtd" >
```

— *Declares the external DTD and specifies the Internet URL where it is available. Note: This is outdated. Currently (Feb 2016): https://github.com/audacity/audacity/blob/master/src/xml/audacityproject.dtd*

```
<project
  xmlns="http://audacity.sourceforge.net/xml/"
  projname="test_data"
  version="1.3.0"
  audacityversion="2.1.2"
  sel0="0.5000000000"
  sel1="3.5000000000"
  vpos="0" h="0.0000000000"
  zoom="154.7142857143"
  rate="44100.0"
  snapto="off"
  selectionformat="hh:mm:s + milliseconds"
  frequencyformat="Hz"
  bandwidthformat="octaves">
```

— *The project element has, among its attributes, the name of the project directory, the DTD version, the Audacity version, the status of the edit environment, including the initial and final instants of the interval that was selected the last time the project was saved, the zoom percentage respect to a duration of about 15 s (depends on the screen resolution), the position (expressed in seconds) of the border of the track control panel with respect to instant $t = 0$, the general sample rate, the "Snap to" function status, the Selection Format, the Frequency Format and the Bandwidth Format.*

```
  <tags>
    <tag name="ARTIST" value="John Smith"/>
    <tag name="TITLE" value="Pure tone"/>
    <tag name="YEAR" value="2013-07-10"/>
    <tag name="Software" value="Cool Edit 2000"/>
    <tag name="Copyright" value="----"/>
    <tag name="COMMENTS" value="Metadata test in WAV"/>
  </tags>
```

— *Here, the typical metadata of the record industry are included.*

```
 <wavetrack
   name="Mixed"
   channel="2"
   linked="0"
   mute="0"
   solo="0"
   height="150"
   minimized="0"
   isSelected="0"
   rate="44100"
   gain="0.501187"
   pan="-0.5">
```

— *The wavetrack element has attributes of track name, channel, mute status, solo status, track window height, minimization status, track selection status, sample rate, gain and panning. It contains 0 or more wavclip elements.*

```
<waveclip
  offset="1.00185941">
```

— *The waveclip element has a single attribute: the time displacement (offset) in seconds of its beginning with respect to the initial instant of the project (t = 0). In this case it starts at instant t = 1.001 859 41 s.*
This in particular has exactly one sequence and one envelope.

```
  <sequence
    maxsamples="262144"
    sampleformat="262159"
    numsamples="44100">
```

— *The sequence element has the following attributes: The maximum number of 32 bit samples that each of its wavebloks can possibly contain; the same number plus 15; the total number of bytes in the sequence (in this case, 44 100, corresponding to 1 s).*
A sequence has a series of audio blocks (waveblocks) that constitute a continuous piece of recording (see figure 3.28).

```
    <waveblock
      start="0">
```

— *Each waveblock element contains a number of bytes that is less than or equal to* 2^{20}*. Its only attribute is the number of initial sample with respect to the initial sample of the waveclip (which is declared in seconds in the waveclip attribute "offset")*
This waveblock contains the reference to an .au file located in the data directory of the project.

```
      <simpleblockfile
        filename="e0000b18.au"
        len="44100"
        min="-0.5"
        max="0.5"
        rms="0.237011"/>
```

— *The simpleblockfile element is one of the possible sub-elements of a waveblock. It is associated with an .au file and it has only attributes. These are the .au file name, the number of samples and its minimum, maximum and RMS values*

```
    </waveblock>

  </sequence>

  <envelope
    numpoints="0"/>
```

— *The envelope element contains 0 or more control points that consist in an ordered pair (time, value). Those points define an amplitude envelope. In this case there are no control points, so the envelope is identically equal to 1 over the whole sequence duration.*

```
</waveclip>
```

```
    <waveclip
      offset="4.00100000">
    — This waveclip starts at  instant t = 4.001 000 00 s.

      <sequence
        maxsamples="262144"
        sampleformat="262159"
        numsamples="44761">
        <waveblock
          start="0">
          <simpleblockfile
            filename="e0000c89.au"
            len="44761"
            min="-0.501237"
            max="0.501229"
            rms="0.293003"/>
        </waveblock>
      </sequence>
      <envelope
        numpoints="0"/>
    </waveclip>
  </wavetrack>

  <wavetrack
    name="Tone 100 Hz"
    channel="2"
    linked="0"
    mute="0"
    solo="0"
    height="150"
    minimized="0"
    isSelected="1"
    rate="48000"
    gain="0.501187"
    pan="0.5">

    <waveclip
      offset="0.00000000">

      <sequence
        maxsamples="262144"
        sampleformat="262159"
        numsamples="336000">

        <waveblock
          start="0">
        — This first audio block starts at the initial instant of the waveclip.
          <pcmaliasblockfile
            summaryfile="e000039a.auf"
            aliasfile="E:\Sounds\Audacity\Tone 100 Hz.wav"
            aliasstart="0"
            aliaslen="262144"
            aliaschannel="0"
            min="-0.501251"
            max="0.501221"
            rms="0.354355"/>
```

— *The pcmaliasblockfile is other of the possible subelements of a waveblock. In this case the waveblock corresponds to a sound that has not been imported to the data directory, but its origin is the file located at E:\Sounds\Audacity\Tone 100 Hz.wav. However, the project does involve an .auf file that contains a summary of the sound (maximum, minimum and rms values once every 65 536 and once every 256 samples).*

```
</waveblock>

<waveblock start="262144">
```

— *This second audio block starts on the sample 262 144 from the initial instant of the waveclip.*

```
<pcmaliasblockfile
   summaryfile="e0000cf3.auf"
   aliasfile="E:\Sonidos\Audacity\Tono 100 Hz.wav"
   aliasstart="262144"
   aliaslen="73856"
   aliaschannel="0"
   min="-0.501251"
   max="0.501251"
   rms="0.266338"/>

</waveblock>

</sequence>

<envelope numpoints="3">
```

— *In this case the envelope has 3 control points (controlpoint elements). Between the second and third control points there is a sudden gain reduction from 1 to 0.527 777 791 023*

```
<controlpoint t="0.000000000000" val="1.000000000000"/>
<controlpoint t="3.044321329640" val="1.000000000000"/>
<controlpoint t="3.044326329640" val="0.527777791023"/>
</envelope>

</waveclip>

</wavetrack>

<labeltrack
   name="Labels 1"
   numlabels="2"
   height="73"
   minimized="0">
```

— *The labeltrack element has the following attributes: track name, number of labels, height of the label track window in pixels and minimization status. It has 0 or more label elements.*

```
<label
   t="1.02585411"
   t1="2.01662050"
   title="White noise"/>
```

— *The label element has the following attributes: initial and final time and title (label's text).*

```
    <label
      t="3.04432133"
      t1="3.04432133"
      title="Gain change"/>
```

— *In this case the initial and final times are identical. This means that it is a label associated with an instant instead of an interval*

```
  </labeltrack>
</project>
```

It is interesting to note that AUP files can be edited or even created outside the Audacity environment. This would allow, for instance, to set with great precision the control points of an envelope, which might be associated with scale changes in the instrument that generated the signal.

3.16.2 AU Format

AU files are located in the data directory of the project. They have a fixed-length header, a variable length annotations field and an audio data segment. Table 3.9 summarizes the structure of an AU file. These structure is based on the .au files of Sun Microsystems. The main differences are that they begin with the string "dns." instead of ".snd" (note that this identifier is the exact reverse of that of Sun's files), that the audio data length is irrelevant and that there is always a single channel.

Although the format is not officially documented, some reverse engineering and some inquiries to Dominic Mazzoni (cofounder of the Audacity project and emeritus developer) were done as to the content of the annotation block.[26] It consists of a *summary*, i.e., a very short version of the wave, intended for graphical purposes at very low zoom levels. It has two parts. The first contains the minimum, maximum, and RMS values of consecutive blocks of 65,536 samples; the second, the minimum, maximum, and RMS value of consecutive blocks of 256 samples.

The summary content is used to plot the signal oscillogram when very long intervals are displayed. In this case, it would make no sense to plot millions of samples on a display whose horizontal resolution is much lower,[27] since many samples would pointlessly collapse into a single vertical line. Instead, the short summary version is plotted for very low zoom, and the long one for intermediate zoom.

The size L_{file} of an .au file can be estimated as the addition of the header length, the audio data segment length and the summary length:

[26] Originally, the annotation block was intended for future inclusion of structured data such as markers or edit points. In Audacity that information is included in the .aup project file (Sun Microsystems 1992).

[27] For instance, a Full HD display has only 1920 pixels in the horizontal direction.

Table 3.9 Structure of an Audacity audio data file, .au

Length (byte)	Content	Comments
4	"dns."	Identifier associated with Audacity's sound files
4	Audio offset A, in bytes	Hexadecimal little endian
4	0xFFFFFFFF	Flag indicating that the audio data size is irrelevant
4	0x06000000	Code indicating that data are in format IEEE-754 (32 bit float)
4	Sample rate in Hz	Hexadecimal little endian
4	Number of channels	It is always 1 for Audacity audio blocks
8	"Audacity"	Program identifier
12	"BlockFile112"	Information type identifier
$A - 44$	Annotation block	Variable length, multiple of 8 byte
$L - A$	Audio data segment	Audio samples in 32 bit floating point IEEE-754 format

$$L_{\text{file}} = 44 + 4N_{\text{audio}} + 12\left[\frac{N_{\text{audio}}}{65{,}536}\right] + 12\left[\frac{N_{\text{audio}}}{256}\right],$$

where [] stands for the integer part.

3.16.3 AUF Format

When one of the original audio sources of an Audacity project is a sound file that has been imported without the option of saving a copy in the data directory (an alias file), Audacity generates one or more *summary files*. Each one has the summary (see the structure in Sect. 3.16.2) of an audio portion of up to 2^{20} bytes. The format of the AUF files is given in Table 3.10. It is an abbreviated version of the AU format in which the header and the audio content has been removed.

3.16.4 External Access to Audio Data

It is possible to access the audio data of an Audacity project via an external software, such as matrix mathematical packages like Scilab, Octave, or MATLAB.

Table 3.10 Structure of an Audacity audio summary file, .auf

Length (byte)	Content	Comments
8	"Audacity"	Program identifier
12	"BlockFile112"	Information type identifier
$A - 20$	Summary	Variable length, multiple of 8 byte

To this end, there are two possibilities. The first one is to export the project from Audacity as a WAV file and then use the **wavread** function to acquire the signal as a column vector (or a two-column matrix in the case of a stereo file) containing the samples normalized to the $[-1, 1]$ range. The second possibility consists in reading the project AUP file, getting the information on the location of each audio block and other attributes such as panning and gain, and finally reconstructing the signal within the mathematical software.

3.17 Recording and Storage Media

Nowadays several recording and storage media for digital audio coexist. To clarify, by *recording* we understand the process of immediate transfer, in real time or with low *latency*,[28] of the digital signal generated by the analog/digital converter to the medium, while by *storage* we understand the long-term preservation of the digital audio signal. The more widely used media are hard disks, flash memories, optical discs[29] such as the CD, the DVD, and the Minidisc, and digital audio tapes (DAT).

3.17.1 Hard Disk

The *hard disk* is, nowadays (2017), still the most widely used mass storage medium for digital information in computers. Hard disks are also used in some digital audio recorders, in particular, multitrack recorders, and in some video cameras. They are also part of portable drives with USB connection intended for backup or for data portability.

They consist in one or more rotating platters made usually of aluminum alloy, glass, or ceramic covered with a magnetic coating, over which the read-write heads "float" by aerodynamic effect (see Fig. 3.29). The "flight" height is well under 1 μm, that is why the surface must be extremely flat, polished, and clean. In order to prevent the deposition of any particle of dust which would impair the heads, all parts except the electronics must be enclosed inside a housing with highly filtered air.

The physical principles of hard disk reading and writing are the same as in a magnetic tape. There are differences only in the mechanical implementation. In tape

[28]*Latency* is the delay that exists between the instant in which the sound reaches the microphone and the instant in which its digital version is transferred to the medium, or, vice versa, the delay between the instant in which the audio is read from the medium where it is stored and the instant in which it is reproduced by a player.

[29]Note that both spellings, "disc" and "disk", are acceptable. However, often optical discs are spelled "disc" while magnetic disks are spelled "disk".

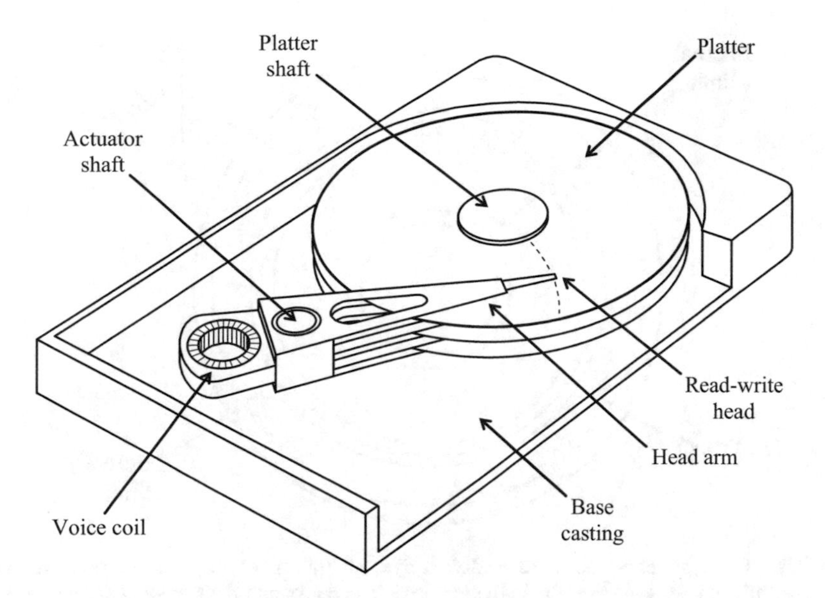

Fig. 3.29 Internal structure schematic of a three-platter six-head hard disk drive. For clarity, part of the base casting has been removed. The pseudo-radial path of the head is shown in *dashed line*

recording, the magnetic tape moves at moderate speed against a static magnetic head with which is in contact. The information is recorded longitudinally on the tape. The tape is only moving during recording and playback operations and random access is not possible.

In a hard disk, the platters move much faster.[30] To minimize erosion, the head is not in contact with the magnetic coating. Information is recorded on concentric circumferences (called tracks), and the head moves radially to reach the desired track, allowing random access. The platter is continuously spinning, even when no write or read operation is being performed. This is because wear is much worse during spinning up and down than when the disk keeps running.

3.17.1.1 Physical Structure of a Hard Disk Drive

In Fig. 3.29, the main mechanical components of a hard disk drive are shown. We first have the disks or *platters*. In order to increase the total disk capacity most hard disk drives have two or more platters, and the platters have both sides magnetically coated, doubling its capacity. Information is distributed on concentric circumferences

[30]In the outer region of 3.5 in disk spinning at 7200 rpm, the linear velocity of the surface with respect to the head reaches 5.33 m/s.

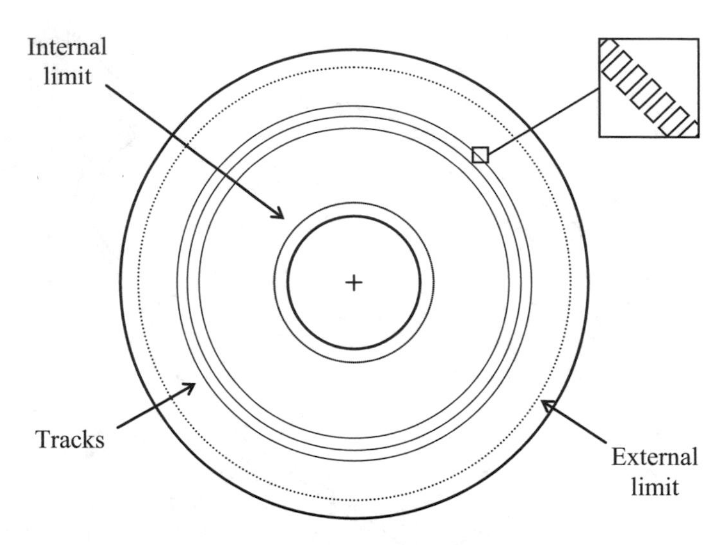

Fig. 3.30 Track structure on a hard disk drive platter. A small track portion has been schematically enlarged to show the individual bits. The tracks and their spacing are out of scale

called *tracks* (see Fig. 3.30). Each individual bit takes a microscopic portion of the track. Because of the large storage capacity of hard disks, the number of tracks is very large (some tens of thousands) and within each track there are a large number of bits.

Each platter is accessed, for writing and reading, by one or more read–write heads attached to the end of the head arms. The arms corresponding to all platters and all sides move together driven by the actuator, which consists on a moving coil immersed in the magnetic field of a permanent magnet (not shown in the figure). Because of the similarity to a loudspeaker, this coil is called *voice coil*.

In the first disks, the actuator was a stepper motor. This type of motor receives electric pulses and for each pulse it spins a fixed angle (a fraction of 2π). Using a gearbox, it was possible to reduce the angle so that each pulse allowed to leap from one track to the next one. This is called *absolute positioning*. It has the virtue of simplicity, since to go to a given track it suffices to apply a number of pulses to the stepper. However, as technology advanced and the track density increased, there were positioning precision issues. Temperature variations, for instance, would cause the platters and the actuators expand or compress differently, causing alignment errors.

In modern hard disk drives, as already described, the actuator is a voice coil that moves within a magnetic field. When an electric current flows through the coil, it creates its own magnetic field that interacts with the existing one causing it to move in either direction according to the electric current direction. As this technique is, in principle, even more imprecise than the stepper actuator, it is complemented with a control loop or *servo system* which compares the desired position with the real one in order to introduce the necessary corrections. This is called *relative positioning*.

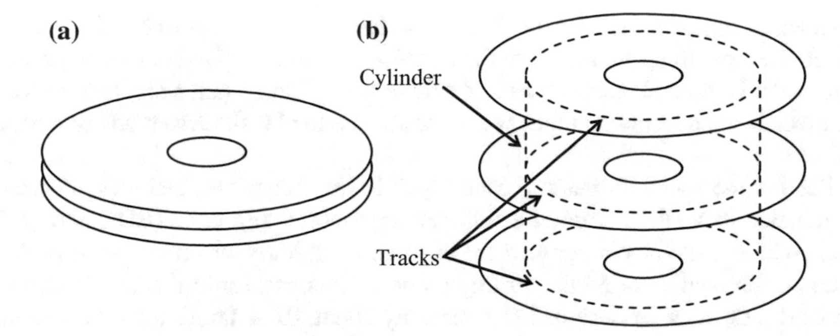

Fig. 3.31 **a** Platters in their normal position. **b** Platters set apart from one another to make visible the tracks corresponding to a cylinder

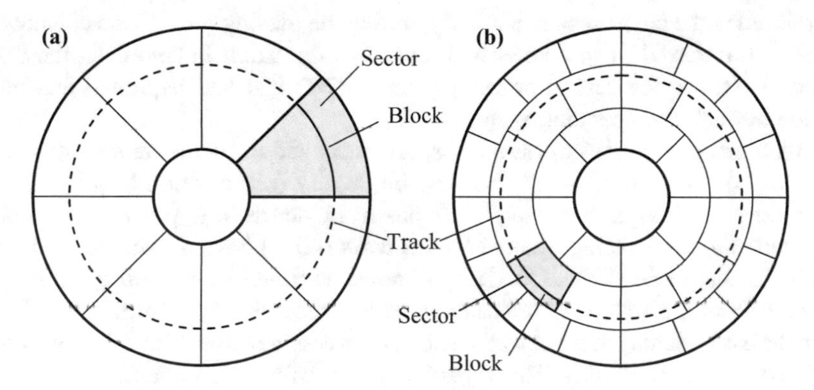

Fig. 3.32 Subdivision of the platter surface in sectors. **a** Uniform number of sectors for all tracks. **b** More sectors in the outer tracks

3.17.1.2 Geometric Aspects of a Hard Disk

An important feature of hard disks is their *geometry*, i.e., the way in which the information is distributed on the platters surface. This information needs to be addressed. This is accomplished specifying the *cylinder*, the *head*, and the *sector* where the desired information is located. A *cylinder* is any set of tracks of equal radius on the different platters. This concept is illustrated in Fig. 3.31. The reason to speak of cylinder instead of *track* is that all heads move together so whatever the actuator's position, all the heads are over the tracks of a single cylinder.

The *head* specification is necessary in order to indicate which of all heads on the cylinder is to be used to write or read. In general, all platters have two heads, one per side.

Finally, a *sector* is an angular portion of the platter (Fig. 3.32) whose tracks have a fixed number of bytes called *block*[31] (in general, 512 or 4096). This subdivision of

[31]The word "sector" is often used to refer either to a sector or a block.

the track information in blocks is due to the fact that the synchronization and error correction difficulties are the same for an individual byte as for a complete group of bytes, so it is more efficient to read or write a whole block than a single byte. Thus, the block is the minimum number of bytes that the hard disk drive reads or writes at a time.

Besides the user information, each block has a header that includes addressing information as well as error-detecting and error-correcting code (EDC and ECC). This code is necessary because there is some probability of error due to material failure or problems during the writing process. This extra information implies that the hard disk must be designed to a capacity about 10 % larger than the specified one.

The information to save in the disk is temporarily stored in a piece of memory in the electronic section of the drive called *buffer*. Once the buffer is full, its content is transferred to the target sector. Similarly, during the reading process the content of a block is transferred to the buffer and, after this operation is complete, the information is sent to the central processing unit (CPU) that had requested that information through the operating system.

In the first generations of hard disks, all tracks had the same number of sectors (Fig. 3.32a), so the outer tracks had less bit density (bits per unit length) than the inner ones. Although this simplified the head addressing process, it implied important waste of storing space. Modern disks (2017) have a variable number of sectors, as shown in Fig. 3.32b. This is known as *zoned bit recording*, *ZBR*.

Hard disks are relatively delicate, since the head size is extremely small and could be easily damaged if subject to shock or excessive vibration. This is why they should be handled carefully. This is particularly true for USB portable units or those in netbooks, and laptops. In return, they allow a fairly high transfer rate and are very reliable for medium-term storage, being capable of keeping the information unaltered for 5 years or more.

All file systems are meant to try to fill with new information the sectors that have been deleted (except in flash drives if it can be avoided). This causes files that are larger than the first empty space available to be split into nonconsecutive sectors. This is called *fragmentation*. After a while, it is possible that files are fragmented in such a way that (1) Access speed is impaired, and (2) The actuator is subject to excessive stress, reducing the lifetime of the disk. This can be corrected performing a disk defragmentation by means of a software application provided with the operating system.

3.17.2 Flash Memory

Flash memory is a very convenient recording medium for digital audio because of its small size, low-power consumption, robustness against shock, vibration or drops, high storage capacity, random access, and high data transfer speed. However, it is not completely reliable for long-term storage, and allows limited

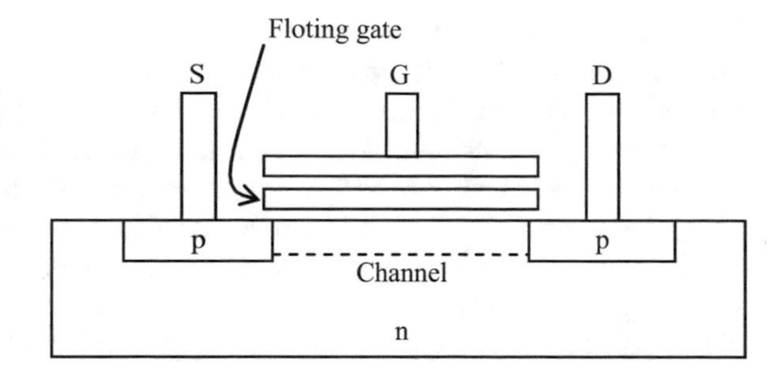

Fig. 3.33 Structure of a FGMOS transistor used in flash memory technology. S is the *source* (it is grounded), D the *drain* (the output) and G the *control gate*; *n* and *p* indicate that silicon has been doped with atoms that provide negative (electrons) and positive (holes) charges

number of write operations at each memory location (high quality memories allow about 10^6 operations). Another drawback is that at the time being (2017) their price per GB is still higher than for the hard disk, even if the latter uses sophisticated electromechanical technology.

Over the last few years flash memory is being used in digital recorders instead of a hard disk. The most popular formats are SD (Secure Digital); SDHC (High Capacity Secure Digital), and CF (Compact Flash). Nowadays, they reach capacities in excess of 256 GB, while the most common models are of 4 GB to 16 GB, allowing to record 6 h to 25 h of CD-quality audio (16 bit stereo with a sample rate of 44 100 Hz).

The transfer speed depends on the *class* specification, a figure expressed in MB/s that is indicated on the memory label. Hence, a class 6 memory will allow data transfer at a rate of 6 MB/s. The most popular memories are class 4, which suffices for CD-quality audio, which requires 176.4 kB/s.

A limited data transfer rate might be an issue in multitrack recorders, since they can record dozens of simultaneous tracks, requiring a high data transfer rate. An application where this is may be useful is in the case of *beamforming*[32] techniques.

Flash memory operation is based on the electric charge storage in a floating conductor, i.e., one that is physically insulated from other conductive parts by an extremely high resistivity dielectric material. Although there may be some electric charge leakage through the insulating material, it is very slow, so the information can stay virtually unaltered for years. However, the long-term stability may be affected by environmental conditions such as high temperature, presence of electromagnetic fields, ionizing radiation, and unexpected supply voltage change (for instance, a blackout).

The basic device of a flash memory is a special type of metal-oxide-semiconductor field effect transistor (MOSFET) called *floating-gate transistor*, FGMOS, whose structure is shown schematically in Fig. 3.33. The floating gate is isolated from the MOSFET channel and from the control gate by two silicon dioxide (SiO_2) insulating

[32]The *Beam-forming* technique consists in using an array with several microphones to receive sound, and apply different weightings (gains) to each one so to create a highly selective directional pattern. Since the directional pattern can be changed during post processing, this technique allows to create images of the acoustic distribution of sound sources (Analog Devices 2013).

layers (not shown). In normal conditions, if a positive voltage is applied to the control gate, the channel turns to be conductive. But if a relatively high voltage (5 V) is applied, the current density through the channel increases, allowing that some high-energy electrons jump over the insulating layer toward the floating gate. This phenomenon is known as *hot-electron injection*. The floating gate is polarized in such a way that it rejects the channel's carriers, turning it permanently nonconductive. The conductive or insulating nature of the channel is associated with a logical 0 or 1. In order to get back to the original state, a −5 V voltage is applied to the control gate, causing the electrons trapped in the floating gate to jump back to the channel, turning it conductive again.

These FGMOS transistors are integrated to form a matrix allowing the individual access by means a complex addressing system.

There are two types of flash memory known as NOR and NAND. Although the implementation details are beyond the scope of this book, let us say that NOR memories allow random access on a byte-per-byte basis, but erasure is only possible on block basis. NAND memories achieve higher information surface density, at the price that read and write access is on a *page* basis (where a page has between 512 bytes and 4096 bytes), while erasure is block based (a block contains 32–128 pages), This is not too much of a problem, since to erase a few bytes it suffices to read the whole block into RAM memory, erase the bytes and rewrite all the pages forming the block back into the flash memory.

This behavior is very similar to the case of a hard disk, where to read or write an individual byte it is necessary to read or write the whole block containing it.

Nowadays (2017), most memories on the market are of the NAND type, since, as mentioned, they allow higher information density and, hence, higher capacity.

In the case of flash memories, defragmentation is not only unnecessary but also detrimental, since it forces a huge number of erasure and write operations that can reduce the lifetime of the device. Moreover, to minimize the number of rewrite operations, the space available after erasure is not used until the full memory has been written.

3.17.3 Optical Discs

Optical disc technology starts in the 80s with the public release of the *compact disc* (CD). Optical discs are made of polycarbonate, a polymer with an excellent dimensional stability that allows precision molding as well. The general philosophy of optical discs is to provide two different light reflectivities. One of them is assigned to the logical 0 and the other to the logical 1. The disc is scanned by a very thin laser beam that, when impinging on the active surface, is reflected with higher or lower intensity. This generates in a phototransistor two different voltage values that can be associated with the logical states.

How the reflectivity change is attained differs from one technology to another. In pressed read-only discs such as audio CD (CD-DA), DVD, and CD-ROM there is a series of pits whose depth is 1/4 the wavelength of the impinging light. When the laser beam strikes either the bottom of a pit or its outside (the *lands*), reflection is maximum because of the aluminum coating which makes the surface reflective. When it strikes the border of a pit, half the light reflects inside and half outside the pit. As there is half a wavelength phase delay ($\lambda/4 + \lambda/4$), there is subtractive interference cancelation, so the intensity of the reflected beam is lower. Hence, pits and lands correspond to a logical 0 and borders to a logical 1. Since for physical reasons it is impossible that there be two consecutive borders, there cannot be two consecutive 1's either. For this reason, it is necessary to recode the information by means of a process called *eight-to-fourteen modulation, EFM.*[33]

In the write-once discs, such as the CD-R and the DVD-R, of widespread use in personal computers for backup purposes, the write technique involves a high power laser that alters the reflective properties of a dye of the family of the *cyanines*, *phtalocyanines*, or *azos*, at a temperature of about 250 °C. Where the dye has burnt, there is lower reflectivity, so when a low-power laser illuminates it there is low reflected intensity, which is taken as a logical 1. While in this case, it would be easier that there were two consecutive 1's, to ensure compatibility with the ordinary CD or DVD the eight-to-fourteen modulation is also used. The cyanines that were used in the first CD-R were chemically unstable and particularly sensitive to ultraviolet light exposure. Dyes used nowadays are far more stable, but the exposure to ultraviolet light is still cause of degradation. A disc with archival purposes should be kept protected from sun light.

Finally, rewritable discs such as the CD-RW use a phase change technique. A polycrystalline reflective material (a silver, indium, antimony, and tellurium alloy) is irradiated by a high power laser beam until it melts at a temperature between 500 and 700 °C. When it solidifies again, it does not return to its original polycrystalline state but reaches an amorphous nonreflective state, which is read as a logical 1. To restore the material to its polycrystalline state, it must be irradiated at medium power so that the alloy reaches 200 °C, a temperature at which it returns to the polycrystalline state.

3.17.3.1 Compact Disc (CD or CD-DA)

We shall study now the digital format of the digital audio compact disc in some detail. The medium consists of a disc made of transparent polycarbonate of a diameter of 120 mm (80 mm in the mini-CD) and a thickness of 1.2 mm (see Fig. 3.34). The disc is placed in the player introducing its central hole in the drive shaft. This operation is automatic in most players, except those in laptop computers. The reading process is performed by means of a laser beam emitted by a laser diode. This beam is concentrated by a lens (not in contact with the disc surface, to

[33]It consists in replacing each byte by a 14 bit word that satisfies two conditions: There cannot be two consecutive 1's or more than ten consecutive 0's.

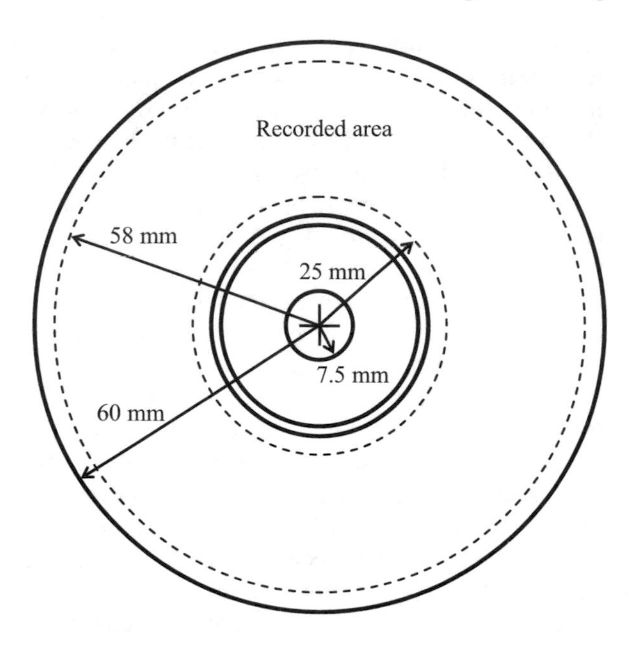

Fig. 3.34 Dimensions of a compact disc. The *dashed* circumferences delimit the recorded area. The circumferences close to the central hole provide thickness increase to separate the recorded area from the reader

increase durability) and finally reaches the reflective layer with a diameter of about 1 μm. The high or low reflectivity is read as a 0 or a 1, respectively.

Unlike the analog vinyl records, playback is from the inner area to the outer area, and the spinning velocity is not constant but it reduces progressively. Another difference is that the spinning velocity differs from a disc to another. This is because the velocity is used to compensate for duration differences, in an attempt to fill the available area as much as possible. The theoretical maximum duration of a recorded program is 74 min 33 s, but it is possible to have longer durations (up to about 80 min). For a given disc, the linear velocity is kept constant (*constant linear velocity*, CLV), instead of the angular velocity. For this purpose there is a servo system that uses synchronization codes within the disc itself.

The linear velocity is about 1.4 m/s in the shortest discs and 1.2 m/s in the longest ones. The relationship between the linear velocity v in m/s and the angular velocity ω in rpm can be obtained as follows. For a given radius r, each turn has a total length

$$L = 2\pi r,$$

so the number of turns per second will be v/L, and the number of turns (revolutions) per minute, 60 times larger. Thus

$$\omega = 60\frac{v}{L} = \frac{60v}{2\pi r}.$$

If we consider an average disc in which the linear velocity may be 1.3 m/s, the angular velocity at the beginning of the disc (i.e., in its internal turn, whose radius is 25 mm) will be

$$\omega = \frac{60 \cdot 1.3}{2 \cdot 3.14 \cdot 0.025} = 497\,\text{rpm},$$

while at the end (in its external turn, of radius 58 mm), it will be

$$\omega = \frac{60 \cdot 1.3}{2 \cdot 3.14 \cdot 0.058} = 214\,\text{rpm}.$$

It is observed that the angular velocity is much higher than in the case of the traditional vynil record (33 1/3 rpm or 45 rpm).

We shall see now how the digital information is encoded into the CD. At the physical level, a structure of ovoidal pits is stamped on the upper side of the disc. That surface is plated with a reflective aluminum layer, and finally coated with a protective lacquer layer. On top of the lacquer is the artwork that identifies the disc and its content (Fig. 3.35b).

The disc scan is performed by the laser beam from the underside (often referred to as the *clean side*). The depth of the pits is adjusted to some value between 0.11 and 0.13 μm, selected such that the near infrared light beam (with a wavelength of 780 nm) reflected on the bottom of a pit is out of phase by π with that one reflected

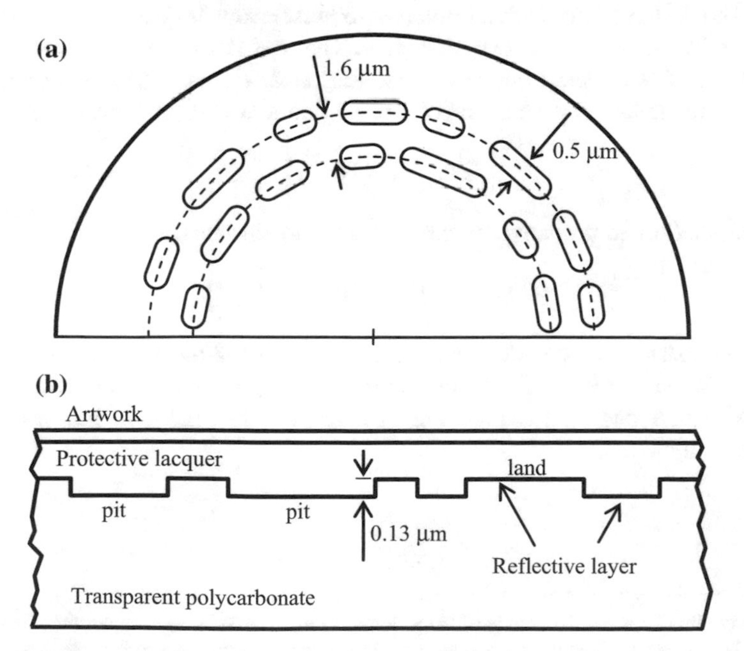

Fig. 3.35 Pits and lands structure in a compact disc. **a** Disc seen from below. **b** Cross-section of a small disc portion along a pit line (Schemes are out of scale.)

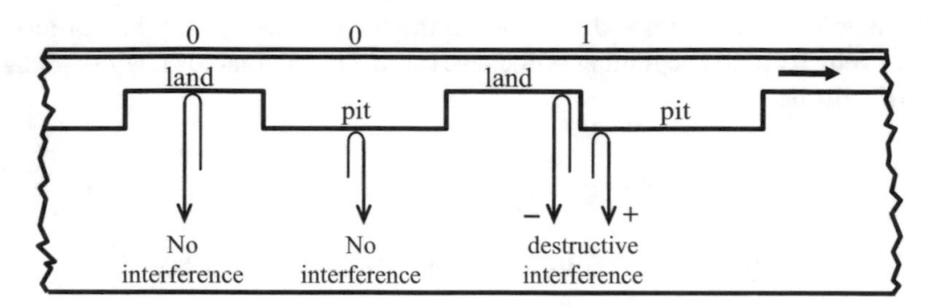

Fig. 3.36 Mechanism to read the raw digital information on a compact disc

on a land. When light strikes the border of a pit, part is reflected inside and part outside the pit, creating a destructive interference (Fig. 3.36). In consequence, every time the border of a pit passes by the photosensor it will receive dim light, and every time a pit or a land passes by the photosensor, it will receive intense light. As mentioned, these are interpreted as 1's and 0's, respectively.

One problem that had to be dealt with in the CD design was that of robustness against data corruption. Unlike a hard disk, which is inside an enclosure free of dust, the CD is a medium prone to be handled negligently by users, so they can easily get dirty or scratched. Since the area occupied by a single bit is microscopic, a small scratch or a fingerprint[34] could render a number of consecutive bits unreadable. The same is true for small manufacturing faults affecting the reflecting surface. This type of error involving a series of contiguous symbols is called a *burst error*. The design of a code that is tolerant to burst errors has been quite a challenge.

The solution calls upon two strategies. The first is to use an *error-correcting code* that is able to detect and correct isolated errors. This is generally attained by means of redundancy. As an example, if we had a system of 2 bit words,

$$00 \quad 01 \quad 10 \quad 11,$$

we could add three properly chosen extra bits, for instance,

$$\underline{000}00 \quad \underline{011}01 \quad \underline{101}10 \quad \underline{110}11,$$

where the extra bits are underlined. Suppose that we allow at most one single bit change. For each original 5 bit word, there are six possibilities: one with no error, and five with a single bit change. For instance, the first one may appear as any of the following:

[34]Anyway, the CD is relatively insensitive to translucent stains such as a fingerprint. The slightly conical laser beam reaches the surface with a diameter of 0.7 mm. Due to the refractive index of polycarbonate (1.585), it changes its angle in such a way that when impinging on the internal reflective surface it has a diameter of about 1 μm. This is why a small stain is not so detrimental.

00000 00001 00010 00100 01000 10000

In total, we have $4 \times 6 = 24$ possibilities. It can be checked that for each one of the original 5 bit words the altered versions cannot come from any of the others, so it is possible to exactly reconstruct the original word, which in turn allows to retrieve the original 2 bit word.

This simple example can be generalized by means of a code originally introduced by Reed and Solomon in 1960 (Pohlman 2002; Watkinson 1988, 2002). Starting from an original word with k symbols,[35] this code adds $2t$ conveniently selected symbols to get a new word with $n = k + 2t$ symbols. This word is resistant to errors in up to t symbols (Pohlman 2002). The code is notated as RS(n, k).

The second strategy consists in redistributing the information in such a way that the symbols originally belonging to a word are now distant from each other. Consequently, the symbols that now are adjacent belong to different original words. Burst errors (the most typical errors in a CD) will then affect very few symbols per original word, allowing error correction according to the Reed–Solomon theory.

The combination of both strategies results in a code that can detect and correct moderate burst errors.[36] It is known as *Cross Interleaved Reed–Solomon Code*, CIRC in honor of its authors.

In the case of the Audio CD, we start with a set of 6 consecutive stereo samples, called a *frame* (or a CD block), i.e., $6 \times 2 \times 2$ bytes = 24 bytes. The first stage consists in interleaving the samples from the original order

$$D_1 \quad I_1 \quad D_2 \quad I_2 \quad D_3 \quad I_3 \quad D_4 \quad I_4 \quad D_5 \quad I_5 \quad D_6 \quad I_6$$

to the new order

$$D_1 \quad D_3 \quad D_5 \quad I_1 \quad I_3 \quad I_5 \quad D_2 \quad D_4 \quad D_6 \quad I_2 \quad I_4 \quad I_6$$

This process is completed with a one-sample delay for the odd samples. The result is coded by a Reed–Solomon encoder RS(28, 24) coder which adds 4 bytes to get a new frame with 28 bytes instead of the original 24 bytes. This encoder is called C2.

In a second stage, a new interleaving is applied so that each one of the 28 bytes will be assigned to a different frame. These frames are separated by a space equivalent to 4 frames. In consequence, each original frame will distribute over a space equivalent to $27 \times 4 + 1 = 109$ frames (see Fig. 3.37).

The third stage involves a new Reed–Solomon encoder RS(32,28) (called C1) which adds another 4 bytes. The new code has in total 8 extra bytes, so it allows correcting errors in up to 4 bytes. The minimum distance between two bytes that

[35]Each symbol may have s bits. For instance, it can be a byte, which has 8 bit.

[36]Of course there is no code capable of correcting massive burst errors. This method is intended for the typical errors such as small stains or scratches.

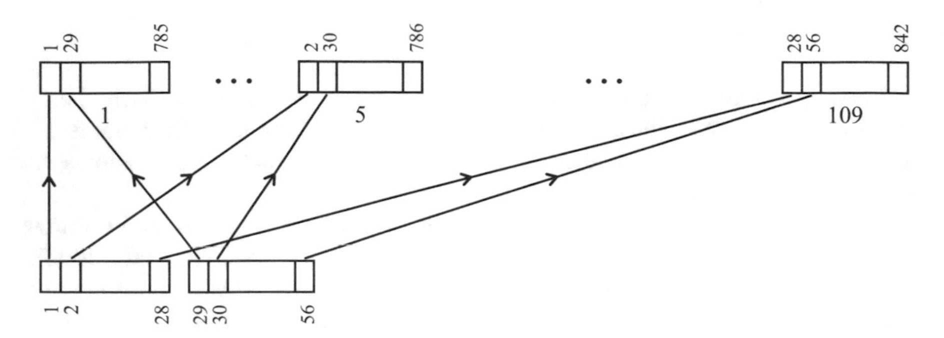

Fig. 3.37 Cross interleaving in a CR after application of a Reed–Solomon RS(28,24) coder. *Bottom* Original frames. *Top* Frames after interleaving

were originally adjacent will be now 32 × 4 bytes, so a burst error affecting up to 4 × 32 × 4 bytes (i.e., 512 bytes or 4096 bit) would change at most 4 bytes of the frame with the double interleaving and the double Reed–Solomon coding, i.e., a change that the code is able to detect and correct.

Two additional elements are added to each 32 byte frame as a result of the CIRC encoding (24 original bytes plus 8 extra bytes for error correction). The first one is a byte known as *subcode* whose bits are called P, Q, R, S, T, U, V, W. It is intended for allocating some metadata. The homologous bits of 98 consecutive frames, for instance, $P_1P_2 \ldots P_{98}$, constitute a 98 bit subcode word. Subcodes P and Q are used for information on track distribution (start, end, and track number), content table, and ISRC catalog code. It has also a cyclic redundancy code CRC for error detecting

The second element is a 24 bit *synchronization word*, selected as a bit pattern that cannot appear as the result of the combination of any of the previously described encoding processes:

$$100000000001000000000010.$$

The aim is to adjust frequently (once every 6/44100 s = 0.13605 ms) the system clock, to prevent a slow drift from causing catastrophic errors.

We have, so far, a code with 256 bit of CIRC-encoded audio, 8 bit of subcode and 24 bit of synchronization word, making a total of 288 bit. This information cannot be directly transferred to the CD for two reasons. First, there is no guarantee that there will not be two or more consecutive 1's and, as we have seen, CD technology does not allow consecutive 1's. Second, there is no guarantee either that there will not be too many consecutive 0's. This would imply too much low frequency content, and this complicates the synchronization process. It is necessary to transform the original sequence of 0's and 1's via what is known as a *group channel code*, i.e., a code that codes groups of bits in other groups of bits that are better suited to be sent through a given transmission channel (in this case, the reflecting surface of the disc), circumventing its limitations.

The channel code used in the CD is the *eight-to-fourteen modulation*, EFM, which substitutes each byte by a 14 bit word that complies with the 2–10 rule (at least two but no more than ten consecutive 0's). This implies that there will be less transitions and, hence, pits, and lands will be larger. There are 16,384 possible combinations of 14 bits, from which 267 satisfy the rule and 256 were included in a look-up table. The synchronization word is not modulated since it already satisfies the rule.

The eight-to-fourteen modulation complies with the 2–10 rule within each valid 14 bit word, but it does not guarantee that the same holds between consecutive words. For instance, one word could end with 1 and the next start with 1. To prevent this and to reduce the overall DC component (a necessary specification for the synchronization system), three new bits are added to each 14 bit word. These three bits are selected dynamically according to the cumulative DC component. Three bits are also added to the synchronization word. It turns out that the channel code of each frame has a length

$$(256 + 8) \times \frac{14 + 3}{8} + 24 + 3 = 588\,\text{bit}$$

The redundancy factor is, thus, 588/192 = 3.0635, i.e., the bit count is three times larger than in the original PCM data. This is the price to pay for a system that is tolerant to burst errors as well as compatible with the characteristics and limitations of the medium.

The CIRC code allows not only to correct many burst errors but also to detect other errors that cannot be corrected because the information loss is too severe. The error detection means that the system knows where the wrong uncorrected bytes are. This implies the possibility of concealing the errors by means of interpolation techniques. This is acceptable for audio playback, but it could significantly alter the signal for applications where the data integrity is crucial.

It is important to take this into account, since one of the most significant aspects of digital audio was the possibility of getting copies that are identical to the original data. The audio CD gives priority to the listening experience over data integrity, so it could not offer any guarantee.

When the CD is to be used for archival purposes, it is advisable to follow some handling and storage recommendations:

(1) Do not touch any of the surfaces directly, since the finger grease could increase the probability of data loss. The acidity of perspiration could, in the long run, damage the lacquer. It is better to hold the disc on the edge.
(2) Do not store the CD's in slim CD cases, since there is so little room that when stacking several CD's the pressure may cause the disc touch the internal surface of the case.
(3) Avoid storing the discs in damp places to prevent fungi and other microorganisms. If possible, use some desiccant such as silica gel in protected bags.

(4) In case it is necessary to clean the CD, use a soft nonabrasive cloth, preferably in radial direction and with minimum pressure. In case it is necessary to use a detergent, use a soft water-based one. Do not use alcohol or solvents. Do not clean the lacquer. It is unnecessary and risky, since the lacquer is very thin and protects the reflecting surface.
(5) Check periodically for data integrity. Since impairment is progressive, the probability of data loss increases with disc age.
(6) For reading or playback use a good quality CD reader in which no dirty or damaged CD's have been played, since the lens could have picked some dirt which in turn could contaminate the CD we intend to play.

3.17.3.2 General Data Compact Disc (CD-ROM)

The CD-ROM is a variant of the CD-DA that allows storing general data, not necessarily audio (Pohlmann 2002). The format is somewhat different, allowing other layers of error control. The information is organized in 98 frames of 24 bytes each (no longer representing 6 samples), forming a 2352 byte data block. The first 12 bytes constitute a synchronism pattern. Then there is a 4 byte block header. The first three bytes indicate, respectively, the minute, the second within that minute, and the number of block within that second; the last byte indicates the *mode*, i.e., the type of data and of error correction.

Note that the reference to minutes and seconds is by analogy with the audio CD, where 1 s is equivalent to 44 100 stereo 16 bit samples, i.e., 176 400 bytes. This number is equivalent to 75 blocks of 2352 bytes.

As regards the mode, there are three possibilities: 0, 1, and 2. Mode 0 corresponds to a block of null data. Mode 2 allocates all the $2352 - 12 - 4 = 2336$ available bytes to user's data. In this case, there is no extra error correction with respect to the normal CD. Mode 1 is the most interesting one for this type of disc, since it reserves 288 bytes for additional error-correcting codes, making this mode a highly reliable one. The price to pay is that there are only $2352 - 12 - 4 - 288 = 2048$ bytes available for user data.

Although we will not go deeper into the details, it is worth commenting that the error rate is extremely low (1 bit in 10^{15} bit). Thus, mode 1 is ideal for storing critical information such as software binary data, which can be rendered useless if a single byte is altered.

3.17.3.3 Write-Once Compact Disc (CD-R)

The discs so far described are manufactured industrially and there are a minimum number of copies (about 1000) to be cost-effective. For that reason, this technology is not a choice for a few backup copies for archival or sharing purposes.

In this case, it is possible to resort to the recordable CD, CD-R, also known as *write-once CD* CD-WO. It is a special type of disc that has a premolded spiral groove 0.6 μm in width with an intergroove pitch of 1.6 μm. The groove is covered by a dye polymer of the family of the *cianines*, *phtalocyanines* or *azos* (Pohlmann 2002). These reflective compounds loose their reflectance when burnt permanently by the action of a high power laser. The pregroove is necessary to assist the servo system devoted to focus the laser beam with precision, making the disc compatible with a regular CD reader. Moreover, the pregroove has a slight wobble of 22 050 Hz when read at normal speed (so-called 1×)

Because of variations in the manufacturing process (including the exact composition of the dye and its thickness), CD-R discs provide, before the first turn (more precisely between, −35 and −13 s with respect to the beginning of the first turn), a space known as *power calibration area*, PCA to be used for calibrating the laser power in order to get the optimal burning power (*optimal power calibration*, OPC). The calibrating process applies different power levels and the result is tested for the most reliable one. Note that there is room for 100 calibrations only. This is important since the CD-R allows multisession writing, and each session requires a full calibration.

Before the −35 s mark, there is a *program memory area*, PMA, with provisional indexing information of the already recorded tracks. When the recording is finished, it is copied to the *table of contents* TOC at the beginning of the disc.

Recordable discs allow recording in single or multiple session. In the first case, the disc is completed at once, including the *lead-in* and *lead-out* areas. Once the recording is complete, it will not be possible to record new data any longer, even if the full capacity of the disc has not been used. In multisession mode, each session (which can contain several tracks) is independent of the others and has its own lead-in and lead-out areas and its own table of contents. However, the old table of contents is copied into the new one. The only table of contents that is read by the reader is the last one. Note that each session uses about 13.5 Mb from the total remaining space. Multisession discs suffer from compatibility problems with readers that support only ordinary prerecorded CD's.

The life time of CD-R discs, if properly handled and preserved, depends on the dye type. Cyanines have an average life time from 10 years to 100 years, while phtalocyanines reach an expected life time of 240 years. In all cases, the disc is supposed to be protected from ultraviolet radiation.

Both for data backup and data sharing, it is preferable to use the data mode instead of the audio CD mode, since the first one offers extra error control and correction, so the information is much more reliable.

3.17.3.4 Rewritable Compact Disc (CD-RW)

The rewritable compact disc CD-RW, also called CD-E, uses, as the CD-R, a premolded spiral groove that works as a guide. The writing medium, or *recording layer*, is a silver-indium-antimony-tellurium alloy which exhibits the property that heated to a temperature between 500 and 700 °C looses its polycrystalline state and reaches a low-reflectance amorphous status. However, if it is later heated to 200 °C, it recovers its original high-reflectance polycrystalline state. The recording layer lies between two dielectric layers that work as thermal insulators. The laser emits in a pulsed way, alternating between the high power needed to write or erase information and the low-power required for reading. This allows a sort of real-time calibration and temperature control, since the information that is being recorded can be monitored.

Unfortunately, the reflectance difference between the polycrystalline and amorphous states is not as high as in the case of prerecorded and write-once CD's, so the compatibility with traditional CD players or readers suffers. There is no problem with readers with the RW specification.

This medium is not as popular as the CD-R, for several reasons. First, because it is more expensive. Second, because it is interesting only when it is necessary to make frequent changes to the recorded information each of which has little or no archival importance. Third, because they have been replaced by the *pen-drive*, a small flash memory device with USB connection, as a temporary mass storage medium or as a sharing device without the risks and safety issues of local area networks.

3.17.3.5 Minidisc (MD)

Although the Minidisc (MD) is seldom used nowadays, it is an interesting case because of the technologies involved. It is a small polycarbonate optical disc with a diameter of 64 mm, protected inside a plastic case. The Minidisc recorder has a mechanism that opens a sliding cover in order to access the disc surface. The disc has a magnetic layer made of a high-coercivity (between 2500 Oe and 5000 Oe) amorphous terbium-iron-gadolinium alloy.[37]

The recording technique is thermo-magneto-optical (Pohlmann 1993). There is a magnetic head on one side of the disc and an optical system on the other side. The magnetic head creates a magnetic field not strong enough to elicit any change in the magnetization state of the magnetic layer. During the writing phase, a high power laser heats the alloy to its Curie temperature (around 180 °C), where the material looses its coercivity (it falls to about 80 Oe) where it is easily magnetized or demagnetized by the magnetic field. The combination of the magnetic field and the laser beam allows a much smaller magnetic mark than would be attained by a magnetic head alone, because of its size. The laser beam allows heating a very small area of the magnetic layer.

The reading phase resorts to the so-called *Kerr effect*, by which a magnetic field rotates the polarization of a laser beam. This polarization change leads to an

[37]*Coercivity* or *coercive force* is the intensity of an inverse magnetic field to be applied to a magnetized ferromagnetic material in order to demagnetize it. The higher the coercivity, the more difficult it is to change its state of magnetization.

intensity change that is detected in a similar fashion as in a conventional CD. In this case the same laser is used, but in low-power mode to prevent demagnetization.

The information encoding on the disc (channel code) obeys the same principles as in the CD, i.e., there is an error-correcting CIRC encoding and an EFM modulation. However, there is a previous stage, a *lossy audio data compression*. This design decision was necessary in order that the Minidisc could store as much audio program as a conventional CD (74 min) on a much smaller area. To save storage space, it was necessary to arrive at a trade-off between quality and information amount. The solution is a technique called *perceptual compression*.

A perceptual compressor makes clever use of the psychoacoustic phenomenon of *masking*, seeking to remove spectral components of the audio signal that would be otherwise masked by other portions of the signal. To this end, the first step is to split the signal into several spectral bands, called sub-bands. Since each sub-band is narrower than the full audio band, it can be undersampled (see Appendix J). Although there are now more signals, each one has a lower sample rate, so the global data rate is, so far, the same.

The next step is to apply some algorithm that implements psychoacoustic masking models in order to get the masked threshold for each sub-band.[38] Considering the threshold as noise, we can estimate the signal-to-noise ratio under that masking condition. For instance, if in a sub-band the masked threshold is 40 dB and the signal level is 70 dB, then we have a signal-to-noise ratio of 30 dB, which can be represented with $n = 30/6 = 5$ bit. We have, thus, saved 11 bit with respect to the original 16 bit.

What would have happened if we had kept the full 16 bit? We would be simply representing sound phenomena below the threshold which are, hence, inaudible.

In the case of the Minidisc, the adopted compression system is called *Adaptive Transform Audio Coding*, ATRAC. It reduces the bit rate from the original PCM corresponding to 44 100 Hz, 2 channels, 16 bit, i.e., $2 \times 16 \times 44\,100 = 1\,411\,200$ bit/s, to a much lower 292 000 bit/s. ATRAC splits the audio signal into three sub-bands, [0, 5.5125 kHz], [5.5125 kHz, 11.025 kHz], and [11.025 kHz, 22.05 kHz], by means of two *quadrature mirror filters*.[39] The first one separates the full audio spectrum into the bands above and below 11.025 kHz and the second one separates the latter into the bands above and below 5.5125 kHz. Each of these three bands is undersampled to its critical sample rate, i.e., 22.05 kHz for the upper band and 11.025 kHz for the lower bands (here the sampling theorem for bandpass signals is applied; see Appendix J).

[38]Masking models provide formulas to find the hearing threshold at any frequency from the levels at the other frequencies.

[39]Quadrature mirror filters are filters that split the spectrum of a digital signal into the upper and the lower regions limited by half the Nyquist frequency. This is done in such a way that if both signals are subsampled to half the sample rate, the residual aliases in both bands cancel each other, allowing a perfect reconstruction of the signal.

Each band is processed with a *modified discrete cosine transform* (MDCT), a form of spectral decomposition particularly suited for overlapped audio frames (in the Minidisc they are 50 % overlapped). The same as in the *discrete Fourier transform*, the spectral resolution varies inversely with the time resolution (time-frequency uncertainty principle). For that reason, the size of each frame is chosen according to the transient behavior of the signal. In audio fragments where the transient response is fast (for instance, impulse noise) and, hence, a good temporal resolution is needed, the analysis window has 32 samples. For the upper band this equals 1.451 ms; for the lower ones, 2.902 ms.

In fragments where the signal varies slowly (for instance, traffic noise), the analysis window is 256 samples long for the upper band and 128 samples long for both lower bands. In this case, it is equivalent to 11.61 ms in all three bands.

This adaptability of some encoder parameters is very interesting because it takes into account not only the spectral masking but also the temporal masking.[40] It is noteworthy that the selection of long and short window size is performed independently on each band. This is because it could happen that the lower band has a fast transient while the others are slowly varying (for instance, a low frequency boom).

Once this conversion from the time domain to the spectral domain is completed, there comes the selective quantization or bit allocation based on the psychoacoustic masking model. The result is that each spectral coefficient is represented with less bits. The number of bits is updated once every 11.61 ms. It is interesting to mention that the quantizing algorithm is not part of the ATRAC specification. This has allowed the evolution of the algorithm over successive versions, with important improvements.

The values so obtained are packaged into 512 bit frames, including a header that informs if the short or long mode has been used, the number of bits and the scale factor for the least significant bit. After this comes the CIRC encoding and the EFM modulation.

While ATRAC compression is generally better than, for instance, MP3, its use for metrological purposes is not recommended, especially because of the availability of uncompressed alternatives based on flash memory.

[40]Temporal masking is the phenomenon by which a strong sound can mask a weaker one that happened some milliseconds earlier. This behavior may seem paradoxical, but it is not. It may be explained through the delay that exists between the instants when the internal ear is excited by the impinging sound and when the actual sound sensation develops in the brain cortex. In this case, this phenomenon is exploited in the following way: if a transient happens at the end of the analysis window, after quantizing a quantization noise appears over the whole window. If the window is long, the noise is clearly noticeable as a breathing sound. If the window is short, it will be temporally masked by the transient.

3.17.3.6 Digital Versatile Disc (DVD)

The (*Digital Versatile Disc*, DVD)[41] is a disc of the same size as the CD (in both 120 and 80 mm versions) but with a higher storage capacity. Each DVD sector has 2048 bytes for user data, which increases to 2418 bytes after adding complementary error control data. Unlike the CD, the DVD may have one or two sides, and each of them can have one or two active layers. The closest one to the surface is semitransparent, while the deepest one is completely reflective. Each layer allows, in principle, a 4.7 GB capacity, so the double-side and double-layer discs reach a capacity of 17 GB. One reason for the large capacity is that the laser wavelength is smaller (635 or 650 nm).

The DVD allows similar versions as the CD: write-once DVD (DVD-R), and rewritable DVD (DVD-RW), which constitutes an interesting medium to store not only a large number of audio files but also large files containing continuous recordings of several hours.

3.17.3.7 Blu-Ray Disc (BD)

The *Blu-ray Disc* (BD) is another optical disc, so called because it uses a 405 nm wavelength *blue* laser[42] which reduces significantly the focus size and, hence, the size of pits and lands. This makes it possible to increase the capacity to 25 GB (or even 33 GB) per layer. Indeed, there are Blu-ray discs of 25, 50, 100, and 128 GB. The physical characteristics of the Blu-ray disc are similar to those of the CD and the DVD, but as the Blu-ray disc is much more sensitive to scratches and fingerprints, the disc has a thin 2 μm protective coating, whose formulation differs from one manufacturer to another, that makes it much more resistant to scratches. One of those formulations consists in colloidal silica-dispersed resin. This material is cured by exposure to ultraviolet light, and it also rejects adherence of grease from fingerprints, a desirable feature since the diameter of the beam on the surface of the disc is a bare 0.13 mm (compare with the 0.7 mm of the CD)

The error correction code of the Blu-ray disc is called *long distance code*. It is similar to a cross interleaved Reed–Solomon with an interleaving distance equivalent to 64 kB (Wikipedia 2013d; Blu-ray Disc Association 2012).

There are also recordable (BD-R) and recordable/erasable (BD-RE) versions with the same capacities as the prerecorded ones. Again it is a very promising medium to store large amounts of audio. However, the technology is still too recent to be able to predict its reliability in terms of long-term archival storage.

---

[41]When it was introduced in 1995, it was generally known as Digital Videodisc. Nevertheless, neither this designation nor the one currently in use today have been official.

[42]The spelling *Blu* has been chosen, seemingly, in order to be able to register the trade mark Blu-ray.

3.17.3.8 Holographic Versatile Disc (HD)

We have, finally, the *Holographic Versatile Disc*, (HVD). It is a disc of the same size of a CD or smaller with a data storage capacity of several TB. It uses a technique called *collinear holography*. It uses two laser beams, red and green, collimated by means of an optical device in order to form a single collinear beam. The red one is used to read the information required by a servosystem controlling the velocity and the alignment of the beam respect to the track, which lies at a deep layer (as in a CD). The green beam reads an almost superficial layer that codes the information by means of laser interference fringes. This medium, which has been publicly demonstrated at technology expositions and even standardized by ECMA (ECMA-377), is not currently available in the market, probably because of its high cost. However it might constitute in the near future a revolutionary medium for the storage of huge amounts of audio information.

3.17.4 Digital Audio Tape (DAT)

We shall briefly mention the system of digital recording on a magnetic tape known as DAT. Because of the bandwidth requirements of digital audio (after error control encoding, about 4 MHz), it is not possible to record using a fixed magnetic head in contact with a low-velocity tape. The DAT system uses, instead, a slightly tilted rotating head which spins at an angular velocity of 2000 rpm performing a helical scan of the tape as the latter moves slowly. The result is a series of parallel tracks crossing the tape from the bottom edge to the top edge with an angle of about 6° with respect to the longitudinal axis of the tape. The separation between consecutive tracks is 13.5 μm. As in the case of the CD, the DAT uses CIRC encoding, but as a channel code it is used eight-to-ten modulation, ETM, instead of eight-to-fourteen.

Nowadays the DAT has virtually fallen into disuse since it has been replaced by solid state recorders that use flash memory cards as a storage medium, since they are superior as regards ease of use and low-power consumption. A serious drawback of the DAT as well as any other tape-based recording or storage system is the impossibility to achieve random access since the medium is intrinsically sequential.

As an advantage it can be mentioned that most portable DAT recorders have been of excellent quality and outstanding specifications, confirmed by laboratory tests. The units currently in use are those recorders purchased before the proliferation of flash memory card recorders.

3.18 File Systems

In most massive data storage media there is a minimum amount of information that can be exchanged at a time, both at writing and reading operations. This is due to medium limitations as well as to error control strategies (which work on data blocks instead of individual bytes). In a hard disk, it has been traditionally a 512 byte sector, but in modern, high capacity drives, it is 4096 bytes. Operating systems can introduce even larger groupings called *clusters* or *allocation units*, each one involving a given number of sectors (for instance, 64 sectors for a high capacity hard disk). This is due to the way files are stored in the disk and the limitations this implies.

Indeed, the access to any information block assumes it can be addressed, and addressing requires space, especially if the files are fragmented, i.e., they consist of a number of nonconsecutive sectors or clusters. This situation arises when files start to be deleted, leaving unoccupied space that has to be filled with parts of the new files to be saved.

In a high capacity disk, the number of sectors is so huge that addressing them requires a prohibitively large space. For example, in a 1 TB disk, there will be about 2×10^9 sectors, each of which requires about $\log_2(2 \times 10^9)$ bits $\cong$ 32 bit, i.e., 4 byte of addressing space. In total, we need $4 \times 2 \times 10^9 = 8$ GB, i.e., 0.8 % of the total available space for an addressing table.

Grouping together several sectors into larger allocation units allows reducing the size of the addressing table. For instance, taking a 64 kB allocation unit, the addressing table will reduce to 125 Mb, i.e., a 0.0125 % of the disk capacity. One disadvantage is that when it is required to save small files there may be waste of space, since any file smaller than 32 kB will occupy one complete cluster. This wasted space is called *slack*.

A more modern approach consists in offering the possibility of reserving contiguous storage area for a file, regardless of whether it is or not occupied. This is done via the concept of *extent* (Buse 2013; McDougall 1999; Wikipedia 2013a), preventing the file fragmentation. This is very useful for large files that must be read sequentially, such as digital audio or video files.

The way in which the available storage space is managed and split into files is known as *file system*. There are many different file systems, being the most popular FAT 32, NTFS, HFS+, EXT3 and EXT4, exFA, and UDF. We shall give a brief description of FAT 32, because it is the simplest one, the most compatible over the variety of operating systems and the one that is implemented in general in flash memory cards such as the SD (Longares 2011; Microsoft Corporation 2000; Kholodov 2010). This is particularly interesting since modern digital recorders are increasingly using this recording medium. We shall also give an overview of the main features of other popular file systems.

3.18.1 FAT 32

FAT32 (32 bit *File Allocation Table*) is an update of the old file systems FAT of the DOS operating system (Disk Operating System) (Microsoft Corporation 2000). The idea of this system is that there is a *file allocation table* formed by successive 32 bit

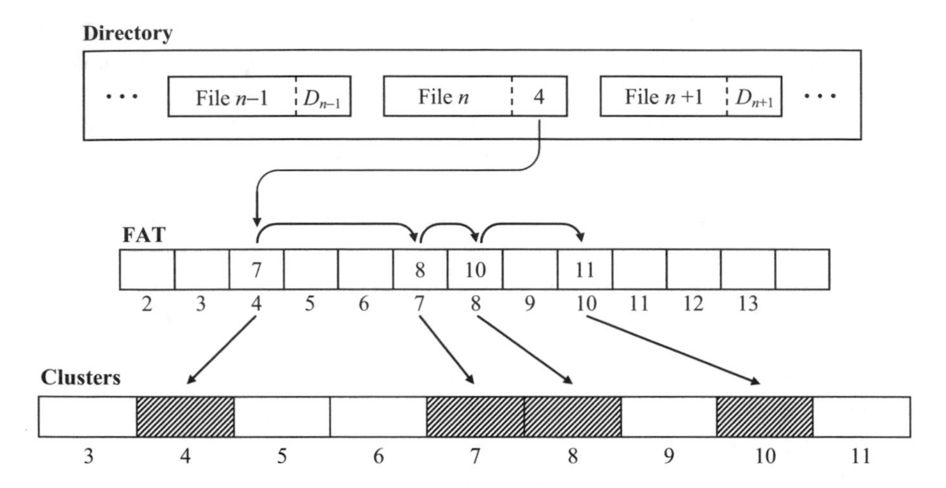

Fig. 3.38 Schematic structure of the FAT system for cluster addressing. One of the fields in the directory entry of file n is the number of the first cluster of the file, $D_n = 4$, which is equal to the number of the corresponding entry of the FAT. This FAT entry contains the following number of cluster and FAT entry. *Note* The directory is also a file and its content is part of one or more clusters. For the sake of clarity here it is presented separately

(4 byte) entries, each one associated with a specific cluster. For instance, entry 157 corresponds to cluster 157.

For every storage unit (hard disc, SD card) each cluster contains a specific number of adjacent sectors. In turn, each file is formed by a number of clusters which may be contiguous or not.

There is a hierarchical organization of the files through a *directory tree*. For the user, a directory is just a container[43] which can contain files and/or other directories (generally referred to as *subdirectories*). At the physical level, a directory is just a file whose content are the *directory entries*, i.e., a few fields that indicate the name of the file or directory (11 ASCII characters), its total size in bytes, the correlative number of the first cluster of the file, some attributes and the date and timestamp of creation, last modification, and last access. Each directory entry is 32 bytes long, except in the case of files with long names, where it is a multiple of 32 bytes.

The correlative number of the first cluster of the file is, indeed, a double pointer since on the one hand it points at the location of the physical beginning of the file, but it also indicates the homologous FAT entry. The content of this FAT entry is, again, a double purpose pointer: it points at the second cluster of the file but also to the homologous FAT entry. This process is illustrated in Fig. 3.38.

As every cluster on the disk has a homologous FAT entry, the preceding is a simple method to locate all the clusters that constitute a file, no matter how

[43]This view is so common that often the directories are called *folders*. We prefer to call them directories since it conveys better the idea of a list with information about elements of a class.

fragmented it is. Since FAT 32 uses 32 bit words to represent the correlative numbers of cluster, the maximum addressable clusters are $2^{32} = 4{,}294{,}967{,}296$, which corresponds, for 32 kB clusters, to 140 TB. However, only 28 bit are really used, so the largest drive that can be addressed has a capacity of 8.8 TB. But the largest addressable *file* is a 4 GB one, since the file size directory field is 32 bit long. This imposes a limit for long-term audio recordings. A stereo file requires 635 MB per recording hour, so the largest continuous recording in FAT 32 is about 6 h 45 min long.

File and directory names allow two versions: short and long. The short one has exactly 11 ASCII characters that correspond to upper case alphanumeric characters and a few additional characters such as +, −, _, ~,) and (. Conventionally, a period is assumed between the first eight and the last three characters (not represented in the name field). The three characters after the period are known as the *extension* of the file, and it usually identifies the type of file (text document, image, audio, etc.). The operating system recognizes the extension and launches the program associated with that extension, but often this is unnecessary, since the type of file can also be derived from the file header.

Long names (up to 255 characters) are an option that was added at a later time, so in order to comply with backward compatibility (i.e., that an older file can be opened with new software), each directory entry includes a short and, if necessary, a long name. The long name entries allow most 2 byte Unicode symbols, except the following ones: \, /, :, *, ?, “, <, > and | since they have special meanings for the operating system. However, how the codes are interpreted may depend upon the code page and the language settings of a specific computer.

An old software will show the short version of the names, which is an *alias* derived from the first characters of the long name according to a series of conventions, such as space and period removal, conversion into upper case and adding ~*n*, where *n* is a correlative number to differentiate file names whose first few characters are coincident. So, “This is a long file.txt” and “This is a longer file.txt” will be listed, respectively, as THISIS~1.TXT and THISIS~2.TXT.

3.18.2 NTFS

NTFS (New Technology File System) is the file system used by Microsoft Windows starting from the NT version (NT, XP, Vista, etc.). Unlike FAT, this file system natively supports long names up to 255 characters. The system respects lower and upper case, but does not make any difference (this means that two files whose names differ only in some characters' case cannot coexist in the same directory). Any character except \, /, :, *, ?, “, <, > and | is acceptable. The following description is based on (Kozierok 2004).

NTFS uses a cluster size between 512 bytes and 4096 bytes. File information is organized as a series of attributes, where even the file content itself is also an attribute. When a file is first created, a record is added to a hidden system file called the *Master File Table* (MFT). This record has a size equal to a cluster or 1024 byte (whichever is larger) and contains some attributes of the file (these attributes are called *resident attributes*.). In the case of very small files, the attribute of content could fit in the record.[44] However, the file content, is in general, larger than a cluster and must continue outside the record as *nonresident attributes*. Non-resident attributes can be stored in other records within the MFT, or outside the MFT, within the so-called *extents*. It is noteworthy that the universe of possible attributes is not restricted to a few created by the system. The user could create new attributes.

In principle, the size of the MFT depends on the files stored in the system, so it could grow considerably. The same as any other file, it could eventually get fragmented. To prevent this to some degree, when a NTFS drive is formatted, a space equal to 1/8 of the disk capacity is reserved for the MFT, so that it can grow without fragmentation. This may seem a gross waste of space, but it is not, since part or all of the file content may be stored within the MFT. The latter happens when the rest of the disk capacity has been occupied. It is also possible that all the reserved space gets eventually occupied by the MFT. In this case, it will continue to grow outside the reserved space where the operating system finds available space. In this case, the MFT will be most likely fragmented. It is important to be aware of this, since the performance of a disk whose capacity has been used in a fraction 7/8 starts to decay.

The file structure is based on a structure used in relational databases called *B + tree*, whose main feature is the possibility to store data in blocks (clusters) in such a way as to attain great retrieval efficiency. To this end each block has pointers to other blocks. Sequential access time as well as insert and erase times are proportional to the logarithm of the data size. The details can be found in the specific literature (Bayer and McCreight 1970, 1972; Wikipedia 2013b, c).

A file's record in the MFT begins with a header that includes pointers to the attributes, including standard information such as date, timestamp and other traditional attributes (such as those present in the FAT file system), the file name (which may include more than one name, such as short and long names and uses Unicode encoding), the data attribute, and the security descriptor attribute (which includes access control such as read, write, execute, and erase permissions). There is also the concept of *audit*, which allows an historical record of different types of events such as directory and file access or session start. A further interesting feature as regards security is the possibility of encryption, available from version 5.0 of NTFS. This is a more reliable security feature than permissions, since by means of tools for low-level access to disk content it is theoretically possible to access any nonencrypted piece of information.

NTFS offers also the possibility of file and directory compression. Although there is a performance decrease because of the use of system resources for

[44]This helps reducing slack and improves the performance, since all the information of the file is accessed at a time.

compression and decompression, in some cases this is compensated by the fact that the file is smaller and less fragmented, reducing the disk access time, which is the true bottleneck in many processes. In line with this, it allows a very fast and basic form of compression, as is the handling of *sparse files*, i.e., files with large "empty" regions (long runs of null bytes). It is the case of audio files with long silence intervals.

This proprietary file system has also a free software version with GPL license (*General Public License*), the NTFS-3G, which allows reading and saving in NTFS drives in Linux, OS X, Android, and other operating systems (Tuxera 2014).

3.18.3 EXT4

EXT4 (*Fourth Extended File System*) is the file system in recent Linux distributions as well as in the Android operating system used in tablet computers and smartphones.

It allows addressing drives up to 2^{60} bytes (i.e., 1000 TB), files up to 16 TB and 64 000 subdirectories per directory. Although the typical hard disks that can be found nowadays (2016) in the market have capacities not greater than a few TB, there is no doubt that the storage capacity will keep growing. To put it in perspective, a 16 TB file is equivalent to 1154 days straight, i.e., 3 years and 2 months of stereo CD-quality recording, suitable for the most ambitious conceivable long-term studies. However, a most likely application for huge files is not long-term recording but high quality multichannel recordings such as might be necessary in multisensor beamforming applications (Analog Devices 2013).

EXT4 also allows the use of *extents*, which, as already commented, constitutes a group of adjacent successive sectors which, unlike clusters, not necessarily have a fixed size. This is especially useful in digital audio, where the access to large, highly fragmented files leads to performance decrease because of the need to reposition many times the read/write head of the hard disk. EXT4 allow extents up to 128 MB.

It is possible to reserve disk space for a file without the need of artificially expanding the file size by means of zero-padding. This is very useful when it is expected that increasingly larger versions of the same file will be saved over time. This space reservation allows reducing fragmentation. It also prevents that other processes that are running at the background use the available space causing that a large operation (such as the disk storage of a large matrix containing the FFT spectra of a large audio file) cannot complete.

On the other hand, EXT4 allows to optimize the distribution of a file on disk by means of *delayed allocation*. It consists in saving a file that is currently being modified in a cache memory instead of the disk until a larger number of blocks have been accumulated, and only then they are transferred to the disk. This reduces fragmentation since larger amounts of information are transferred at a time, allowing the system to choose a better allocation. There is some risk of data loss in case of a blackout or a system crash.

Time measurement in any file system is affected by *granularity*, i.e., the effect of time resolution. In FAT 32, the timestamp resolution is 10 ms. In EXT4, it reaches the nanosecond. For audio files that mean that the creation time can be provided with a resolution far better than an individual sample. Note that this does not mean that the timestamp is accurate to that resolution, since it depends on the accuracy of the system clock. EXT4 allows to extend the limit of representable dates up to year 2242. This is an improvement respect older systems that represent dates with a signed 32 bit word representing seconds from 1970, yielding a limit of year 2038.

Finally, this file system provides mechanisms to accelerate file system check operations (Mathur et al. 2007; Wikipedia 2013e).

3.18.4 HFS+

HFS+ (*Hierarchical file System Plus*) is the file system used by Macintosh's operating system OS X (Wikipedia 2013f). This system uses 512 bytes sectors grouped in allocation blocks with 32 bit addressing, so the maximum number of addressable blocks is 2^{32}. The effective number of blocks depends on drive capacity. The maximum file size is 2^{63} bytes. In practice, it is limited only by the available disk space.

There is an *allocation file*, also called *bitmap* (not to be confused with the BMP image file format) which associates a bit with each allocation block, indicating if it is free (0) or occupied (1). The size of the allocation file depends on the number of blocks, but it is rounded up in order that its size is an integer number of sectors (Developer Connection 2004).

There is a *catalog file* structured as a *B*-tree*. The B*-tree is a way to organize the information into *nodes*, each of which has a size of 4 kB. Nodes contain a descriptor, a group of records that contain information and a series of *offsets* that address the records. The descriptor contains information on the type of node, its level in the tree, the number of records in the node, and the forward and backward links to other nodes. The aim of a B*-tree is to make more efficient the retrieval, erasure, and substitution of information. Several structures within HFS+ are organized as B*-trees.

Files in HFS+ are *forked*, so that for each file there is a *resource fork* and a *data fork*. The resource fork is headed by a resource map. For instance, for an executable file the resources might describe the menus, dialogs, icons, memory usage, the code itself, etc., while the data fork may be empty. Document files, on the contrary, have an empty resource fork (Developer Connection 2007) and nonempty data fork.

File and folder (directory) names can have up to 255 Unicode characters in HFS + , and dates are represented by a 32 bit unsigned integer that represents the number of seconds since 1904, allowing dates up to year 2040.

Files have attributes which can be an alias, custom icons for files and folders, whether it is on the desktop or not, whether it is a system file or not, etc. This file system supports extents, i.e., contiguous space assigned to a fork.

3.18.5 UDF

UDF (*Universal Disk Format*) is the file system used by the DVD and other optical media. It is defined by the Optical Standard Technology Association (OSTA 2005), and it is in turn is a derivation of the specification ECMA-167 by the European Computer Manufacturers Association (ECMA). File names may use different codes such as ASCII, Unicode and ECMA-094, and can reach 255 characters. The directory structure depth allows paths up to 1023 bytes. The file size may reach 2^{60} bytes. It allows to record dates of creation, archiving, modification, and attribute change and access.

Structurally, it has sectors whose size depends on the specific medium, which is organized in blocks and extents up to $2^{30} - 1$ bytes. Depending on whether the medium allows writing, the system allows random write and read. The system has three variants. The first one, called *plain*, is applicable to discs that allow random read and write access, such as the DVD-RW. The second one, known as *incremental writing*, is applicable to write-once discs such as the CR-R and DVD-R. It uses a *virtual allocation table*. In this case there must be a precompilation of the information to be incrementally recorded on the disk. However, it is allowed to write during different incremental sessions, replacing the table during each new session. It is possible to delete information, but only in a logical sense, i.e., the new table will not contain an entry addressing the deleted content any longer. However, since each session creates an updated table, this erasure actually implies a loss of space, since the old table cannot be physically erased. In a CD-R, each rewrite implies the loss of 13 Mb. The third variant, *spare*, is applicable to rewritable disks such as the DVD-RW. It is designed to minimize the rewrite operations on the same physical portion of the disk in order to prevent medium wear. Therefore, in this case the rewrite operation is performed preferentially over a clean area. The corresponding table features a record of errors due to wear (OSTA 2005; Wikipedia 2013g).

3.19 Long-Term Preservation

A significant issue with digital audio (as well as any other type of digital information) is long-term preservation with several objectives: (1) Amortize the cost and effort implied in any measurement or recording; (2) Document past measurement campaigns; (3) Preserve audible testimonials of the soundscape of a given place and time.

The physical principles of vinyl record or analog tape recording playback are widely understood and easy to implement. The first sound recordings by Thomas Alva Edison in 1877 on a tin foil have been preserved and restored (National Park Service, n.d.). On the contrary, modern digital techniques such as complex error-correcting codes, spectral representation, compressed codes with arcane strategies, file atomization, etc., may represent quite a challenge for those who intend to retrieve the information in the future (Pogue 2011).

An obvious complication is that most commonly available storage media are very delicate because of the microscopic size of their blocks of information, making them prone to massive data loss.

Furthermore, digital information is *virtual*, i.e., it is not inextricably associated with the storage medium. This implies that the preservation of the physical object, the container, not necessarily guarantees data preservation. While along history much information has been preserved and subsequently deciphered, nowadays far more information is generated but a vanishingly small fraction is preserved in sufficiently rugged and dependable physical media. In most cases, the information is stored in devices with a relatively short lifespan where data can be easily corrupted or even completely deleted, such as hard disks, optical media, or flash memories.[45]

Several approaches have been identified for long-term preservation of digital information (Peres 2007; Thibodeau 2002; Rothenberg 1995, 1999). The first and most simple one consists in doing nothing, letting the future "digital archeologists" deal with the problem on their own. This practically guarantees that, eventually, the information will be definitively lost, since the cost of the detectivesque retrieval services would be so high that, except in very special cases, it would not be justified.

The second method is the *museum approach*, which requires to preserve not only the storage media but also the hardware necessary to retrieve the information. This includes computers and peripherals as well as working copies of the appropriate software. A serious difficulty is *planned obsolescence*, which plagues many industry products such as computers. It is intended for ensuring the cyclic replacement so that manufacturers can enjoy higher income. The consequence of planned obsolescence is that equipment will stop working properly after a few years because of malfunction of one or more parts, either mechanical or electronic. The possibility of repair will be more and more complicated because of the increasing difficulty to purchase spare parts for obsolete equipment and the lack of technology disclosure to ensure the possibility of replicating them at a future time.

A third technique is *emulation* and *virtualization* of obsolete equipment, i.e., some software that emulates a hardware in which the software needed to read the encoded information would run. Of course, this does not relieve us from the need to

---

[45]Text documents or graphical output enjoy the privilege of being printable in high quality alkaline paper, which in the absence of accidents or natural disasters ensures preservation during decades or even centuries.

preserve the digital information itself, both the specific data of interest and the emulating software code.

Migration is another long-term preservation technique. There are two possibilities. In the first case the digital information is preserved as is, so it is necessary to have software to decode it or transfer it to a new format. In the second, the information is migrated to a more current format in such a way that there are no changes in the essential information. Here it is important to clarify what is *essential* and what not in each case. For instance, it is possible to preserve a pure text file without any format, or it is possible to preserve also its layout, or it is even possible to preserve structural information (such as chapters, sections, cross references). It may seem that in the long run some accessory information becomes irrelevant, but for a historical study, for instance, such information might prove essential. Similarly, in the case of audio one may be interested only in the audio signal itself, but also in several types of metadata (which not always are compatible across formats).

Another technique is to use permanent (read only) media that have an estimated longevity of decades or even centuries. It is difficult to find in the consumer market such media, since that is precisely what consumption is about, an ephemeral product lifespan that ensures periodical replacement. High quality CD-R with phtalocyanine dye and gold reflective layer are supposed to last in excess of 100 years. Note that, because of the already commented issues of software and hardware obsolescence, this does not purport that the CD-R is meant to be stored for a century!

Finally, an important resource is redundancy and geographic spread of the information, in order to render the preservation less dependent on a particular copy. It consists in keeping several copies, if possible at different locations, thus reducing the risk of loss by mishandling, accident, sinister, natural disaster, or robbery. This technique has an antecedent, the publication of books, particularly by the thousands, since the massive destruction of all copies is unlikely.

As can be noted, these techniques are not exclusionary, and indeed some of them are based on others. We can see that as long as technology does not develop a satisfactory and cost-effective solution to this problem,[46] the best provisional solution seems to encompass frequent migration combined with the use of the best storage media available, open encoding standards, redundancy, and geographic spread. It is possible to take advantage of the so-called *cloud storage* services, i.e., storage in Internet servers located elsewhere. However, the security of such systems is not fully guaranteed, as regards both hardware integrity (hard disks and solid state drives) and information confidentiality, so it is advisable to keep local copies and encrypt files.[47]

---

[46]This problem seems to circumscribe to academy, since industrial societies discourage attachment to history, its objects and teachings, so there are few individuals who feel the need for preservation and, consequently, there is not enough pressure on the industry to manufacture high quality media.

[47]It is also arguable whether a specific company offering such service will remain in the market in the long run.

Although it is not feasible to give here a complete specification, a serious attempt to preserve audio signals over decades requires designing a protocol where the following items are stipulated:

(1) Medium-term planning of the actions to take
(2) One or more responsible people
(3) Repository selection (location)
(4) Economic investment planning (equipment, hosting services, personnel)
(5) Exhaustive documentation of all actions taken, including description and specification of equipment and technologies applied; format standards; redundancy and geographic distribution management; periodic checking of data integrity; file rewrite to prevent aging or weakening of magnetic, electrostatic, or optical marks on the storage devices; detailed description of the sounds, including geographic location and time stamp of the recording, its motives and reference to documents where they have been studied, analyzed, or reported.

3.20 Conclusion

We have covered the digital audio recording techniques, including digitization, storage media, the most widely used file systems and the binary format of some sound files. The latter allows *interoperability*, i.e., the possibility of being able to read, create or modify a file by means of software other than the one commonly used to create or handle it. In particular, it allows accessing some features such as cues and labels, which is useful for the automation of some analysis functions, for instance, accessing to calibration information, as well as assigning GPS coordinates or attaching videos, images, or descriptive text.

Problems

(1) Consider an audio signal whose peak voltage is 1 V and assume the total broad-band noise *above* 20 kHz has an RMS value of 0.7 mV. Find the minimum order of the anti-aliasing filter to be used at the input of an A/D converter working at a sample rate of 44 100 Hz and a resolution of 16 bit. Hint: Once filtered, the residual noise should be below the least significant bit (LSB). Assume that the peak factor (ratio of the peak value to the RMS value) of the noise is 3. The slope of a filter of order n is $-6n$ dB/oct.
(2) Contemporary orchestral music has a long-term spectrum whose spectral density grows at a rate of 12 dB/oct, then is approximately constant from 100 Hz to 500 Hz, then falls at −6 dB/oct up to 2000 Hz and finally falls at −12 dB/oct. Its peak factor is typically 10. Find the minimum order of an anti-aliasing filter for an A/D converter with a sample rate of 44 100 Hz and a resolution of 16 bit. It should be flat within a tolerance of −1 dB at 20 kHz, and it should reduce the signal above Nyquist limit below the least significant bit. Hint: Starting with a provisional spectral density equal to 1 between

100 Hz and 500 Hz compute the RMS value in the audio range, then find the peak value and apply a constant such that the peak value is coded as the maximum digital value. Then compute the RMS above 20 kHz.

(3) (a) In order to test the effects of aliasing use Scilab to create a signal equal to the superposition of two tones of 200 and 8000 Hz with $F_s = 40\ 000$ Hz. Simulate an undersampling at a sample rate of 10 000 Hz without anti-aliasing filter by picking every fourth sample. Use the function **sound** to listen to both sounds (make sure the signal is normalized to a peak amplitude smaller than 1 to prevent distortion). While the 200 Hz tone remains unchanged, the 8000 Hz has reflected over the new Nyquist frequency. Can you tell what the alias frequency is?
(b) Repeat part (a) with the superposition of a 2 kHz tone and a Gaussian noise. Compute the RMS value of both signals and subtract (on an energy, i.e., square, basis) the theoretical RMS of the tone. Both results are very similar. This shows that the energy that was originally distributed over the spectral range from 0 to 20 kHz has been folded onto the range from 0 to 5 kHz.

(4) Write a Scilab script to test the quantization noise. To this end, generate a quasi-analog sine wave of amplitude 1 and many samples per cycle (for instance, 1000). To simulate 2^n steps just multiply by 2^{n-1}, use the function **floor** to get the integer part and divide back by 2^{n-1}. Then subtract the original sine wave to isolate the quantization noise. (a) Plot it versus time. (b) Compute the RMS value of the noise for different n's. Check if it obeys the theoretical formula. (c) Apply the function **fft** with $N = 4096$ to perform a spectrum analysis and plot its absolute value from 1 to $N/2$.

(5) Write a Scilab script to test the effect of dither. (a) Set $F_s = 44\ 100$, $N = 4096$ and generate a sine wave of amplitude 1.1 with a frequency equal to a multiple of F_s/N. In order to simulate the sampling process near 1 LSB use the function **floor**. (b) Apply the function **fft** with size N to get the spectrum and plot its absolute value from 1 to $N/2$. (c) Repeat (a) and (b) adding different amounts of random noise with an RMS value in the range from 0 to 1.

(6) Generate a brief sound (for instance one cycle of a sine wave) and save it to a wav file using the function **wavwrite**. Then open the file using a hexadecimal editor (such as HxD) and analyze its content and its compliance to the format.

(7) Connect a microphone to the audio input of a computer. If there is a recording application installed, use it to record some sounds. Otherwise, download and install Audacity and use it to record. Assuming that a calibration tone that is −18 dB from a sine wave of maximum amplitude corresponds to an acoustic signal of 94 dB, use Scilab to find the equivalent level of the sound you have recorder. Hint: Save or export the signal as a wav file and then open it from Scilab using the **wavread** function.

(8) If you have access to a sound level meter and a digital recorder, generate a file using a calibration tone. There are three ways to do so: (a) Many sound level meters have an internal calibration tone equivalent to 94 dB, (b) You can use

an acoustic calibrator; (c) You can generate a 1 kHz sine wave and play it using the function **sound**; set the volume such that the sound level meter measures 94 dB and record that sound (you must wear hearing protectors since 94 dB is quite loud). Once you get the calibrating tone, go out and record some traffic noise in a new file. Then repeat problem (7).

(9) Go to the FLAC website (https://xiph.org/flac/) and download the reference codec. Use it to compress the files of problems (7) and (8) and then decode them. Check with a file comparison tool (for instance, HxD) that the recovered files are identical to the original ones.

(10) Using Audacity save one of the wav files as a project. Analyze the .aup file with any text processor and check its conformance to the format. Open any of the associated .au files with any hexadecimal editor (such as HxD) and analyze its content.

References[48]

Analog Devices (2013) Microphone beamforming simulation tool. Simulation software. http://wiki.analog.com/resources/tools-software/microphone/beamforming-simulation-tool

Bayer R, McCreight E (1970) Organization and maintenance of large ordered indexes. Mathematical information sciences report no. 20. Mathematical and information sciences. Laboratory Boeing Scientific Research Laboratories. July 1970. http://infolab.usc.edu/csci585/Spring2010/den_ar/indexing.pdf

Bayer R, McCreight E (1972) Organization and maintenance of large ordered indexes. Acta Informatica 1 Fasc. 3, pp 173–189 http://www.liacs.nl/~graaf/STUDENTENSEMINARIUM/OMLO.pdf

Blu-ray Disc Association (2012) White paper blu-ray disc™ format—general, 3rd edn. December, 2012. Accessed 26 Feb 2016, http://www.blu-raydisc.com/Assets/Downloadablefile/White_Paper_General_3rd_Dec%202012_20121210.pdf

Boser B, Wooley B (1988) The design of sigma-delta modulation analog-to-digital converters. IEEE J Solid-State Circ 23(6), December 1988 http://www.uio.no/studier/emner/matnat/ifi/INF4420/v10/undervisningsmateriale/Boser_Wooley.pdf. Accessed 26 Feb 2016

Brüel & Kjær (2016) User manual—hand held analyzers types 2250 and 2270. Brüel & Kjær BE 1713–32. Nærum, Denmark http://www.bksv.com/doc/be1713.pdf. Accessed 26 Feb 2016

Buse JW (2013) Intro to extents. Linux.org, 2013/07/09. Accessed 2014/04/20, http://www.linux.org/threads/intro-to-extents.4131/. Accessed 26 Feb 2016

de la Rosa JM (2011) Sigma-delta modulators: tutorial overview, design guide, and state-of-the-art survey. IEEE Trans Circ Syst—I: Regul Papers 58(1), http://ieeexplore.ieee.org/stamp/stamp.jsp?tp=&arnumber=5672380

Developer Connection (2004) Technical note TN1150. HFS plus volume format. Apple Inc., 2004. http://dubeiko.com/development/FileSystems/HFSPLUS/tn1150.html. Accessed 26 Feb 2016

Developer Connection (2007) Inside Macintosh: files. Apple Inc., 2007. http://www.dubeyko.com/development/FileSystems/HFS/inside_macintosh/inside_macintosh.htm. Accessed 26 Feb 2016

[48]Now and then, besides a documentary reference, a Wikipedia entry is included since it provides a more accessible description of some item as well as further documentary references.

ECMA-167 Volume and file structure for write-once and rewritable media using non-sequential recording for information interchange. 1997. Accessed 2013/07/23, http://www.ecma-international.org/publications/files/ECMA-ST/Ecma-167.pdf. Accessed 26 Feb 2016

ECMA-377 Information interchange on holographic versatile disc (HVD) recordable cartridges capacity: 200 Gbytes per Cartridge. 2007. http://www.ecma-international.org/publications/files/ECMA-ST/ECMA-377.pdf. Accessed 26 Feb 2016

Gray RM (1987) Oversampled sigma-delta modulation. IEEE Trans Commun, Com-35, No. 5, May 1987 http://ieeexplore.ieee.org/stamp/stamp.jsp?tp=&arnumber=1096814

IEC 61672-1 (2002) Electroacoustics—sound level meters—part 1: specifications. International Electrotehcnical Commission

ISO 1996-2 (2007) Acoustics—description, measurement and assessment of environmental noise—part 2: determination of environmental noise levels. Geneva, Switzerland

Kholodov I (2010) CIS-24 the microcomputer environment. Bristol Community College. Fall River, Massachussets, USA. http://www.c-jump.com/CIS24/Slides/FAT/lecture.html. Accessed 26 Feb 2016

Kozierok CM (2004) New technology file system. The PC Guide. 2004. http://www.pcguide.com/ref/hdd/file/ntfs/index.htm. Accessed 26 Feb 2016

Longares J (2011) Sistemas de archivos FAT16 / FAT32 para las tarjetas SD / MMC". Octubre de 2011. Accessed 20 Jul 2013, http://www.javierlongares.com/arte-en-8-bits/sistemas-de-archivos-fat16-fat32-para-las-tarjetas-sd-mmc/

MacKay D (2005) Information theory, inference, and learning algorithms. Cambridge University Press. http://www.inference.phy.cam.ac.uk/itprnn/book.pdf. Accessed 08 Oct 2016

Mathur A, Cao M, Dilger A (2007) EXT4: the next generation of the EXT3 file system. LOGIN, June 06 Volume 31 (2007/05/27 10:22 AM). Page 25

Marengo R, Roveri E, Guerrero JMR, Treffiló M (2011) Análisis comparativo de codificadores de audio sin pérdidas y una herramienta gráfica para su selección y predicción de su desempeño". Mecánica Computacional Vol XXX, págs. 3167–3186. In: En Oscar Möller, Signorelli JW, Storti MA (eds.) Rosario, Argentina, 1–4 Noviembre 2011

McDougall R (1999) Getting to know the solaris filesystem, Part 1. Sun World, May 1999. http://sunsite.uakom.sk/sunworldonline/swol-05-1999/swol-05-filesystem.html. Accessed 26 Feb 2016

Microsoft Corporation (1994) Multimedia standards update. New multimedia data types and data techniques. Revision 3.0 1994/04/15. http://www-mmsp.ece.mcgill.ca/documents/audioformats/wave/Docs/RIFFNEW.pdf

Microsoft Corporation (2000) Microsoft extensible firmware initiative—FAT32 file system specification—FAT: general overview of on-disk format. version 1.03, 2000/12/06. http://staff.washington.edu/dittrich/misc/fatgen103.pdf. Accessed 26 Feb 2016

Miyara F (2001a) ¿Ruido o señal? La otra información. En defensa del registro digital del ruido urbano. Cuarta Jornada Regional sobre Ruido Urbano, Montevideo, Uruguay, 2001. http://www.fceia.unr.edu.ar/acustica/biblio/reg-dig.pdf. Accessed 26 Feb 2016

Miyara F (2001b) Grabación digital: ¿DAT o MiniDisc? Tecnopolitan. Año 2 Nº 15, 2001. http://www.fceia.unr.edu.ar/acustica/biblio/md-vs-dat.pdf. Accessed 26 Feb 2016

Miyara F, Accolti E, Pasch V, Cabanellas S, Yanitelli M, Miechi, P, Marengo F, Mignini E (2010) Suitability of a consumer digital recorder for use in acoustical measurements Internoise 2010. http://www.fceia.unr.edu.ar/acustica/biblio/consumer_recorders_for_noise_measurement_INTERNOISE_2010_993.pdf. Accessed 26 Feb 2016

National Park Service (no date) Listen to Edison sound recordings http://www.nps.gov/edis/photosmultimedia/the-recording-archives.htm. Accessed 26 Feb 2016

OSTA (2005) Universal disk format® specification. Revision 2.60. March 2005. http://www.osta.org/specs/pdf/udf260.pdf. Accessed 26 Feb 2016

Peres MR (ed) (2007) Focal encyclopedia of photography. Digital imaging, theory and applications, history, and science, 4th edn. Focal Press—Elsevier, Burlington

Pogue D (2011) Seeing forever: storing bits isn't the same as preserving them Scientific American, April 2011. http://www.scientificamerican.com/article/seeing-forever/. Accessed 26 Feb 2016

Pohlmann K (1993) Advanced digital audio. SAMS—Prentice Hall, Carmel
Pohlmann K (2002) Principles of digital audio (fourth edition). McGraw-Hill Video/audio Professional, New York
Proakis JG, Manolakis DG (1996) Digital signal processing. Principles, algorithms and applications (third edition). Prentice-Hall International, Upper Saddle River
Raynaudo E, Treffiló M (2013) Registro digital de ruido ambiental con transferencia inalámbrica de datos y post-procesamiento basado en software y hardware libre. Proyecto final de la Carrera de Ingeniería Electrónica, FCEIA-UNR. Rosario, Argentina
Robinson T (1994) Simple lossless and near-lossless waveform compression. Technical report CUED/F-INFENG/TR.156, Cambridge University, 1994. http://www.ee.columbia.edu/~dpwe/papers/Robin94-shorten.pdf. Accessed 20 Apr 2014, http://citeseer.nj.nec.com/robinson94shorten.html Accessed 20 Apr 2014
Rothenberg J (1995) Ensuring the longevity of digital documents. Sci Am 272(1):42–47
Rothenberg J (1999) Ensuring the longevity of digital information. Council on library and information resources. http://www.clir.org/pubs/archives/ensuring.pdf. Accessed 22 Apr 2014
Roveri E (2011) Estudio y comparación de codificadores de audio sin pérdidas. Proyecto final Ingeniería Electrónica. Univ. Nac. Rosario, FCEIA. Noviembre de 2011
Salomon D (2007) Data compression. The complete reference. Springer, London
Sun Microsystems (1992) Header file for Audio, .au. Sun Microsystems, Inc. http://pubs.opengroup.org/external/auformat.html. Accessed 28 Feb 2016
Thibodeau K (2002) Overview of technological approaches to digital preservation and challenges in coming years. In: Conference proceedings: the state of digital preservation: an international perspective. http://www.clir.org/pubs/reports/pub107/thibodeau.html. Accessed 26 Feb 2016
Tuxera (2014) "NTFS-3G + Ntfsprogs". http://www.tuxera.com/community/ntfs-3g-download/. Accessed 27 Feb 2016
Watkinson J (1988) The art of digital audio. Focal Press, London
Watkinson J (2002) An introduction to digital audio. Focal Press, Burlington
Wikipedia (2013a) "Extent (file systems)". Wikipedia: the free encyclopedia. Wikimedia Foundation, Inc., 2013/02/27. Accessed 2013/07/21. http://en.wikipedia.org/wiki/Extent_(file_systems)
Wikipedia (2013b) "B+trees". Wikipedia: the free encyclopedia. Wikimedia Foundation, Inc., 2013/05/30 Accessed 2013/07/22. http://en.wikipedia.org/wiki/B%2B_tree
Wikipedia (2013c) "B trees". Wikipedia: the free encyclopedia. Wikimedia Foundation, Inc., 2013/07/15. Accessed 2013/07/22. http://en.wikipedia.org/wiki/B-tree
Wikipedia (2013d) "Blu-ray disc". Wikipedia: the free encyclopedia. Wikimedia Foundation, Inc., 2013/07/21. Accessed 2013/07/22. https://en.wikipedia.org/wiki/Blu-ray_Disc
Wikipedia (2013e) "Ext4". Wikipedia: the free encyclopedia. Wikimedia Foundation, Inc., 2013/06/26. Accessed 2013/07/22. http://en.wikipedia.org/wiki/Ext4
Wikipedia (2013f) "HFS+". Wikipedia: the free encyclopedia. Wikimedia Foundation, Inc., 2013/05/06. Accessed 2013/07/23. http://es.wikipedia.org/wiki/HFS%2B
Wikipedia (2013g) Universal disc format. Wikipedia: the free encyclopedia. Wikimedia Foundation, Inc., 2013/06/12. Accessed 2013/07/23. http://en.wikipedia.org/wiki/Universal_Disk_Format
Zwicker E, Fastl H (1999) Psychoacoustics—facts and models. Springer, Berlin

Chapter 4
Digital Audio Editing

4.1 Introduction

When an audio signal is acquired by means of a digital recorder or a portable computer with a sound card, we get in general a .wav file stored in some medium, removable or not, such as an SDHC flash memory card or the laptop's hard disk.

Depending on the measurement protocol we may have different files corresponding to the different events of interest, or a single large file that includes all of them. Those events might be short intervals corresponding to specific predominant sources, intervals of a specified duration according to some protocol (for instance, 5 min) without regard of the specific sources, calibration signals, oral comments by the operator or even auxiliary signals such as synchronization signals (in the manner of the *clapperboard*[1] used at the beginning of movie or video footage) or signals coming from vehicle counters, GPS devices, etc.

In any case it will be necessary to subject the sound files to a variety of processes, such as computing sound levels, exceedance levels, and spectrum analysis, among others. In general, it is previously necessary to perform some editing work on the signal with the following objectives:

(1) Apply calibration constants that give sense to the abstract figures with which samples are digitally represented.
(2) Isolate or circumscribe short signal fragments with characteristic features, for instance, a given type of source.
(3) Demarcate and identify events in order to classify or categorize them.
(4) Remove irrelevant portions of signal.
(5) Transcribe oral comments by the operator as well as handwritten notes to a label or note in the file itself, associated with a particular instant or interval.

[1]The *clapperboard* is a small slate whose top edge is articulated, allowing to identify the scene or footage and at the same time to synchronize the image and the audio through the crack sound produced by striking together the hinged parts.

F. Miyara, *Software-Based Acoustical Measurements*, Modern Acoustics and Signal Processing, DOI 10.1007/978-3-319-55871-4_4

(6) Synchronize two or more simultaneous recordings.
(7) Associate specific cues with the standard time.
(8) Apply specific filters to remove unwanted or irrelevant components (for instance, some parasitic tone generated during the recording process).
(9) Change the sample rate or the bit resolution.
(10) Mix or combine different signals.

These and many more operations can be carried out by a means of a *digital audio editor* often called also *digital audio workstation*, DAW, especially if they offer a bundle of advanced features. While there are many digital audio editors, most of them are intended for musical production for the record industry. We can mention, for instance, Adobe Audition, Ardour, Audacity, Cool Edit, Cubase, Expstudio Audio Editor, Free Audio Editor, Goldwave, Music Editor Free, Nuendo, Power Sound Editor, Pro Tools, Sound Forge, Traverso DAW, Wave Editor, Wave Pad, Wavosaur, Sound Engine, Wavesurfer, among many others (list as of 2014). Some of them are free of charge, others allow a trial period for free, (Tan n.d.), and others offer demo versions not fully functional.

4.2 Audacity[2]

One editor that stands clearly alone is Audacity because it is free ("free" as in "freedom"), open-source[3] and cross-platform[4] software, distributed under GPL license[5] and in permanent expansion through an increasing number of enthusiastic collaborators all around the world. It can be downloaded from http://www.audacityteam.org/. In addition to a beautiful graphical user interface it offers a host of powerful functions that make it competitive with commercial software. For instance, it allows for an unlimited number of audio tracks and label tracks, and it includes many effects, filters, generators, and analysis tools, as well as the possibility of adding new features as *plugins* that can be implemented in C, or in an interpreted (but highly efficient) language called Nyquist,[6] directly by the user.

It is always risky to include in a book too many details on a program that is frequently updated, since even the next version could exhibit changes that would

[2]Audacity® software is copyright© 1999–2016 Audacity Team. The name Audacity® is a registered trademark of Dominic Mazzoni.

[3]*Open source* means that it is possible to download the source code in C++ for its study, analysis and possible improvement.

[4]There are versions for GNU Linux, Windows, and Macintosh.

[5]GPL is the acronym for *General Public License*, a type of license promoted by the Free Software Foundation as a means to favor the sharing of software as well as the possibility to study and modify it, with the only condition that any distribution of derived work must be issued under the same GPL license. More details can be found in http://www.fsf.org/es.

[6]*Nyquist* was originally designed for the generation and management of virtual musical instruments, but it also offers great potential for signal generation and analysis.

render the explanation obsolete. The decision to include, however, some of the features and functions of the current version (2.1.2) at the moment of this writing (2016), obeys three reasons. The first one is that several of the selected features are fairly common among many audio editors including several recent versions of Audacity. The second reason is that should there be important changes in future versions, all the previous versions are available for download in the Audacity project site (http://www.audacityteam.org/), so the reader could download the one described here (2.1.2). Finally, it is of the author's interest to make this program known because it is free software and it has already reached an interesting level of maturity.

NOTE: What follows is not meant to be a tutorial but an explanation of several aspects to consider in any digital audio editor to be used within a measurement system. The explanation of several of Audacity's functions is but a complement. For a complete explanation of all the features and functions it is suggested to read the manual that is accessible from the Help menu.

We shall refer to some of the possibilities that offers Audacity in the last version available at the moment of this writing (2.1.2). Before starting, consider that it is possible to change the language selecting *Edit/Preferences.../Interface*.[7] Once there, the Language selector offers more then 50 languages. From now on we will assume that English has been selected.

4.2.1 Opening an Existing File

Although .wav is not Audacity's native format, it is possible to *import* files of this type either with *File/Open...* or *File/Import/Audio...* A third way is to drag a file from any file explorer and drop it into the Audacity interface. In Fig. 4.1 there is an example of a waveform so imported. As can be observed, the signal is shown normalized between ± 1. When, as in this case, a large amount of audio is shown, the program is actually plotting the file *summary* instead of the signal. As explained in 3.16.2, the summary is a condensed version that contains the minimum, the maximum and the RMS value every 256 samples and every 65 536 samples. The 256 sample version is used for medium zoom and the 65 536 sample version for very low zoom (i.e. very long intervals are represented). In the oscillogram, negative and positive peaks are shown in dark blue and $\pm$ the RMS value in light blue (dark and light gray in the monochrome figure).

In all cases, the project adopts the original sample rate, but the numeric format is converted from 16 bit to 32 bit float, as can be seen in the control panel to the left of the track window. It is possible to keep the original bit resolution, but it is necessary

[7]We adopt the convention of separating with / the items selected in a menu or sub-menu from those selected in any sub-menu.

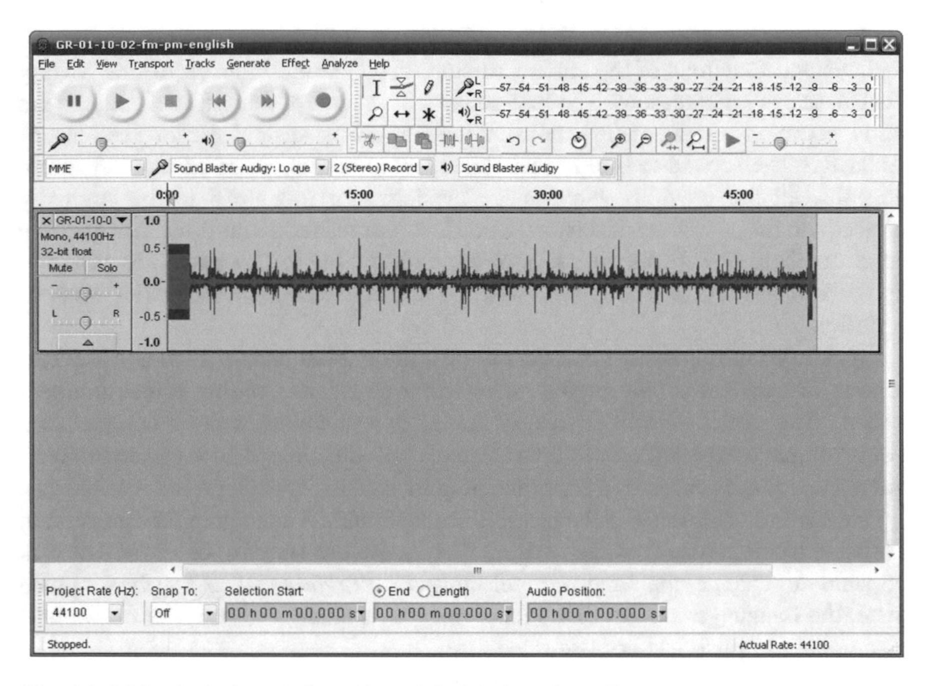

Fig. 4.1 Main Audacity window (the original is in color). There are menu areas, transport controls for recording, playing, pausing and stopping, interactive tools, track tools, vumeters, faders, and selection tools

to change the default sample format in the preferences editor (*Edit/Preferences.../ Quality*).

This decision may be appropriate if the editing process will not introduce waveform changes (for instance, setting apart a portion of the waveform to be analyzed later, adding cues or labels, joining two segments, removing irrelevant audio).

When the signal is to be subjected to signal processing, such as scaling or filtering, it is better to work with 32 bit, since cumulative truncation errors will arise in a much less significant digit. In order to illustrate this concept, suppose that we multiply the number 3.2456 four times by 0.65321 (a typical operation in digital filters). The exact value is

$$3.2456 \times 0.65321^4 = 0.385\,974\,948\,890\,394\,580\,168\,251\,860\,56$$

If each intermediate result is truncated to 10 significant digits after each operation, we shall have

$$3.2456 \times 0.65321^4 \cong 0.385\,974\,94\underline{7\,9}$$

where the incorrect digits have been underlined. If we now truncate to 5 significant digits after each operation,

$$3.2456 \times 0.65321^4 \cong 0.3859\underline{5}$$

If we compare with the exact result, we can see that in the first case we have 8 exact digits, while in the second case we have only 4. If we truncate the first approximation to 5 digits, all digits will be correct, unlike the second case, where only 4 digits are correct.

This situation, illustrated in decimal code, also holds for binary code. That is why it is better to work with 32 bit than with 16 bit.

Now, according to Sect. 3.7, prior to the final truncation to 16 bit (conceptually, a requantization with less bits) it will be necessary to add some *dither* in order to decorrelate.

Audacity adds dither by default before requantizing, that is, why it is not convenient to convert to 32 bit a signal that is not meant to be processed, since the dither implies an unnecessary change in the signal. Dithering can be disabled in the menu item *Edit/Preferences/Quality/Dither*.

4.2.2 Recording Sounds

Another way to load a signal in order to work with it is to record it directly. To this end, the transducer signal, which in our case is the calibrated output of a sound level meter, must be connected to the line input of a sound card, either internal or external (external sound cards are preferable since they are of better quality).

Before beginning to record, it is necessary to make sure that the signal at the sound card input is not so small to be affected by the digitization noise, nor so large to cause saturation.

The level adjustment is performed by sliding the fader at the right of the microphone icon while observing the corresponding vumeter (volume meter). If no activity is noted, it may be necessary to open the context menu (by right-clicking on the vumeter) and check *Enable Meter.*[8] In the case of monophonic signals, it is necessary to select such condition in the device tool bar. Once the level has been properly adjusted, recording is started by pushing the red button in the transport toolbar. Audacity automatically creates a new track.

4.2.3 Generating Signals

Finally, we can also introduce signals to work with by generating them from the *Generate* menu. Possibilities include tones, triangular and square waves, noise and

[8]When the mouse pointer is over any active element of the Audacity interface, a brief context help pops up. This is useful to discover new functions or remember where they are.

chirps, which are useful as test signals. While there are not many functions available yet, it is possible to create new ones using the Nyquist language (though we will not go any further, there is plenty of information in several Internet sites) (Dannenberg 2016). An interesting feature is that one track may be played back while another one is recording. This is useful for acoustical tests, where a signal (for instance, a tone sweep) is played back and the system response is recorded in other track.

4.2.4 Adding New Tracks

Once a track has been loaded or created, it is possible to add new tracks in several ways. The first one is to import another .wav file, which will automatically create a new track with the new signal.[9] Note that Audacity allows to import files with different sample rates. However, each project has a global sample rate that is set by default in the preferences editor (*Edit/Preference/Quality*). This default value is overridden by the sample rate of the first imported track. It can also be changed from the Audacity selection toolbar, normally at the bottom of the main window.

The possibility of combining signals with different original sample rates may be interesting for electroacoustic music or for soundscape simulation, but it is not advisable in a metrological context, since they will be in general signals of different qualities, being the bottleneck that one with the poorest sample rate.

Each track has a default name (Audio track) that can be changed by clicking on it to open a context menu, one of whose items is *Name*... Clicking on it a dialog box opens where the new name can be entered.

4.2.5 Saving a Project

Once the intended sound waves are loaded, one can proceed to save the project. An .aup project file is then created, as well as a directory of the same name ending with _data that contains, in a subdirectory tree, one or more .au or .auf files, according to the importing mode. There are two modes: 1) Saving a real copy of the audio file, in which case one or more .au files will be created according to the format described in Sect. 3.16.2; 2) Linking the original .wav file without creating any copy, in which case one or more .auf summary files are created according to Sect. 3.16.3. In the latter case a *dependency* is created. This is simply a link to the original file that appears in the project file. Audacity warns the user before saving, since even if the original file will not be modified by Audacity, it might be modified by means of an

[9]Bear in mind that in this case only the import function can be used. Attempting to use the open file function will open the waveform in a new, independent window.

external application, which could cause inconsistencies that might affect the project's integrity. The dependencies of a project can be checked from the menu item *File/Check dependencies...*

Anyway, there is a limit on the editing actions that can be performed without causing the program to make a real copy of the audio files. If the edit involves an effective waveform change, such as replacing some part or filtering it, when saving the project the corresponding files will be generated within the project's data directory and the dependency will be removed. An exception is when the edit is of a type that can be recorded in the project file, such as a change in the gain, panning or envelope.

4.2.6 Selection

One of the preliminary operations necessary for any edit event is to *select* some portion or region of the signal. In Audacity this can be done in two ways. The first is to visually identify the region of interest, point with the mouse at the beginning and drag to the end while pressing the main button. The result is shown in Fig. 4.2. The second is to identify the initial and final instants and enter them in the corresponding boxes in the selection tool bar, usually at the bottom of the window.[10] It is also possible to define the initial time and the length of the selection by choosing "Length" instead of "End" as the meaning of the second box.

Even if the first method has been used, the limits of the selection can be edited entering new instant values in the corresponding boxes. Finally, it is possible to make changes with the mouse by getting close to either limit until a hand replaces the arrow pointer. Then one can drag the limit to a new position on the time axis.

A selection may be saved momentarily by means of *Edit/Region Save* and then can be restored by means of *Edit/Region Restore*. Note that saving a selection does not save or copy to the clipboard the audio content of the selection but only the limits of the region. In case there are two or more tracks, it is possible to copy the selection made in one track to other track by pressing *Shift* and clicking on the track control panel at the left of the track.

An interesting function related to selection is zero-crossing detection. Once selected a region, its limits can be slightly adjusted to the closest zero crossing by means of *Edit/Find zero crossings* (or with the keyboard shortcut Z). This is useful to prevent jump discontinuities that could alter the spectral content or the RMS value.

[10]This position may change since in Audacity most of the tool bars can be relocated and allow *docking*. This means that if picked with the mouse from their handle at the left they can be freely moved to other docking area, where, once released, they dock permanently until a new relocation. Docking areas are recognized because a blue triangle appears pointing at the new possible location.

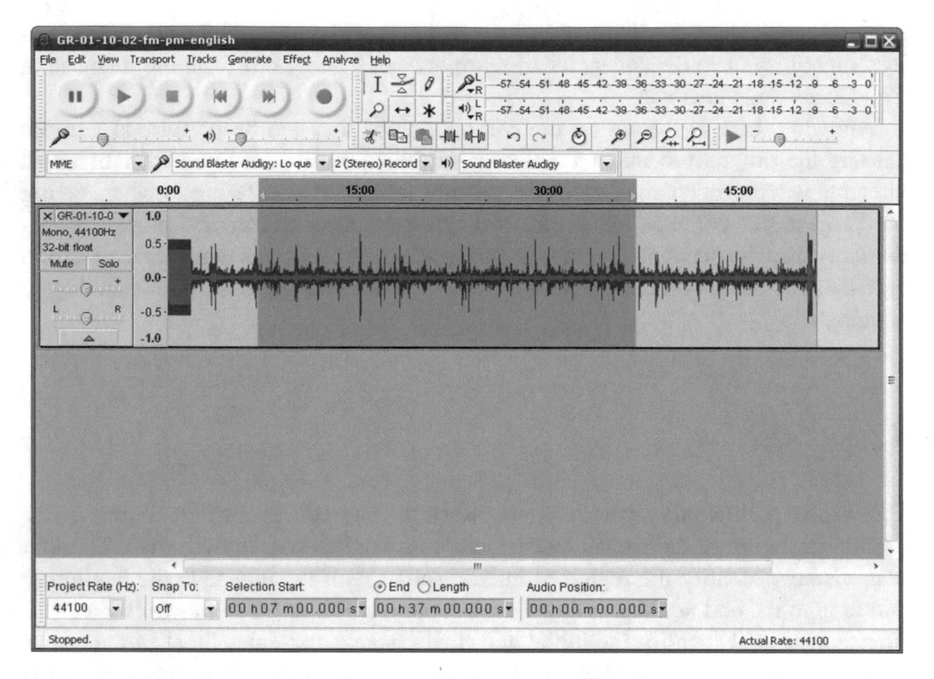

Fig. 4.2 Selection of part of a sound track in Audacity

4.2.7 Labels

Label tracks are a very useful tool to attach text information to specific time instants or intervals. It is possible to create as many label tracks as desired by means of *Tracks/Add new/Label track*. The default name of label tracks ("Label track") can be changed as in the case of audio tracks. Once a label track has been created, there are two ways to add labels. The first one is to select a region (time interval) and click on *Tracks/Add Label At Selection*. The label is an initially empty box where it is possible to write the desired text. The box will widen to accommodate the text.

The second way is to click on *Tracks/Edit labels* to open a table where all labels are listed and can be edited. It is possible to write a text as well as to edit the initial and final instants of the region to which the label refers. It is also possible to insert new labels. The labels created in this way are automatically located at 0, so the initial and final times should be edited. Figure 4.3 shows two labels added to the project of the previous figures.

Labels can be used as part of an analysis protocol that could be automated letting some software code read the XML project file (.aup) to detect key words such as "calibration," "date," "time," "GPS," or "synchro," and then extract values to be used in analysis or documentation. There could also be some structured events such as vehicle count as well as a cognitive classification performed by the operator upon listening to the track (for instance a dog's bark, a shout, some object falling down).

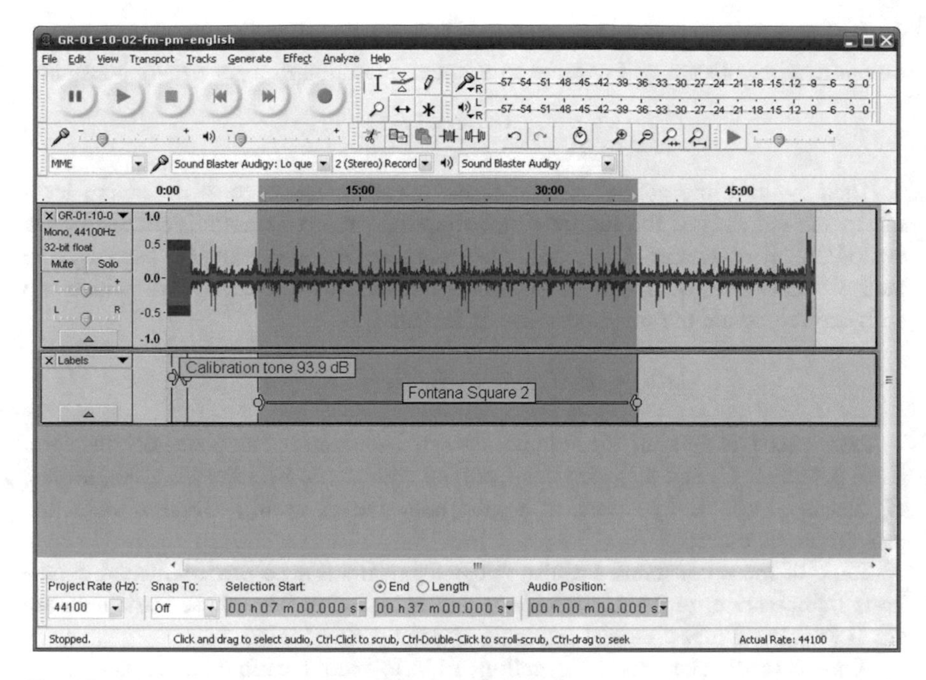

Fig. 4.3 Selection of a region of an audio track in Audacity

Since the structure of an XML file is designed precisely to facilitate automated parsing, the information can be retrieved by some software routine capable of detecting key words that define some information categories and within those categories, their attributes or associated parameters.

4.2.8 Selection in the Presence of Labels

Labels also facilitate repeating selections, since when dragging a selection boundary close to one of the labeled region limits, it attracts the selection boundary and makes it coincide with the region limit (this is sometimes referred to as *sticky* or *magnetic selection*). A vertical yellow line (called *snap guide*) appears over the limit to inform the user that the limit has attracted the selection boundary.

To extend the selection to other tracks it suffices to press Shift and point with the mouse inside the track control panel.

4.2.9 Calibration Tone

When a transducer (microphone) is used as a signal generator, the whole process that eventually leads to a sound file is mediated by a series of components each of which modifies the signal. The first one is the microphone itself, whose input–output relationship we will assume as a linear one through a *sensitivity* constant:

$$v(t) = S_{\mathrm{m}} p(t). \tag{4.1}$$

Then we have the gain G_{c} of the signal conditioning stage of the sound level meter, the gain G_{d} of the digitizer's preamplifier, the conversion constant of the analog/digital converter $K_{\mathrm{A/D}}$, and the normalization constant K_{n} used by the analysis software. Thus, the dimensionless numeric value $x(n)$ with which the software represents the pressure value at instant t_n is

$$x(n) = K_{\mathrm{n}} K_{\mathrm{A/D}} G_{\mathrm{d}} G_{\mathrm{c}} S_{\mathrm{m}} p(t_n) = K\, p(t_n). \tag{4.2}$$

In practice it is difficult, if not impossible, to keep track of all constants that form K. In particular G_{d} and $K_{\mathrm{A/D}}$ are not specified and cannot be adjusted at all, unlike, G_{c} that is adjustable by means of a calibration preset, or K_{n}, which is under the programmer's control.

Some of these constants are also subject to drifts due to environmental conditions (temperature, pressure, humidity) and aging. However, these variations are small for typical measurement intervals of some tens of minutes.

A more practical method for getting K is to record calibration signal whose sound pressure level or, preferably, equivalent level is known with very low uncertainty. This signal could be, in principle, any signal that can be measured by the sound level meter while simultaneously recording it. In practice, a calibration tone from an acoustic calibrator is used, since its uncertainty is lower then that of the sound level meter. The calibrator generates a high purity sinusoidal tone of known sound pressure level and frequency inside an acoustic coupler to which the microphone is coupled.

There are two possibilities. The first one is to take some characteristic instantaneous value such as the peak value of the recorded calibration tone. The second one is to take the effective value (RMS value) extended to an interval of a few seconds. The latter is preferable since it reduces the influence of acoustic and electric noise.[11]

It can be calculated by approximating numerically the integral from Eq. (1.1) by means of

$$X_{\mathrm{ef}} = \sqrt{\frac{1}{N_2 - N_1 + 1} \sum_{n=N_1}^{N_2} x^2(n)}, \tag{4.3}$$

where N_1 and N_2 are the indices of the initial and final samples of the selected region of the calibration tone. We shall call such value X_{cal}.

[11]The acoustic calibrator has some degree of acoustic insulation, but the ambient noise, even attenuated, can alter instantaneous values such as the peak value.

Assuming that the nominal sound pressure level inside the coupler is L_{cal}, which corresponds to a sound pressure

$$P_{cal} = P_{ref} 10^{L_{cal}/20}, \tag{4.4}$$

where $P_{ref} = 20\ \mu\text{Pa}$, we can estimate the constant K as

$$K = \frac{X_{cal}}{P_{cal}} = \frac{X_{cal}}{P_{ref} 10^{L_{cal}/20}}. \tag{4.5}$$

We can now apply that constant in combination with Eq. (4.2) to get the sound pressure that corresponds to instant t_n:

$$p(t_n) = \frac{1}{K} x(n). \tag{4.6}$$

Naturally, this equation also holds for RMS values, as can be easily derived from (4.3):

$$P_{ef} = \frac{1}{K} X_{ef}. \tag{4.7}$$

If, as is customary, we prefer to work with sound pressure level instead of the instantaneous value of sound pressure, then

$$L_{eq} = 20 \log \frac{X_{ef}/K}{P_{ref}} = 20 \log X_{ef} - 20 \log X_{cal} + L_{cal}. \tag{4.8}$$

This means that we can convert a numerically computed RMS value X_{ef} into the corresponding equivalent sound pressure level L_{eq}. All that is needed is a recorded portion of the calibration signal whose equivalent level is known.

We can implement this by means of external software (for example any mathematical package, such as Scilab), but it is also possible to do it directly in Audacity following this procedure:

(1) Open the window *Analyze/Contrast...*;
(2) Identify the calibration tone, whose oscillogram is characterized by a constant amplitude;
(3) Select an interval several seconds long and apply the function *Edit/Find Zero Crossings*. This will reduce the error in the RMS value, since the error is 0 when the integrating time equals an integer number of half-cycles (see Fig. 3);
(4) In the contrast window opened in (1), click on the *Measure selection* button corresponding to *background*.

The result is a level in dB that equals $20 \log X_{ef}$. It is a level referred to 1, i.e., the maximum normalized signal, hence it is always ≤ 0. In the example of Fig. 4.1 the result is −8.2 dB.

It is very important not to fall into the error of thinking that this value is the sound pressure level. As a matter of fact, the true sound pressure level of the calibrator was 93.9 dB.

If we repeat operations (2) to (4) with the selection corresponding to the other label, but clicking on the *Measure selection* button corresponding to *foreground*, we get −30.2 dB. Applying Eq. (4.8), we have:

$$L_{eq} = -30.2\,\mathrm{dB} - (-8.2\,\mathrm{dB}) + 93.9\,\mathrm{dB} = 71.9\,\mathrm{dB},$$

so the equivalent sound pressure level in the labeled interval is 71.9 dB. The box labeled *Difference* computes automatically the difference between both values. It suffices to add the calibration tone level to get the equivalent level.

Two final observations. First, for long measuring intervals (several hours), it is convenient to repeat the calibration periodically (for instance, once every 30 min). Although the differences are usually very small, this is to take into account possible calibration drifts, particularly if there have been temperature or humidity variations.

Second, it is convenient, immediately after recording the calibration tone, to orally record the instrument's reading as an acoustic documentation of the calibration value.

4.2.10 FFT Filters

Although we will refer in detail to FFT filters in the chapter on spectrum analysis, we will see here the possibilities that Audacity offers. Conceptually, FFT filters allow to implement an arbitrary frequency response, for instance A and C weighting filters or octave and one-third octave filters. Once a portion of the signal has been selected this function can be applied by means of *Effect/Equalization*. A window as illustrated in Fig. 4.4 opens.

This window allows to draw the frequency response by grabbing and dragging points of the default plane response, and also to select a *preset* filtering curve by opening the menu *Select curve*. The thick curve (blue in the original screenshot) with empty circles represents the desired specification of the filter. The thin curve (green in the original screenshot) represents the actual frequency response. In the low frequency range the response differs somewhat from the desired one. It fits better if the filter length is increased, since the FFT increases the frequency resolution, which affects mainly the low frequency end. The two lateral sliders allow to adjust the upper and lower bounds of the plot.

The filter shown in Fig. 4.4 corresponds to an A filter (see also Fig. 5 and Table 1.1), which is not among the presets presently bundled with the program. By means of *Save/Manage Curves*... one can open a dialog window where it is possible to import previously saved curves. They should have been saved with an .xml extension according to the following structure:

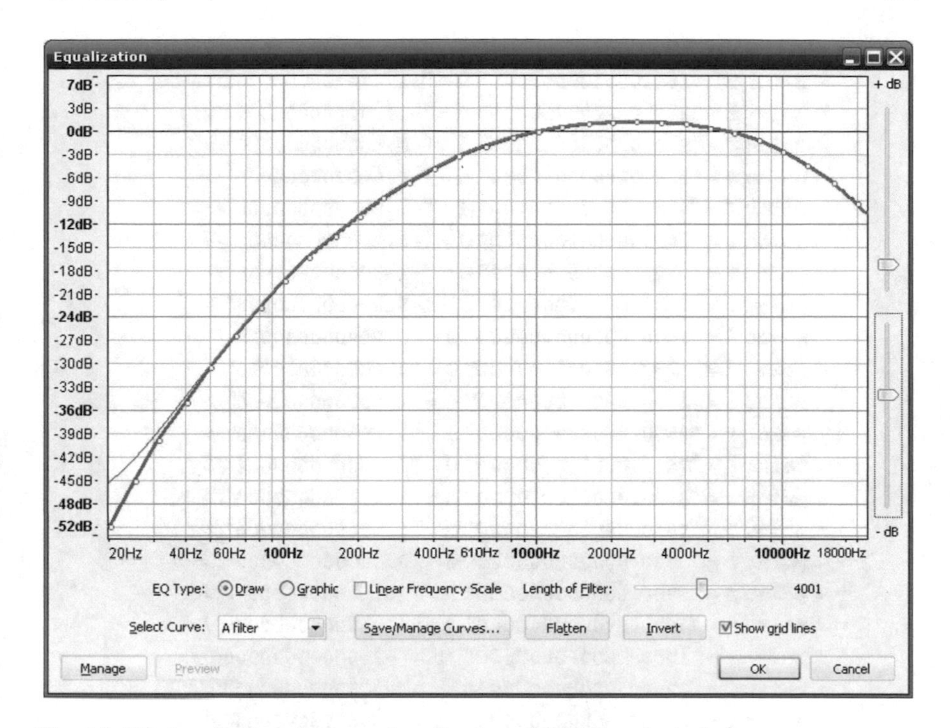

Fig. 4.4 Window for the selection and application of FFT filters in Audacity

```
<equalizationeffect>
     <curve name="name">
          <point f="f1" d="d1"/>
          <point f="f2" d="d2"/>
          ...
          <point f="fn" d="dn"/>
     </curve>
</equalizationeffect>
```

where *name* is the filter's name, f_1, …, f_n are the numerical values of the frequencies in Hz and d_1, …, d_n are the corresponding numerical values of the frequency response in dB. As a particular example, the A filter is defined as:

```
<equalizationeffect>
     < curve name = ”A filter”>
          < point f = ”20.000000000000” d = ”-50.500000000000”/>
          < point f = ”25.000000000000” d = ”-44.700000000000”/>
          < point f = ”31.000000000000” d = ”-39.400000000000”/>
          < point f = ”40.000000000000” d = ”-34.600000000000”/>
          < point f = ”50.000000000000” d = ”-30.200000000000”/>
          < point f = ”63.000000000000” d = ”-26.200000000000”/>
          < point f = ”80.000000000000” d = ”-22.500000000000”/>
          < point f = ”100.000000000000” d = ”-19.100000000000”/>
```

```
        < point f = ”125.000000000000” d = ”-16.100000000000”/>
        < point f = ”160.000000000000” d = ”-13.400000000000”/>
        < point f = ”200.000000000000” d = ”-10.900000000000”/>
        < point f = ”250.000000000000” d = ”-8.600000000000”/>
        < point f = ”315.000000000000” d = ”-6.600000000000”/>
        < point f = ”400.000000000000” d = ”-4.800000000000”/>
        < point f = ”500.000000000000” d = ”-3.200000000000”/>
        < point f = ”630.000000000000” d = ”-1.900000000000”/>
        < point f = ”800.000000000000” d = ”-0.800000000000”/>
        < point f = ”1000.000000000000” d = ”0.000000000000”/>
        < point f = ”1250.000000000000” d = ”0.600000000000”/>
        < point f = ”1600.000000000000” d = ”1.000000000000”/>
        < point f = ”2000.000000000000” d = ”1.200000000000”/>
        < point f = ”2500.000000000000” d = ”1.300000000000”/>
        < point f = ”3150.000000000000” d = ”1.200000000000”/>
        < point f = ”4000.000000000000” d = ”1.000000000000”/>
        < point f = ”5000.000000000000” d = ”0.500000000000”/>
        < point f = ”6300.000000000000” d = ”-0.100000000000”/>
        < point f = ”8000.000000000000” d = ”-1.100000000000”/>
        < point f = ”10000.000000000000” d = ”-2.500000000000”/>
        < point f = ”12500.000000000000” d = ”-4.300000000000”/>
        < point f = ”16000.000000000000” d = ”-6.600000000000”/>
        < point f = ”20000.000000000000” d = ”-9.300000000000”/>
        < point f = ”25000.000000000000” d = ”-12.300000000000”/>
        < point f = ”31500.000000000000” d = ”-15.700000000000”/>
        < point f = ”40000.000000000000” d = ”-19.400000000000”/>
        < point f = ”48000.000000000000” d = ”-22.300000000000”/>
        < point f = ”50000.000000000000” d = ”-23.000000000000”/>
        < point f = ”63000.000000000000” d = ”-26.800000000000”/>
        < point f = ”80000.000000000000” d = ”-30.900000000000”/>
        < point f = ”100000.000000000000” d = ”-34.700000000000”/>
    </curve>
</equalizationeffect>
```

In a similar fashion other filters such as the C filter, octave band and one-third octave band filters, critical band filters for use in psychoacoustics, as well as many others that could be useful for some specific application, can be designed either graphically or numerically. In a near future, implementations of time-varying filters are expected. They could prove interesting to separate certain sources (Boggino 2014; Bergero 2014).

4.2.11 Spectrogram and Spectrum Analysis

The spectrogram and the spectrum analysis plot are two essential tools in any audio editor. The spectrogram shows the spectral amplitude, as a function of time (abscissa) and frequency (ordinate), through a color (or a grayscale) map. The color map is a purely conventional scale ranging from dark (cold) to brilliant (warm). Internally it is actually a five-color scale which is interpolated on an RGB (red-green-blue) basis.

In Fig. 4.5 the spectrogram corresponding to Fig. 4.1 is shown. To make the spectrogram visible it is necessary to click on the small arrow in the track control panel and select *Spectrum* from the menu.

Although some practice is required, the spectrogram is very useful to visually detect particular acoustic events such as bird chirps, human speech, vehicle passings. In the figure, the calibration tone can be readily seen, represented by an intense line centered on 1 kHz as well as some weaker harmonics at 2, 3 and 5 kHz. The rest of the spectrogram shows traffic noise events, with emphasis at the low frequency end. Note that the default scale ranges from 0 to 8 kHz, but it is possible to increase or decrease the zoom level by clicking on the scale while pressing or not the Shift key. It is also possible to select a vertical spectral range dragging with the mouse between two frequencies. The second frequency might be outside the original range.

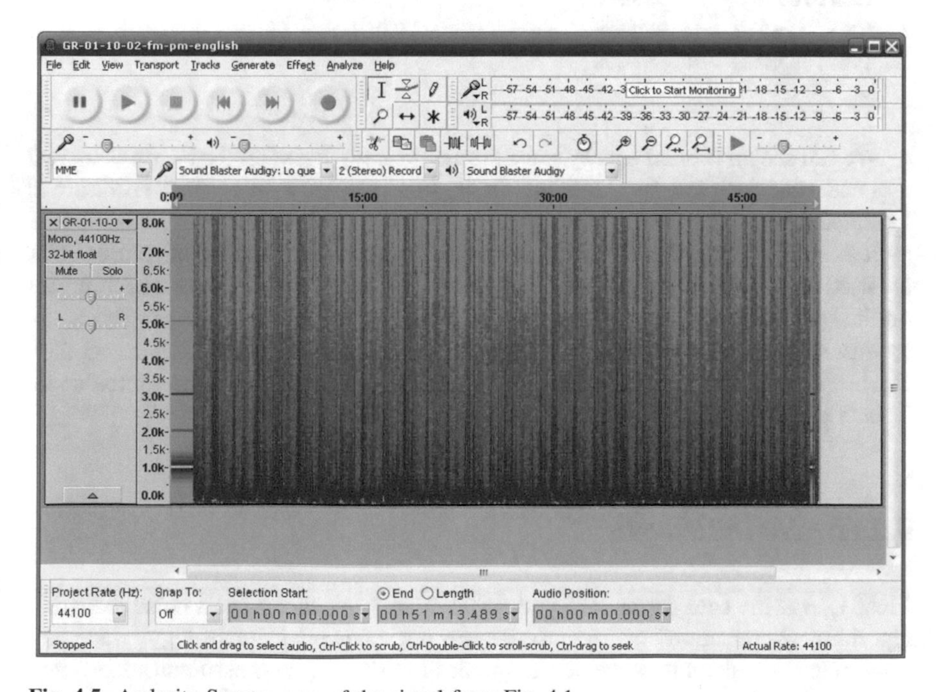

Fig. 4.5 Audacity Spectrogram of the signal from Fig. 4.1

In general, a range up to 5 kHz allows to see all relevant details, for instance the harmonics of tonal sounds, such as human voices, sounds of animals, motors, vehicles, etc. In order to appreciate characteristic patterns of short events, such as bird chirps, speech, car horns, mechanical noise, impulse noise, it may be necessary to appropriately select the vertical and horizontal zoom level. It is also possible to drag down the bottom border of the track to enlarge the track height so to get a more detailed view.

The other spectral tool is the spectrum analysis window. Once a region to analyze has been selected, the spectrum can be plotted by means of *Analyze/Plot spectrum*. Figure 4.6 shows the spectral window showing the calibration tone spectrum.

An interesting possibility of the spectrum plot window is to export the spectrum data as a text file containing all the spectral lines as a frequency–volume table. The following is an extract of what happens near 1000 Hz (the nominal frequency of the calibration tone)

```
Frequency (Hz) volume (dB)
968.994141     -97.818932
979.760742     -65.876839
990.527344     -30.771830
1001.293945    -14.120676
1012.060547  -6.435168
1022.827148  -5.694033
1033.593750  -11.738699
1044.360352  -26.038492
1055. 126953 -54.824665
1065.893555  -98.254608
```

We can see that sound energy concentrates strongly between the bands 1001.293 945 and 1033.593 750 Hz. Later on it shall be explained how an FFT spectrum is interpreted.

It is also possible to change the spectrum size, i.e., the number of samples that are used to compute it (in general, a power of 2). This parameter has a direct impact on the spectrum resolution. The larger the window length, the more detailed is the spectrum. In the current version the supported lengths range from 128 to 65 536. However, this limit is a parameter that could be easily edited in the source code. This is possible because Audacity is open-source software.

4.2.12 Noise Reduction

Finally, as is the case for most digital audio editors, Audacity has a noise reduction algorithm. Applications of this process are not frequent in metrology, since any noise reduction alters to some degree the desired signal (the measurand). However, for the sake of completeness we shall briefly introduce this tool.

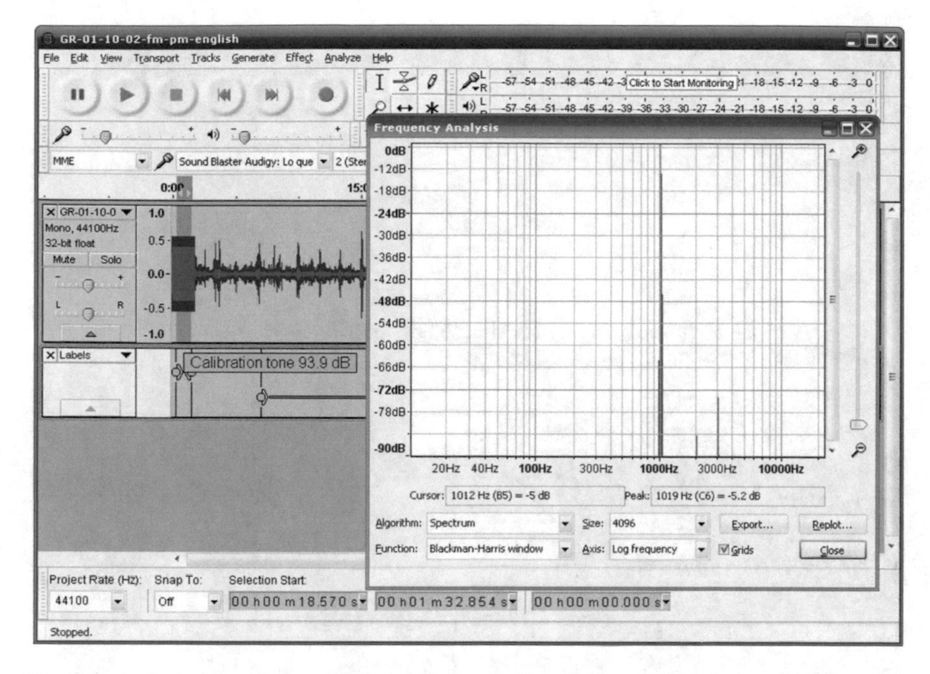

Fig. 4.6 Audacity Spectrum analysis window showing the spectrum of the calibration tone included in the signal of Fig. 4.1

First of all, bear in mind that the aim of a noise reduction algorithm is to get rid of some systematic noise component. However, given an interval including signal + noise, the algorithm has no way to discriminate signal and noise. That is why it is necessary to provide some interval where only the type of noise to reduce is present. Figure 4.7 shows the window that opens when *Effect/Noise Reduction…* is selected.

The procedure consists in the following: (1) Select first a noise-only interval and press *Get Noise Profile*. The algorithm computes the average spectrum and keeps it for later use; then the window closes. (2) Select the interval to remove noise from, press again *Get Noise Profile*, adjust parameters if necessary, and press OK to remove noise.

Essentially, the noise removal algorithm implements a multiband noise gate. At both stages, a 2048 sample FFT is used, providing a 1024 band spectrum.[12] The

[12]We shall see later that since the FFT is a complex spectrum, in order to get the real harmonics it is necessary to add pairs of conjugate lines. That is why a 4096 point FFT yields only 2048 spectral lines.

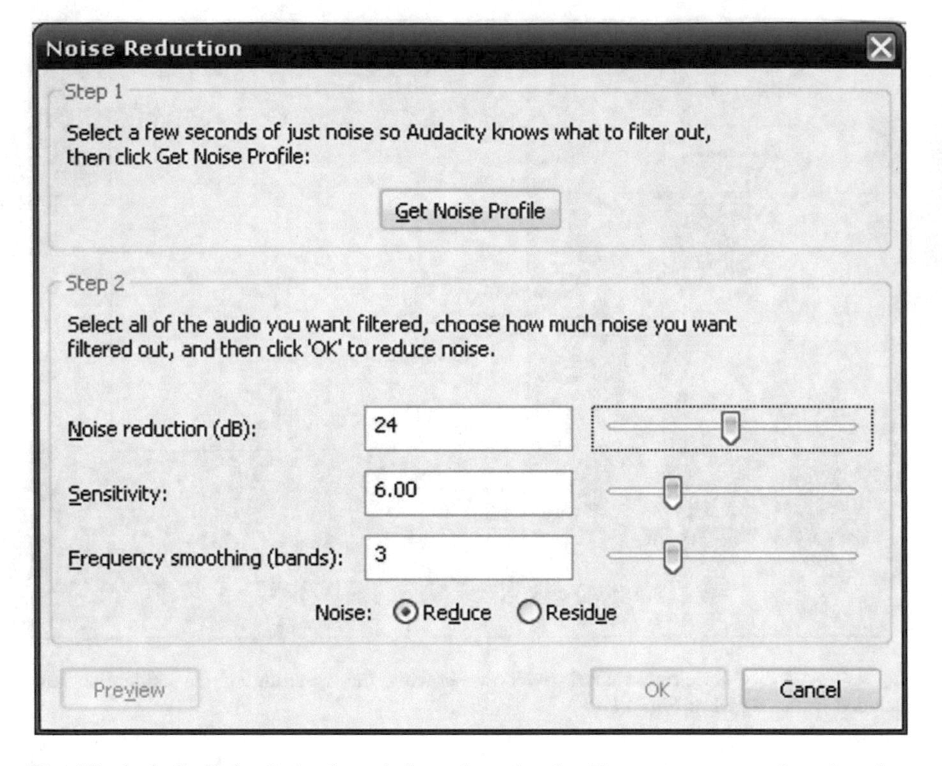

Fig. 4.7 Audacity Noise Reduction window where the algorithm parameters can be selected

noise profile is basically a high percentile (a value close to the maximum) of the spectrum during the noise-only interval.

To reduce the noise, the 1024 band spectrum is computed for consecutive (or, rather, overlapped) 2048 sample windows. For each window the algorithm finds out, on a separate spectral line basis, whether or not the signal exceeds certain threshold based on the noise profile and the *sensitivity* parameter (which indicates how many decibels above the noise profile is the threshold). If the answer is *yes*, nothing is done. If it is *no* (meaning that the signal is very likely just noise), it is reduced by the number of decibels indicated by the *Noise reduction* (dB) parameter.

In order to prevent *artifacts*[13] two precautions are taken. First, a single spectral line is never removed or reduced alone. This is because except in very particular situations, a single tone is represented by more than one spectral line,[14] so if only one of them is removed, a noticeable amplitude modulation may appear. The

[13]An *artifact*, in the context of audio signals is any audible disturbance caused as a consequence of some process. For example, clicks and parasitic tones.

[14]A tone may be represented by a single spectral line if its amplitude is constant and its frequency is also constant and equal to the frequency of a spectral line, i.e. kF_s/N (where k is an integer, N the window length and F_s the sample rate).

bandwidth to set the degree of similarity in noise reduction for adjacent spectral lines is indicated by the *Frequency smoothing* parameter.

The second precaution is to prevent an abrupt change over time, i.e. that a noise component be suddenly removed, since that would cause a perceivable transient. This used to be controlled by the *Attack/decay time* parameter. This parameter has been optimized and removed from the user's control.

Once the bands where noise prevails have been reduced in the spectral domain, the inverse FFT is taken to get the time-domain signal. Since the windows are too short to describe the entire signal, they must be somehow joined. The way to do this without creating jump discontinuities is to overlap and crossfade windows (see FFT filters) (Mazzoni n.d.; Crook 2011).

Problems

Note: These problems require Audacity installed on your computer.

(1) (a) Connect a microphone to the audio input of a computer. Open Audacity and start recording the ambient sound during about 30 s. If the recording is stereo go to *Edit/Preferences/Devices* and select 1 channel. You can also convert the stereo recording into mono selecting the menu *Tracks/Stereo track to mono*. (b) Select a portion of the signal and apply a label going to *Track/Apply label to selection*. (b) Go to *Analyze/Contrast*... to measure the RMS value (expressed in dB relative to a square wave of maximum amplitude). If −18 dB is equivalent to 94 dB, find the sound pressure level of the labeled region.

(2) Connect the audio output of a sound level meter to a digital recorder and start recording. Make sure the output is unweighted. First record a 94.0 dB calibration signal during several seconds. Then record the ambient noise. Open the file with Audacity. (a) Locate the calibration tone and select a portion. Go to *Edit/Find zero crossings* in order to adjust the ends so that an integer number of cycles is selected and add a label with the text "calibration tone." Open *Analyze/Contrast*... and measure selection as background. Then select a portion of ambient noise, label it as "noise" and measure selection as Foreground. Use the difference to find the sound pressure level.

(3) Create the specification of an FFT filter to implement the A-weighting (see Sect. 4.3.10), save it with an .xml extension and import it from *Effects/Equalization/Save/Manage Curves.../Import*. Once imported, apply it to a selection of an ambient noise recording. Once more, use *Analyze/Contrast*... to measure the RMS value of the A-weighted signal. Use the calibration data to find the A-weighted sound pressure level.

(4) Using a digital recorder, make some recordings of: (a) Traffic noise; (b) People speaking; (c) Bird chirps; (d) Dogs barking; (e) Some music and (f) Any other interesting sound you find there. Then download the files to the computer, open them in Audacity and select the spectrogram view. Visually analyze the different sounds you have recorded while listening to them, trying to detect features that are common to all sounds in a given category. Notes: (i) If you cannot find those sounds in your area, you can download many sounds of different categories from several Internet sites (for instance https://freesound.org/).

(ii) While you certainly will not obtain calibrated recordings, even a regular digital recorder will have enough fidelity to attain the purpose of this practice.

(5) Open some audio files containing discrete tones, for instance some piano music or single-speaker speech. Use the FFT spectrum analyzer (*Analyze/Plot spectrum*) to measure the frequency of the different harmonics. Note that the cursor sticks to the peaks, facilitating the measurement of frequencies. Try different window functions and sizes.

References

Bergero F, Miyara F (2014) Diseño de una interfaz gráfica para ingreso de una respuesta en frecuencia variable en Audacity. Facultad de Ciencias Exactas, Ingeniería y Agrimensura, Universidad Nacional de Rosario. 2014. Unpublished internal document

Boggino L, Miyara F. (2014) Especificación de un filtro variable en el tiempo mediante una herramienta de dibujo en Audacity y su implementación. Universidad Nacional de Rosario, FCEIA, 2014. Unpublished internal document

Crook J (2011) How noise removal works. Audacity Team, 2011. http://wiki.audacityteam.org/wiki/How_Noise_Removal_Works. Accessed 27 Jul 2013

Dannenberg RB (2016) Nyquist reference manual. Carnegie Mellon University. School of Computer Science. Pittsburgh. http://www.cs.cmu.edu/~rbd/doc/nyquist/. Accessed 29 Feb 2016

Mazzoni D (no date) Audacity: a digital audio editor. NoiseRemoval.cpp. https://audacity.googlecode.com/svn/audacity-src/trunk/src/effects/NoiseRemoval.cpp. Accessed 26 Feb 2016

Mazzoni D, Licameli P (no date) "Audacity: a digital audio editor. NoiseReduction.cpp". https://audacity.googlecode.com/svn/audacity-src/trunk/src/effects/NoiseRemoval.cpp. Accessed 26 Feb 2016

Tan K (no date) 25 Free digital audio editors you should know. http://www.hongkiat.com/blog/25-free-digital-audio-editors/. Accessed 24 Jul 2013

Chapter 5
Transducers

5.1 Microphones

The quintessential transducer in use for acoustical measurements is the instrumentation microphone. This transducer converts sound pressure $p(t)$ into voltage $v(t)$ through an ideally linear relationship:

$$v(t) = S_{\mathrm{m}} p(t), \tag{5.1}$$

where S_{m} is the sensitivity constant. In a real microphone, S_{m} depends on frequency and on the type and orientation of the sound field.

Although there are several physical principles on which microphones can be based, instrumentation microphones are always of the condenser type, since they have, in theory, a flat frequency response in a wider frequency range.[1] In Fig. 5.1 the schematic of a condenser microphone is shown.

The operating principle is as follows. There is a rigid *back plate* with several openings and a stretched membrane or diaphragm, both of them electrically conductive. A condenser is thus formed, whose capacitance is

$$C_{\mathrm{o}} = \varepsilon_{\mathrm{o}} \frac{\pi a^2}{X_{\mathrm{o}}}, \tag{5.2}$$

where $\varepsilon_{\mathrm{o}} = 8.84 \times 10^{-12}$ F/m is the dielectric constant of air, a is the diaphragm radius, and X_{o} is the separation at rest between the diaphragm and the back plate.

[1]Dynamic (moving coil) microphones have limitations at low frequency due to Faraday's law (the induced voltage depends on the velocity and hence on the frequency) and at high frequency due to the large inertia of the coil mass (the acceleration and, so, the velocity, decrease with frequency). In order to get a flat frequency response, sophisticated acoustic filters to correct the natural response are needed.

F. Miyara, *Software-Based Acoustical Measurements*, Modern Acoustics and Signal Processing, DOI 10.1007/978-3-319-55871-4_5

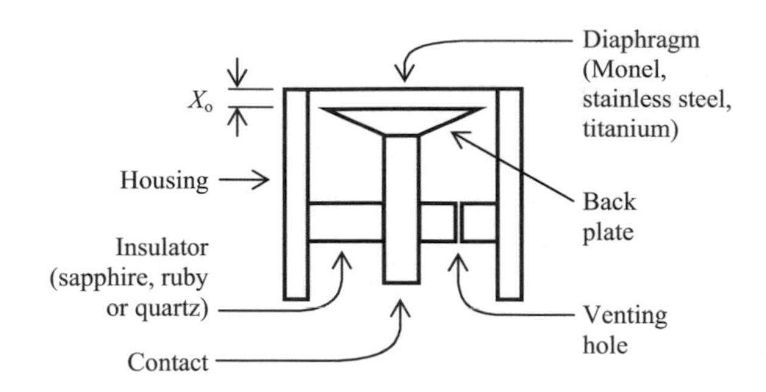

Fig. 5.1 Schematic of an instrumentation condenser microphone

Suppose now that both plates of the condenser have fixed and opposite electric charges $\pm Q_o$. Then between both plates there will be a voltage E_o, called *polarization voltage*:

$$E_o = \frac{Q_o}{C_o}. \tag{5.3}$$

When a sound wave impinges on the diaphragm, this will be set into vibration with an amplitude proportional to the outer sound pressure amplitude (the internal side is at an approximately constant pressure), causing the distance between both plates to vary. Hence, according to Eq. (5.2), the capacitance will vary and so will the voltage:

$$E = \frac{Q_o}{C}. \tag{5.4}$$

It can be shown that, with good approximation, the variable component of the voltage is given by

$$v(t) = \frac{E_o}{X_o} \frac{1}{\pi a^2 \left(\frac{8\tau}{\pi a^4} + \frac{\gamma P_o}{V_o} \right)} p(t), \tag{5.5}$$

where τ is the linear tension (force per unit length) applied to the diaphragm, $\gamma = 1.4$ is the ratio between constant-pressure and constant-volume specific heats of the air, and V_o the volume of the microphone enclosure (Miyara 2007, 2013).[2] If the linear tension applied to the diaphragm is high, the second term of the denominator is negligible so the approximation can be simplified to

$$v(t) = \frac{E_o}{X_o} \frac{a^2}{8\tau} p(t). \tag{5.6}$$

[2]This volume V_o is not equal to $\pi a^2 X_o$ because the zone between the back plate and the diaphragm is communicated by means of several openings with the volume behind the back plate.

The sensitivity is, therefore,

$$S_{\mathrm{m}} = \frac{E_{\mathrm{o}}}{X_{\mathrm{o}}}\frac{a^2}{8\tau} \tag{5.7}$$

In an instrumentation microphone a high sensitivity is desired in order to have a high signal-to-noise ratio. The larger the signal-to-noise ratio the smaller the minimum level that the instrument can reliably measure.

Although the diaphragm's linear tension τ should be low to get a high sensitivity, this has a negative impact on the frequency response, which gets reduced. This is an unacceptable tradeoff for an instrumentation microphone, so large linear tensions of the order of 1000 N/m are applied. For this reason it is necessary to use materials with a high yield point, such as Monel (a nickel (67 %) and copper (33 %) alloy which is extremely resistant to corrosion), special stainless steel or titanium.

In order to get a large sensitivity it is necessary, therefore, to increase the polarization voltage E_{o} and reduce X_{o}. A voltage of 200 V and a distance of around 20 μm between the diaphragm and the back plate are used. The thickness of the diaphragm is less than 10 μm. It can be shown that at a sound pressure level of 110 dB the mean displacement is of the order of 1 μm. It is also possible to increase the diaphragm radius, but except in some laboratory applications, today the most widely used microphones are the half-inch (1/2″, approximately 13 mm) and the one-quarter-inch (1/4″) ones.

5.2 Polarization

In the preceding discussion it was assumed that somehow a charge had been previously applied to both plates of the condenser. In practice, there are two ways to accomplish this. The first one is to apply an external polarization voltage through a resistor with a sufficiently high resistance so that the signal is not severely attenuated by the voltage divider effect at the load. The condenser will get charged to that voltage acquiring the necessary electric charge.

The second way is by means of a special material known as *electret*. It is a polymer that is normally an insulator, but when heated to high temperature it becomes electrically conductive. While in that state, it may be electrically charged if used as a plate of a condenser. Upon cooling, the electric charge is permanently trapped within the material, providing the polarization charge Q_{o} that the microphone needs to operate. This electric charge induces an opposite charge on the other plate by electrostatic attraction, creating the electric field necessary for the microphone operation.

Even if electret microphones are more expensive, they are a good choice for their use in handheld sound level meters, since this avoids the need of using a DC–DC switching converter to rise the battery voltage to 200 V.

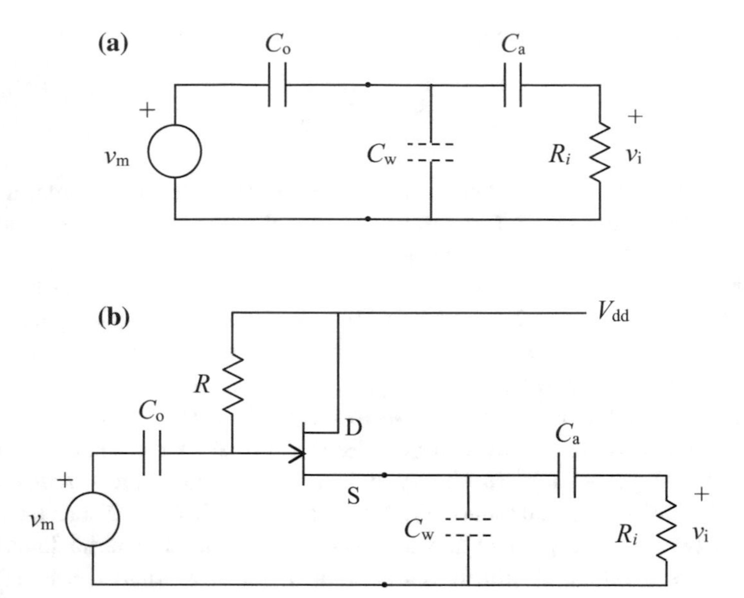

Fig. 5.2 **a** Electrical model of the connection of a condenser microphone to an amplifier. **b** The microphone with a FET preamplifier connected as a voltage follower. In both cases C_w is the wire capacitance, C_a is used to block the DC component and R_i is the input resistance of the remote amplifier. V_{dd} polarizes both the microphone and the FET

5.3 Preamplifier

As a circuit element, the microphone is a voltage source with a series capacitor in the range of some tens of pF. This means that at 20 Hz the internal impedance is of the order of 100 MΩ. Should we connect it directly to the instrument's input amplifier, the voltage divider between the microphone impedance and its load (the amplifier's input resistance R_i) would attenuate the signal in an inordinate way, rendering it useless. Moreover, the capacitance C_w of the wire between the microphone and the amplifier would cause additional high-frequency attenuation.

To prevent this, a preamplifier formed by one or two field effect transistors (FET) is used. These FET's are used as high input impedance voltage followers (unity gain) that work as an impedance adapter by reducing the extremely high impedance of the microphone capacitor to a few hundred ohms (Fig. 5.2b). The preamplifier is located very close to the microphone, enclosed in the tubular stem that connects the microphone with the meter body (which acts as an electrostatic shield).

5.4 Sound Fields

The space and time distribution of the sound pressure is usually referred to as *sound field*. Although the complexity of any real-life sound field is very high in almost all cases, from the metrological point of view it may be frequently approximated by one of four limiting cases with well-defined properties: free field, pressure field, diffuse field, and stationary field.

5.4.1 Free Field

A *free field* is any sound field whose propagation direction is clearly determined. A condition for the presence of a free field is that there are no reflecting surfaces near the sound source or the receiver. Special cases of free field are plane, cylindrical, or spherical waves, but at a considerable distance from the source any free field exhibits an approximately plane behavior. In general, free field takes place outdoors, but it is also found inside an *anechoic chamber*, i.e., a room with all its surfaces covered by highly absorbent materials or structures so that the reflected sound is negligible.

5.4.2 Diffuse Field

A *diffuse field*, also known as *random field* is any sound field such that all propagation directions are equally probable. In general, we have a diffuse field in large closed rooms whose dimensions are large compared to the wavelength range of interest, particularly those with irregular shape or that contain a variety of sound reflective objects. In general, a diffuse field has also a continuous spectrum.

5.4.3 Pressure Field

A *pressure field* is any sound field that oscillates in phase and with the same amplitude at every location of a given spatial region. In general, this condition takes place in small enclosures where the maximum linear dimension is much smaller than the minimum wavelength of interest. In order to have at most a 1 % amplitude variation within an enclosure (which is equivalent to a difference of 0.1 dB, i.e., the resolution of most sound level meters), it is sufficient that its maximum dimension be about $\lambda/40$.

A typical situation is within an acoustic coupler, for instance, an acoustic calibrator where a sound source and the microphone under test share a small closed volume. Another case is that of an artificial ear for earphone testing.

5.4.4 Stationary Field

A *stationary field* is any sound field that oscillates in phase or opposite phase in all locations of a region. It happens only for sinusoidal tones in small enclosures, even if the conditions for a pressure field do not hold. An example is the *Kundt tube* or *impedance tube* used to measure the normal-incidence sound absorption coefficient.

5.5 Microphones and Sound Fields

A given microphone behaves differently in sound fields of different type. To see why, take into account that unless the microphone had vanishingly small dimensions compared to the minimum wavelength of interest, its very presence is capable of altering the sound field. Indeed, reflection and diffraction phenomena take place on the microphone and preamplifier body, adding to the original field. There is also an effect of the radiation acoustic impedance, since when the microphone diaphragm vibrates by the action of an external wave, it also radiates an additional pressure that, again, adds to the sound field that existed in the absence of the microphone.

Acknowledging this physical reality and the metrological need to measure different sound fields, manufacturers design microphones that detect and convert the real sound pressure to which they are exposed (including their own effect) but include corrections to the frequency response that make them suitable for the different sound fields. These corrections modify their pressure response to make it variable with frequency,[3] hence compensating for the effect of the microphone.

So we have, in the first place, the *pressure-field microphones*, which present a flat frequency response to the pressure field actually present on its diaphragm, including any effect caused by the microphone itself. They are the ideal choice to measure pressure and stationary fields.

Then we have the *free-field microphones*, in which the frequency response has been modified to compensate for the effect of the microphone itself, so that it is flat when measuring free fields. However, this is only true when the microphone axis aims at the sound source.

Finally, we have the *diffuse-field microphones*, designed to have a flat frequency response in an environment where there is a diffuse field.

---

[3]In professional audio there are other types of microphones whose intrinsic response (previous to any correction) is not proportional to the diaphragm pressure. They have the back of their case partially or completely open so that the diaphragm is exposed to the external sound field on both sides. When the opening is complete the microphone is called *pressure-gradient microphone* since its output voltage is not proportional to the sound pressure but to the spatial gradient of pressure. Therefore, in a direction along which the sound pressure varies considerably (for instance pointing toward the source) the microphone will have high sensitivity, while in a direction in which it does not change (for instance, perpendicular to the wave propagation) it will have zero sensitivity. They have a polar pattern that resembles the shape of number eight, so they are often called figure-of-eight microphones. If the opening is partial, the output voltage is a weighted superposition of the output of a pressure microphone and a pressure-gradient microphone. Some examples are cardioid, supercardioid, and hypecardioid microphones.

Some sound level meters use digital signal processing to change the frequency response in order to adapt it to different sound fields simulating a different type of microphone.

5.6 Frequency Response

The frequency response of the different microphone types is designed to be flat when used within a sound field that corresponds to the microphone type and, in some cases, in compliance with some specified orientation. Figure 5.3a shows the sound pressure really present on the diaphragm for frontal incidence with respect to the sound pressure in the absence of the microphone. Figure 5.3b gives the correction needed to get a flat response as in Fig. 5.3c. For free-field microphones, this correction is built in and is attained by adding acoustic damping.

Figure 5.3b explains why when a free-field microphone is subject to a pressure field (as when inserted into an acoustic calibrator) the frequency response will exhibit some high-frequency loss. For the typical case of a 1/2″ free-field microphone, at 1 kHz there is a −0.15 dB loss, so if the calibration tone is nominally 94 dB, the instrument's reading should be 93.85 dB. An instrument so calibrated will read correctly a free field of 94 dB, since the real pressure on the diaphragm will be 94.15 dB, which will be compensated by the −0.15 dB free-field correction.

Free-field microphones have a flat frequency response only for frontal incidence. If the sound wave arrives from a different angle, then the frequency response changes. The upper panel of Fig. 5.4 shows the sound pressure level on the diaphragm of a typical 1/2″ microphone relative to the unaltered sound pressure level in the absence of the microphone, for different directions relative to the axis of the microphone. The lower panel shows the frequency response for a *free-field*

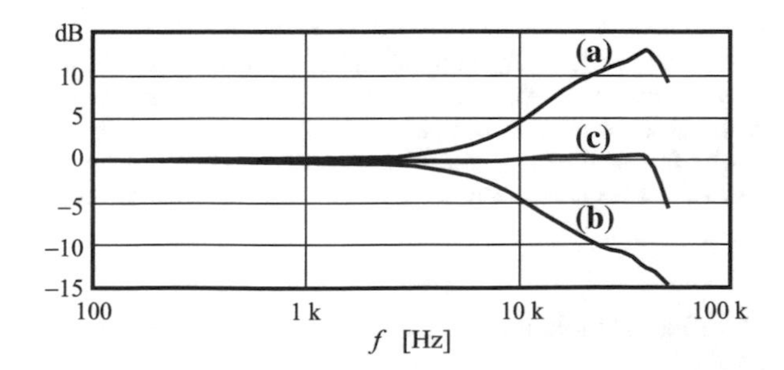

Fig. 5.3 **a** Sound pressure level on the diaphragm of a 1/2″ microphone with its protective grid when exposed to a frontal incidence free field, relative to the sound pressure level in the absence of the microphone. **b** Compensated pressure response of a microphone designed to have a flat free-field response. **c** Total response. The compensation in (**b**) combined with the sound pressure level increment in (**a**) yields a reasonably flat free-field response. Copyright © Brüel & Kjær (1996)

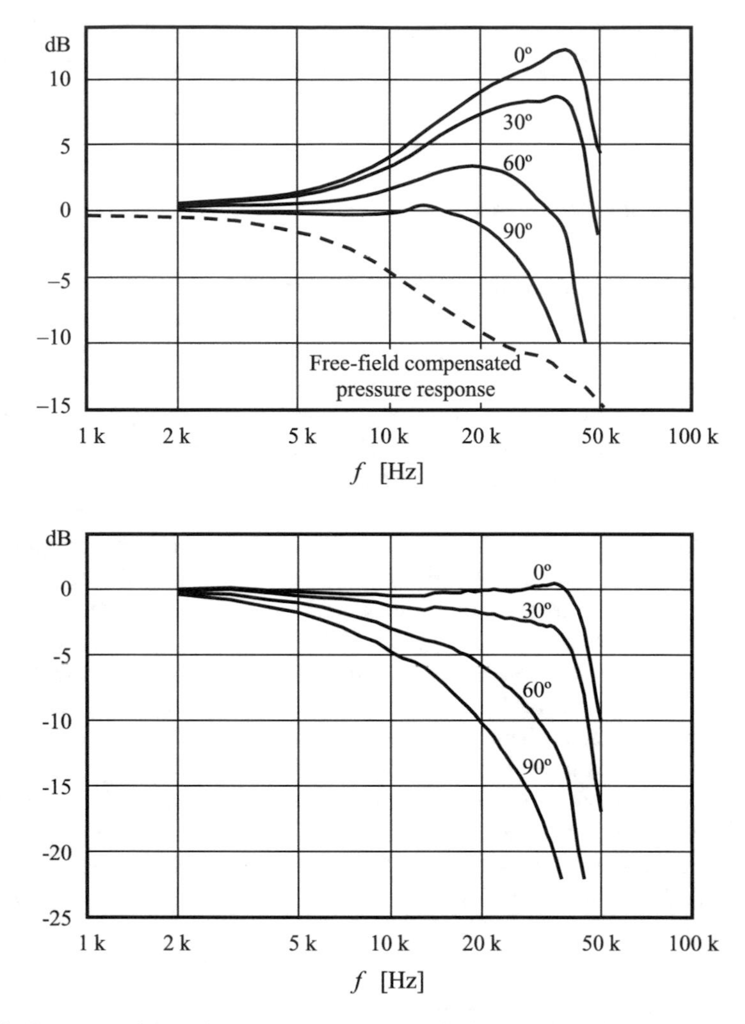

Fig. 5.4 *Upper panel* Sound pressure level on the diaphragm of a 1/2″ microphone with its protective grid when exposed to a free field with different incidence angles, relative to the sound pressure level in the absence of the microphone. The *dashed line* indicates the compensated pressure response of a free-field microphone. The 0° response corresponds to Fig. 5.3a (Beranek 1993). *Lower panel* Actual frequency response of a 1/2″ free-field microphone for each incidence angle

microphone for different incidence angles. It is obtained by adding the correction mentioned above.

We can see that there is an important attenuation at high frequencies. At 10 kHz the loss is of 1 dB for 30°, 3 dB for 60°, and 5 dB for 90°. This shows that the free-field frequency response may be significantly different according to the orientation of the microphone with respect to the sound field.

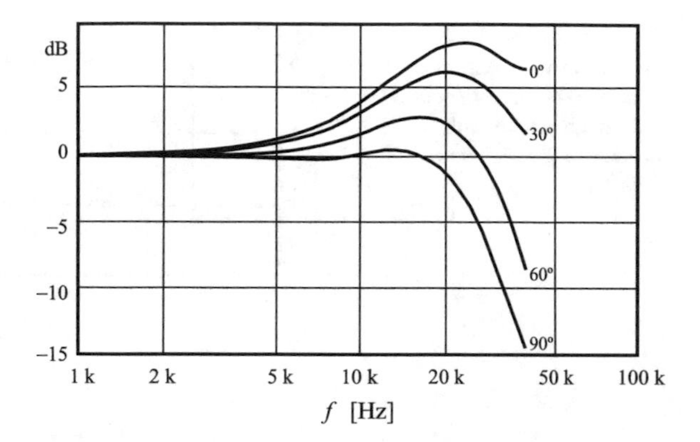

Fig. 5.5 Sound pressure level on the diaphragm of a typical 1/2″ microphone *without* its protective grid when exposed to a free field with different incidence angles, relative to the sound pressure level in the absence of the microphone. Copyright © Brüel & Kjær (1996)

In Fig. 5.5 the corrections corresponding to the case in which the protective grid has been removed are shown. We can see that the curves are more regular. This is because of the significant effect, particularly at high frequency, of the grid's complex shape with its slits of different sizes. However, as the diaphragm is extremely delicate, it is not advisable to remove the grid, except in a highly controlled laboratory situation where it may be convenient to take advantage of the smoother response.

An interesting situation takes place when measuring vehicle noise, particularly with light traffic formed by the pass-by of isolated vehicles. Indeed, the direction of the noise of a single vehicle changes during the pass-by, so if one wanted to measure the true vehicle noise it would be necessary to compensate for that effect with a time-variable correction of the frequency response. If the velocity is approximately known, the correction can be made with a time-variable FFT filter. Another possibility is to point the microphone vertically upwards. In that case the frequency response is not flat (since the angle exceeds 90°) but it is less dependent on the direction.

Although it is frequently necessary to measure indoor noise, where it can be safely assumed that the sound field is roughly diffuse, diffuse-field microphones are unusual. Most sound level meters and analyzers come with a free-field microphone. Fortunately, the correction to transform a free-field response into a diffuse-field one reduces to a single correction curve. This is true because in a diffuse field there is no prevailing direction, so in all directions the reading is the same. This correction is obtained by integrating the corrections for all angles, following a procedure that is normalized in standard IEC 61183 (1994). In some cases, a ring-shaped plastic accessory is provided with the microphone. It performs a diffractive acoustic filtering that is equivalent to the diffuse-field correction curve. In other cases the sound level meter has an optional DSP filter that implements the correction curve, transforming the instrument's response into a diffuse-field one. In both cases the frequency response is flat for a diffuse field (as would be the response of a pressure microphone so small that it does not alter the sound field at the relevant frequencies).

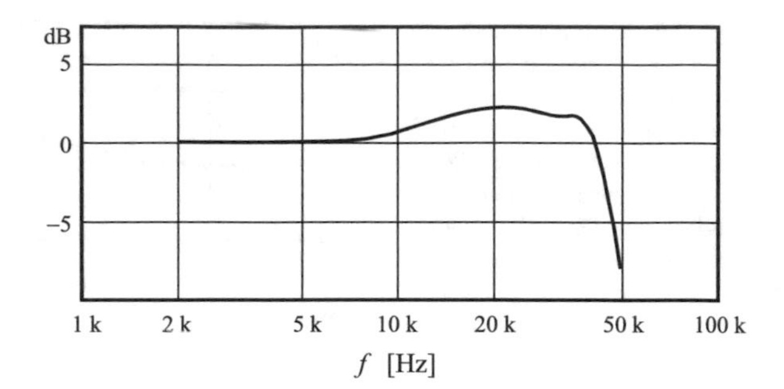

Fig. 5.6 Diffuse-field correction with respect to the pressure response of a typical 1/2″ microphone with protective grid (Beranek 1993)

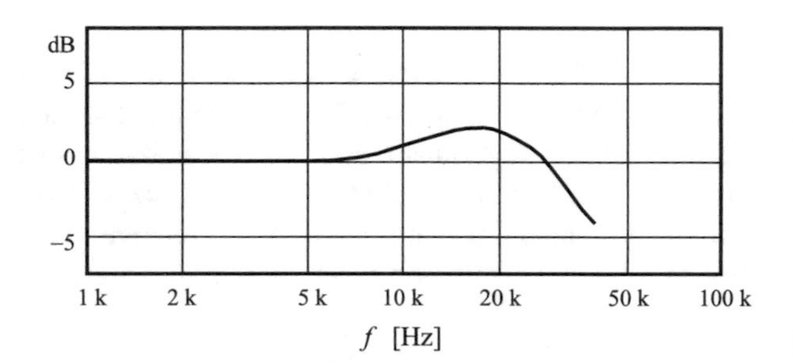

Fig. 5.7 Diffuse field correction with respect to the pressure response of a typical 1/2″ microphone with the protective grid removed. Copyright © Brüel & Kjær (1996)

Figure 5.6 shows the correction with respect to the pressure response to get a diffuse-field response. Bear in mind that this correction is not the one to be applied to the measurements made with a free-field microphone, since it has already a frontal-incidence free-field correction that should be subtracted before proceeding to apply the diffuse-field correction. If, for instance, we want to know the free-field/diffuse-field correction for 10 kHz, we see from Fig. 5.7 that the pressure-field/diffuse-field correction is about 1 dB and, from Fig. 5.3a, we know that the pressure response is artificially attenuated by about 4 dB, so the actual diffuse field is 1 dB − (−4 dB) = 5 dB above the reading.

Figure 5.7 corresponds to the case in which the protective grid has been removed, though the correction must be applied to the pressure response with the protective grid also removed.

5.7 Directional Response Pattern

The frequency response and the directional response of a microphone are intimately related. Another way to present this information is to use polar diagrams such as shown in Fig. 5.8.

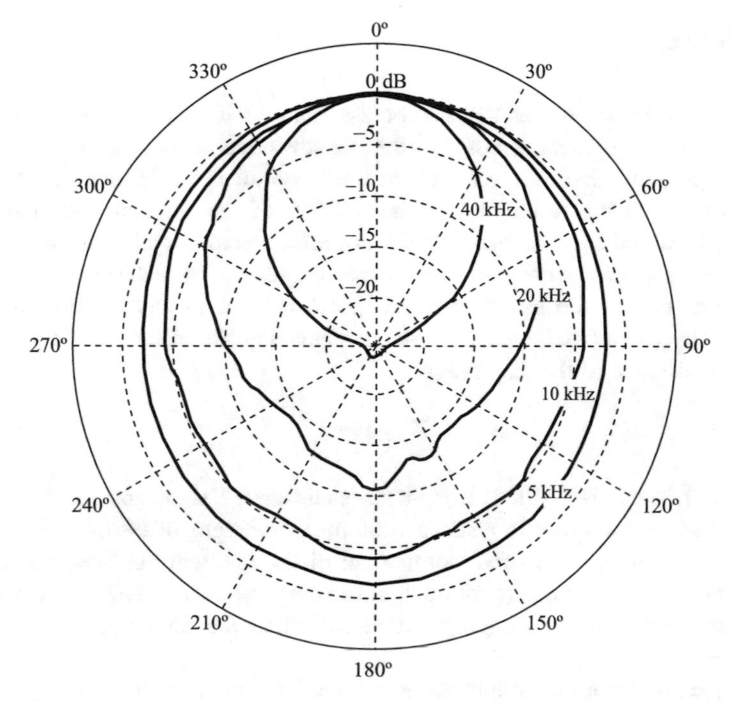

Fig. 5.8 Directional response of a free-field microphone for different frequencies, normalized to the frontal response sensitivity. The angle is measured with respect to the frontal direction. Copyright © Brüel & Kjær (1996)

We can see that in the range up to 20 kHz the error is smaller than 3 dB within ±30° from the microphone's axis, but for larger deviations the errors can be significant.

The preceding discussion does not take into account the sound level meter body. Although it has usually an aerodynamic shape designed to minimize its effect on the sound field at the microphone, it can influence readings considerably. The technical specifications of sound level meters usually include these effects. In order to reduce the effects of the instrument a microphone extension cable can be used, though in most normal situations this is an unnecessary precaution since the effect on the frequency response is usually below ±0.2 dB at most frequencies (Brüel & Kjær 2001).[4] This is much lower than the typical measurement uncertainty due to the instrument and noise variability.

[4]Besides, microphone extension cables are often very expensive (probably without a strong reason) if compared with a cable of similar quality for other applications.

5.8 Noise

There are two noise sources that can be associated with a condenser microphone. The first one is the *intrinsic noise* of the transducer itself and the second one, the preamplifier noise due to several phenomena within the FET and resistors. The intrinsic noise is the result of the transduction of the thermal vibration of the diaphragm, caused by the random motion of air molecules against the diaphragm.[5] The power spectral density of this noise, expressed in $\mathrm{Pa}^2/\mathrm{Hz}$, has a similar behavior to that of the electric thermal noise present in any resistor. It can be computed by means of Johnson's formula, replacing the electric resistance by the acoustic resistance of the diaphragm:

$$\overline{p^2} = 4kTR_\mathrm{a}, \tag{5.8}$$

where $k = 1.38 \times 10^{-23}$ J/K is Boltzmann's constant, T is the absolute temperature in K and R_a is the acoustic resistance of the diaphragm in $\mathrm{Ns/m}^5$. The acoustic resistance is defined as the real component of the quotient between pressure and volume velocity (volume per unit time) at the diaphragm. The acoustic resistance is mainly the result of the damping that is added to the diaphragm to control its resonance.

There are low noise microphones in which the acoustic resistance is reduced at the expense of increased high-frequency resonance and ringing. In order to get a flat frequency response it is necessary to insert an electronic equalizing network. This equalization may also be implemented by DSP software.

Since for constant damping the acoustic resistance is proportional to the diaphragm area, the spectral density of the noise is also proportional to the area. In consequence, the effective noise pressure, given by

$$P_\mathrm{ef} = \sqrt{\int_0^{f_\mathrm{máx}} \overline{p^2}(f)\mathrm{d}f} = \sqrt{4kTR_\mathrm{a}f_\mathrm{máx}} \tag{5.9}$$

happens to be proportional to the square root of the area, i.e., proportional to the radius. Since the sensitivity is, according to (5.7), proportional to the squared radius, it turns out that the signal-to-noise ratio is directly proportional to the radius. Therefore, a 1″ microphone will have twice the signal-to-noise ratio of a 1/2″ microphone. In decibels, it will have a 6 dB higher signal-to-noise ratio.

The pressure thermal noise has a flat spectrum, i.e., it is white noise. For this reason, the power spectral density of the equivalent electric noise (after the transduction process) behaves as the square of the pressure frequency response of the

[5]This is sometimes incorrectly called *Brownian motion*. The true Brownian motion does not refer to a relatively large object such as a microphone's diaphragm nor to the air molecules, but to the motion that the air molecules cause to dust particles in suspension. These motions can be often appreciated with a magnifying glass.

microphone (see Fig. 5.3a). However, the usable signal is available *after* the preamplifier, so the preamplifier noise must be taken also into account. It consists of some components of white noise due to the resistors and the inversely polarized FET junction, as well as a pink noise (1/*f*) component, with emphasis in low frequency. The result is a noise where the preamplifier component predominates at low frequency and the microphone component predominates at medium frequencies. At high frequencies, usually above the normal operating range, the microphone noise falls because its frequency response has a rapid roll-off due to diaphragm inertia.

The microphone noise is often specified by means of a hypothetical external sound pressure level that would cause the same output in an ideal noiseless microphone of equal sensitivity. Both Z (flat) and A weightings are used to express the residual noise. Because of the low-frequency noise of the preamplifier, the A-weighted acoustic equivalent noise is smaller. Microphone noise is often specified in one-third octave bands.

5.9 Distortion

Distortion in condenser microphones is due to two factors. First, the polarization charge Q_o, which was assumed to be constant, actually varies over time, and the amplitude of variation increases with the current through the condenser. This current is the consequence of the voltage variation (the signal itself) and circulates through the external elements such as the stray capacitance between the back plate and the case (which is connected to the diaphragm), the amplifier's input resistance (which is very high, hundreds of MΩ), and input capacitance.

The second factor is the fact that the formula (5.5) is only an approximation, since it ignores higher order terms that for small amplitudes are negligible but for a large signal are not. These originate in the fact that the membrane deformation ceases to be linear with sound pressure at very high levels.

Both factors cause second and third harmonics distortion that grow as a power of the pressure amplitude, so that in a graph whose abscissa is the sound pressure level

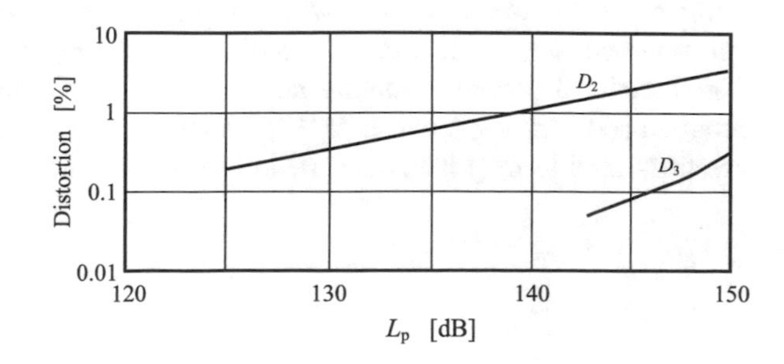

Fig. 5.9 Second and third harmonic percent distortion, D_2 and D_3, in a typical condenser microphone (after Brüel and Kjær 1996)

in linear dB scale and its ordinate is the distortion in logarithmic scale, the plot is linear (see Fig. 5.9). These distortion components reach 1 % for sound pressure levels of the order of 140 dB. Distortion is lower for smaller microphones since the membrane deformation is smaller, all the other parameters left the same.

The preamplifier contributes very little to the distortion, except at the highest end of the dynamic range (way above the useful range), where it causes a slight increment.

5.10 Micromachined Microphones

There is a new microphone technology that has been already in use in cell phones but is rapidly evolving toward instrumentation applications. We refer to MEMS Microphones (Micro-Electro-Mechanical Systems), a technology that combines integrated circuit technology with micromechanical devices (Lee 2009)

As an example, we have the miniature condenser microphones ADMP421 and ADMP521 manufactured by Analog Devices (Analog Devices 2012). A very interesting feature is that they provide a digital output as a *pulse density modulation* (PDM) stream. It is generated by a sigma-delta modulator with fourth order noise shaping (equivalent to four integrations). The recommended clock frequency is 2.4 MHz, a 64 times oversampling with respect to a frequency of 37.5 kHz, which guarantees a maximum frequency of 16 kHz. One problem is that because of its tiny size the low-frequency cutoff is about 100 Hz. This limits its application range.

This type of microphone has an intrinsic equivalent noise of the order of 30 dBA and a sensitivity of −26 dBFS. The dBFS symbol stands for *decibel referred to full scale* (full-scale decibel). This means that with a 94 dB tone the output peak will be 26 dB below the maximum representable digital value (Lewis 2012a). It is specified as omnidirectional, a feature attributable to its small size (4 mm × 3 mm × 1 mm), turning it into a pressure microphone that does not alter significantly the sound field over the spectral range of interest.

A promising application field is *beamforming*, where a microphone array is used to create a sensor with a programmable directional pattern. According to the direction from which the sound is coming we get a particular delay pattern for the different microphone signals. If we combine all signals so that the delay is exactly compensated, we get maximum combined sensitivity for that particular direction. The directional pattern is lobular and the main lobe's direction can be, therefore, programmed (Analog Devices 2013; Lewis 2012b). This technique is being successfully used to map the sound radiation of machines and vehicles.

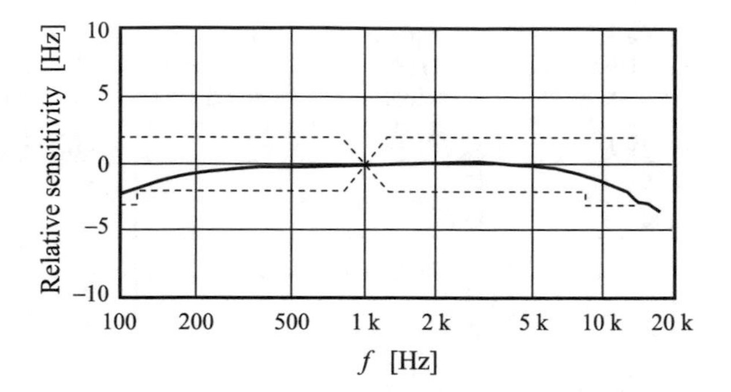

Fig. 5.10 Frequency response of micromachined (MEMS) microphone ADMP421 relative to the 1 kHz response (Lewis 2012a)

5.10.1 Frequency Response

The frequency response specification for MEMS microphones is shown in the example of Fig. 5.10, where the full line represents a particular measured response and the dashed lines represent the upper and lower guaranteed limits.

5.11 Audiometric Earphones

The microphone, as already seen, converts an acoustic signal (sound pressure) into an electric signal (voltage). In many measurements the opposite transduction is needed. An interesting and frequent case is that of psychoacoustic measurements, where the human ear is involved. In this case we use *audiometric earphones*. The problem to be solved is to find the relationship between the electric signal applied to the earphones and the sound pressure they generate. However, this, that would seem to be quite a simple and basic specification, is troublesome for several reasons.

In the first place, the desired relationship depends on the shape of the individual ear, and on the way the earphones are worn, including the force they exert on the head. It depends also on the type of sound field we intend to use as a reference. For instance, we may be interested in generating at the ear entrance a similar sound field as we would get when exposed to a free field measured at the ear position with the subject absent.[6] We may otherwise be interested in getting a specified sound pressure level at the tympanum. It will differ from the other case because of resonance phenomena in the ear canal.

We will analyze the case of the Sennheiser HDA 200 headset, a pair of earphones suitable for high-frequency audiometry (i.e., where the test tones extend to 16 000 Hz), for which it has been possible to gather reasonably complete

---

[6]Note that both sound fields will differ because the ear, in combination with the head and torso causes diffraction, reflection, and absorption phenomena that alter the original free field. Moreover, these effects depend strongly on frequency and sound direction.

Table 5.1 Specified frequency response of the audiometric headset Sennheiser HDA 200, measured in a Brüel & Kjær Type 4153 acoustic coupler that complies with IEC 60318-1 (1994). Also provided is the sensitivity in Pa/V and the headset noise attenuation (Sennheiser, n.d.)

f_k (Hz)	$L_{A,C}$ (dB) @ $V_{ef} = 0.5$ V	$S_{A,C}(f_k)$ (Pa/V)	Atten (dB)
125	112.5	16.9	14.3
250	113.0	17.9	16.9
500	112.0	15.9	22.5
750	111.0	14.2	–
1000	108.5	10.6	28.6
2000	104.0	6.3	32.0
3000	104.0	6.3	–
4000	104.0	6.3	45.7
5000	106.5	8.5	–
6000	107.5	9.5	–
8000	105.5	7.5	43.8
9000	105.0	7.1	–
10 000	102.5	5.3	–
11 200	102.0	5.0	–
12 500	103.0	5.7	–
14 000	98.5	3.4	–
16 000	100.0	4.0	–

information, even if it is scattered over several documents.[7] It is a *circumaural headset*, which means that it covers completely the ears, providing acoustic insulation in excess of 28 dB above 1000 Hz

Let us note that because of the diversity of ear shapes, the specification must be based on measurements with a standard *acoustic coupler*, which is specified in International Standard IEC 60318-1 (1994). Table 5.1 presents the information provided by the manufacturer, where $L_{A,C}$ is the sound pressure level produced by the audiometric earphone (A) within the coupler (C) when excited by an RMS voltage of 0.5 V.

Also included is the earphone sensitivity referred to the coupler, computed for each audiometric frequency f_k according to the formula

$$S_{A,C}(f_k) = \frac{p_{A,C}}{v_A} = 10^{\frac{L_{A,C}(f_k)}{20}} \frac{P_{ref}}{0.5\,\mathrm{V}}, \tag{5.10}$$

where v_A is the effective (RMS) voltage applied to the headset (in this case, 0.5 V).

We define the *free-field sensitivity* of the earphones as

[7]The HDA 200 has been discontinued by Sennheiser, and replaced by the new model HDA 300. Comparable data for the HDA 300 is available from the manufacturer and can be downloaded from http://en-de.sennheiser.com/downloads/download/file/4789/HDA300_RETSPL.pdf.

$$S_{\mathrm{A,FF}} = \frac{p_{\mathrm{FF}}}{v_{\mathrm{A,FF}}}, \tag{5.11}$$

where p_{FF} is the sound pressure of a free field (in the absence of the subject) and $v_{\mathrm{A,FF}}$ is the voltage that has to be applied to the earphones to produce a sound that is perceived as equally loud as the free field (IEC 60645-2 1993).[8]

In order to convert the sensitivity values referred to a coupler, $S_{\mathrm{A,C}}(f_k)$, into the earphone free-field (FF) sensitivity, $S_{\mathrm{A,FF}}(f_k)$, we can use the result of tests by the German PTB laboratory (Richter 2003), which gives the difference between the *sensitivity levels* referred to free field, G_{F}, and to an acoustic coupler, G_{C}:

$$G_{\mathrm{F}}(f_k) - G_{\mathrm{C}}(f_k) = 20\log\frac{S_{\mathrm{A,FF}}}{S_{\mathrm{o}}} - 20\log\frac{S_{\mathrm{A,C}}}{S_{\mathrm{o}}} = 20\log\frac{S_{\mathrm{A,FF}}}{S_{\mathrm{A,C}}}, \tag{5.12}$$

where S_{o} is an arbitrary reference sensitivity (usually 1 Pa/V, though in this case it is irrelevant). Such a correction is given in Table 5.2, for the normalized one-third octave bands between 125 and 8000 Hz and has been included also in ISO 389-8 (2004) Standard. Solving for the free-field sensitivity, we get

$$S_{\mathrm{A,FF}} = 10^{\frac{G_{\mathrm{F}}(f_k) - G_{\mathrm{C}}(f_k)}{20}} S_{\mathrm{A,C}}. \tag{5.13}$$

Table 5.2 Correction between the coupler sensitivity and the free-field sensitivity for the HDA 200 (Richter 2003; see also ISO 389-8 (2004) Annex C)

f_k (Hz)	$G_{\mathrm{F}}(f_k) - G_{\mathrm{C}}(f_k)$ (dB)
125	−5.0
160	−4.5
200	−4.5
250	−4.5
315	−5.0
400	−5.5
500	−2.5
630	−2.5
800	−3.0
1000	−3.5
1250	−2.0
1600	−5.5
2000	−5.0
2500	−6.0
3150	−7.0
4000	−13.0
5000	−14.5
6300	−11.0
8000	−8.5

[8]The operational definition requires to get the average of at least 10 otologically normal subjects.

In order to get the diffuse-field (DF) sensitivity, $S_{\mathrm{A,DF}}$ (defined similarly as $S_{\mathrm{A,FF}}$), since there is no equivalent to ISO 389-8 (2004) Annex C for diffuse field, we must resort to International Standard ISO 11904-1 (2002), which provides the correction between the free-field sound pressure level and the sound pressure level at the tympanum, ΔL_{FF}, and the correction between the diffuse-field sound pressure level and the sound pressure level at the tympanum, ΔL_{DF}, defined as

$$\Delta L_{\mathrm{FF}}(f_k) = L_{\mathrm{T}}(f_k) - L_{\mathrm{FF}}(f_k). \tag{5.14}$$

$$\Delta L_{\mathrm{DF}}(f_k) = L_{\mathrm{T}}(f_k) - L_{\mathrm{DF}}(f_k). \tag{5.15}$$

These are the values to add to the sound pressure level measured at free field or diffuse field (depending on the case) with the subject absent, to get the sound pressure level that would be measured at the tympanum by means of a tiny microphone probe placed close to the tympanum. These corrections are summarized in Table 5.3.

Taking the level at the tympanum as the common quantity to both situations, it is possible to get the difference between the sound pressure levels of a diffuse field and a free field that are aurally equivalent at a given frequency:

Table 5.3 Corrections between the sound pressure level in the absence of the subject and the sound pressure level at the tympanum for diffuse field and free field. Adapted from Table 1 of ISO 11904-1 (2002), refer to the source document for the original text. Copyright remains with ISO

f_k (Hz)	$\Delta L_{\mathrm{DF}}(f_k)$ (dB)	$\Delta L_{\mathrm{FF}}(f_k)$ (dB)
100	0.0	0.0
125	0.2	0.2
160	0.4	0.4
200	0.6	0.6
250	0.8	0.8
315	1.1	1.1
400	1.5	1.5
500	2.1	2.0
630	2.8	2.3
800	3.3	3.1
1000	4.1	2.7
1250	5.5	2.9
1600	7.7	5.8
2000	11.0	12.4
2500	15.3	15.7
3150	15.7	14.9
4000	12.9	13.2
5000	10.6	8.9
6300	9.4	3.1
8000	9.5	−1.4
10,000	6.8	−3.8
12,500	3.8	−0.1
16,000	0.7	−0.4

$$L_{\mathrm{DF}}(f_k) - L_{\mathrm{FF}}(f_k) = \Delta L_{\mathrm{FF}}(f_k) - \Delta L_{\mathrm{DF}}(f_k). \tag{5.16}$$

From (5.13) and (5.16) we can find the diffuse field sensitivity, $S_{\mathrm{A,DF}}$, of the earphone:

$$S_{\mathrm{A,DF}} = 10^{\frac{\Delta L_{\mathrm{FF}}(f_k) - \Delta L_{\mathrm{DF}}(f_k)}{20}} 10^{\frac{G_{\mathrm{F}}(f_k) - G_{\mathrm{C}}(f_k)}{20}} S_{\mathrm{A,C}}. \tag{5.17}$$

Finally, if we are interested in the tympanum sensitivity, $S_{\mathrm{A,T}}$, if suffices to apply only the free-field correction from Table 5.3 to Eq. (5.13):

$$S_{\mathrm{A,T}} = 10^{\frac{\Delta L_{\mathrm{FF}}(f_k)}{20}} 10^{\frac{G_{\mathrm{F}}(f_k) - G_{\mathrm{C}}(f_k)}{20}} S_{\mathrm{A,C}}. \tag{5.18}$$

Formulas (5.13), (5.17), and (5.18) are adequate to characterize the audiometric earphones for most relevant situations. It does not characterize it for non-frontal free fields (for instance, one having some azimuth and/or elevation). That problem must be dealt with using *head related transfer functions*, HRTF. However, this topic is too specific and falls out of the scope of this book.

EXAMPLE: Assume that we need to expose a subject to a sensation that is equivalent to the one they would experience if immersed in a diffuse field of 4 kHz and 60 dB, using the HDA 200 headset. Find the RMS voltage to apply.

We use formula (5.17) to get the diffuse field sensitivity:

$$S_{\mathrm{A,DF}} = 10^{\frac{13.2 - 12.9}{20}} \times 10^{\frac{-13}{20}} \times 6.3\,\mathrm{Pa/V} = 1.46\,\mathrm{Pa/V}.$$

The desired sound pressure level, 60 dB, corresponds to an effective pressure

$$P_{\mathrm{ef}} = 20 \times 10^{-6} 10^{60/20} = 0.02\mathrm{Pa},$$

so the required RMS voltage will be

$$v_A = \frac{P_{\mathrm{ef}}}{S_{\mathrm{A,DF}}} = \frac{0.02}{1.46}\,\mathrm{V} = 13.7\,\mathrm{mV}.$$

If we are working with a computer and a sound card as a signal source, we may characterize, by means of a simple test, the constant between a normalized digital quantity ($-1 \leq x \leq 1$) and the earphone output of the sound card (measured with a true-RMS millivolt-meter). Then it is easy to digitally generate a sine wave that corresponds to the desired 13.7 mV.

Depending on the type of output stage of the sound card, it may be necessary to take into account the voltage divider effect that the signal could suffer because of the loading of the headset impedance (nominally 40 Ω). If the output is a line output, it may have an output impedance as high as 100 Ω. In this case, a software correction may be applied to compensate for the load effect.

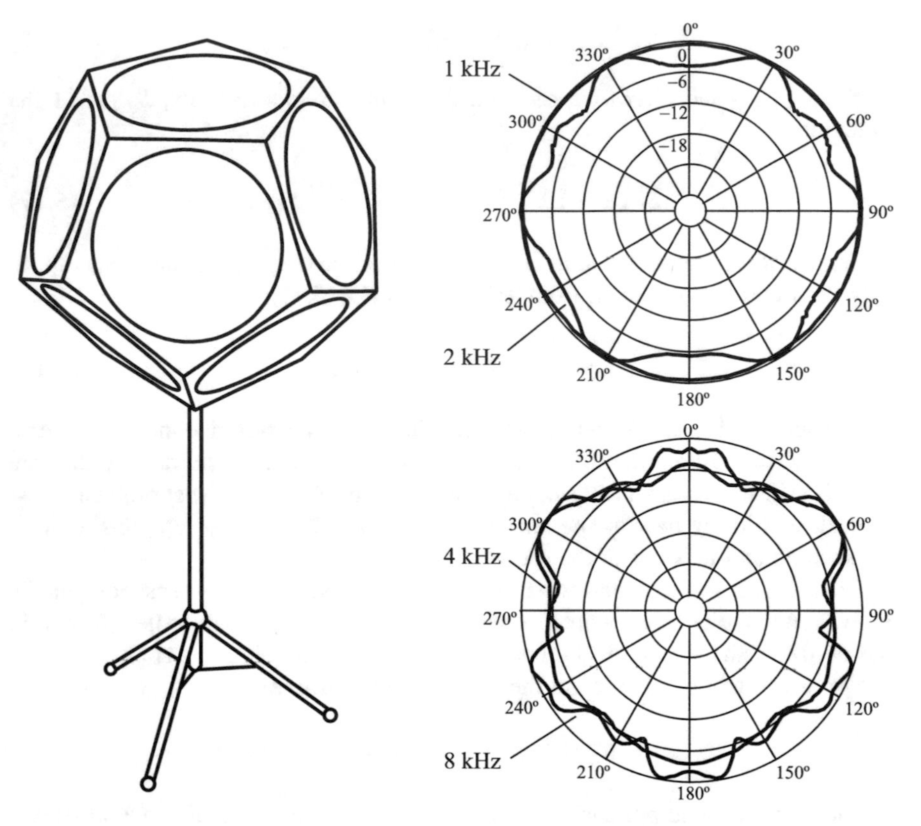

Fig. 5.11 *Left* omnidirectional source (dodecahedron) *Right* typical polar response for frequencies 1 kHz, 2 kHz, 4 kHz, and 8 kHz measured at a 1 m distance

5.12 Omnidirectional Sources

Omnidirectional sources are another type of transducer of interest in acoustics. They are mainly used in the measurement of architectural acoustics parameters, either at the laboratory or in situ. They have generally the shape of a regular dodecahedron, each of whose faces has a loudspeaker. In its traditional version, all the loudspeakers are fed by a single amplifier, so that all of them radiate in phase. The regular distribution of the faces ensures the maximum directional homogeneity of the radiated field. It might be feasible to use an icosahedron instead of a dodecahedron. The problem is that the proportion between the area of a circle inside a triangle and the triangle is smaller than in the case on a dodecahedron, so the latter is a better approximation to a pulsating sphere all of whose points radiate in phase. Figure 5.11 shows an example of a typical omnidirectional source.

According to the International Standard ISO 3382-1 (2009), omnidirectional sources must comply with a maximum directivity variation between sliding circular arcs of 30° such as indicated in Fig. 5.12a. On the other hand, International

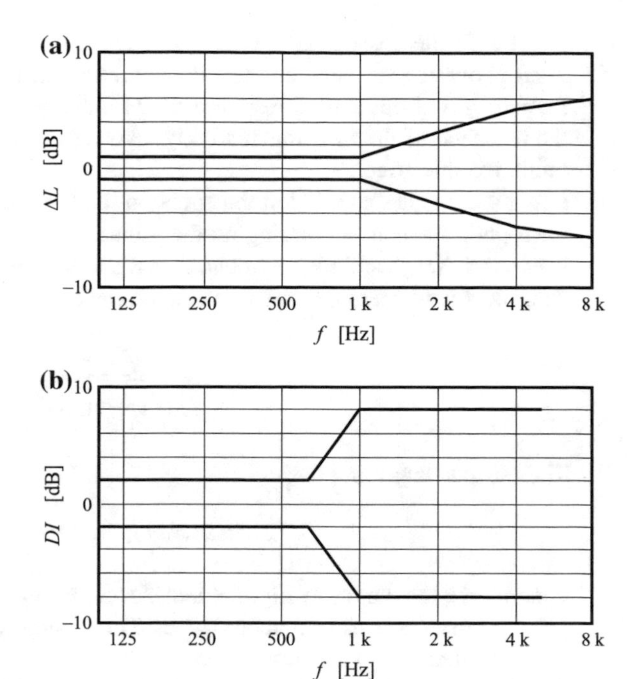

Fig. 5.12 **a** Tolerable limits for the directivity variation for sliding arcs of 30° for an omnidirectional source according to International Standard ISO 3382-1 (2009). **b** Tolerable limits of the directivity index *DI* for an omnidirectional source according to International Standard ISO 140-3 (1995)

Standard ISO 140-3 (1995) gives an omnidirectionality criterion based on the directivity index, DI, calculated as:

$$DI = L_{360^\circ} - L_{30^\circ}, \tag{5.19}$$

where L_{360° is the energy-averaged sound pressure level along a 360° circular arc centered at the source, while L_{30° is the energy-averaged sound pressure level along a 30° arc contained in the same 360° arc. In order that the source be considered as omnidirectional the directivity index must be within the limits shown in Fig. 5.12b for the worst-case plane. However, if the source is polyhedral, such as in the case of the dodecahedron, any single plane suffices.

The deviations from the values here stated cannot be corrected by software, since in this case there is not a single frequency response to compensate, but one for each specific direction.

Problems

(1) A measuring microphone presents a sensitivity of −26 dB referred to 1 V/Pa. (a) Find the sensitivity in V/Pa. (b) If its diameter is 13 mm, the distance between the diaphragm and the back plate is 10 μm, its polarization voltage is 200 V, find its linear tension.

(2) A sound level meter with a 1/2″ free-field microphone measuring a 10 kHz frontal incidence tone reads 68 dB. (a) What is the true sound pressure level on the microphone's diaphragm? (b) What is the sound pressure at the same

position in the absence of the microphone? (c) What is the new reading if the microphone is reoriented so that the sound field arrives 60° off axis? Note: The frequency and directional response of the microphone may change slightly in the presence of the instrument's body, even if its aerodynamic design seeks to minimize this effect.

(3) Find by computer simulation the effect on the frequency response of a free-field microphone when measuring traffic noise. Hint: If the minimum distance between vehicles and the microphone is d and the maximum sound level is P_{max} then the sound pressure when a vehicle is at a distance x is

$$p(x) = \frac{P_{max}}{\sqrt{1 + (x/d)^2}}$$

The off-axis angle is

$$\alpha = \arctan(x/d)$$

Parametrize the curves of Fig. 5.4 and, for each frequency, interpolate over α to get the correction as a function of α. Then integrate numerically the square pressure affected by this correction.

(4) An intensity probe is a device intended to measure sound intensity. The most frequent approach is to use two matched pressure microphones facing each other usually at a small distance d. The intensity in a given direction can be computed as the product of the sound pressure $p(x, t)$ and the particle velocity $u(x, t)$ in that direction (not to be confused with c, the wave velocity). The pressure can be computed as the average of the pressures at both microphones. The particle velocity obeys the following equation:

$$\frac{\partial u}{\partial t} = -\frac{1}{\rho_o}\frac{\partial p}{\partial x},$$

where ρ_o is the air's density. The velocity is obtained integrating the second member with respect to time, and $\partial p/\partial x$ can be approximated as the difference between the pressure measured by both microphones divided by the separation d. Suppose the output of both microphones is digitally recorded to a stereo wav file. (a) Write a Scilab script that computes the instantaneous intensity, (b) Compute the mean intensity, (c) Identify the sources of error and try to quantify them. Does the wavelength affect? (d) Simulate the output of the two microphones when excited by a sinusoidal plane wave varying the frequency and the direction of the probe with respect to the propagation direction, (e) Apply the previously developed algorithm to compute the intensity and compare with the theoretical result. Assume the microphones are compensated so that they measure the sound pressure that would exist in their absence.

(5) A microphone has a sensitivity of −26 dB referred to 1 V/Pa and its noise floor is specified as 20 dB. (a) Find the RMS value of the noise voltage. (b) Assuming the noise is of the $1/f$ type (pink noise) compute the A-weighted noise floor.

(6) An audiometric headset HDA 200 is to be used to conduct a psychoacoustic experiment. (a) Specify the frequency response of an equalizer that when applied to a signal that has been recorded in free field ensures that the subjects perceive the sound as if they were listening in free field. (b) Since the electrical load of the headset may vary with frequency, generating digital constant-amplitude tones may lead to a voltage amplitude that changes with frequency. Assuming you measure with a true-RMS precision multimeter the voltage for different frequencies, modify the specification of part (a) to take into account the load effect.

References

Analog Devices (2012) Ultralow noise microphone with bottom port and PDM digital out-put. Datasheet. http://www.analog.com/static/imported-files/data_sheets/ADMP521.pdf

Analog Devices (2013) Microphone beamforming simulation tool. Simulation software. http://wiki.analog.com/resources/tools-software/microphone/beamforming-simulation-tool

Brüel & Kjær (1996) Microphone handbook—Vol. 1: theory. Technical documentation Brüel & Kjær BE 1447–11. Nærum, Denmark. http://www.bksv.es/doc/be1447.pdf. Accessed 26 Feb 2016

Brüel & Kjær (2001) Integrating sound level meter type 2239A. Technical documentation BB 1219-11. Nærum, Denmark

Beranek Leo (1993) Acoustical measurements. Acoustical Society of America, Cambridge

IEC 60645-2:1993 Audiometers—Part 2: equipment for speech audiometry. IEC, Geneva, 1993/11/11

IEC 60318-1 ed2.0 (1994) Electroacoustics—simulators of human head and ear—Part 1: ear simulator for the measurement of supra-aural and circumaural earphones. IEC, Geneva, 2009/08/31

IEC 61183 ed1.0 (1994) Electroacoustics—random-incidence and diffuse-field calibration of sound level meters

ISO 140-3 (1995) Acoustics. Measurement of sound insulation in building element. Part 3: laboratory measurements of airborne sound insulation of building elements

ISO/DIS 389-8 (2004) Acoustics—reference zero for the calibration of audiometric equipment—Part 8: reference equivalent threshold sound pressure levels for pure tones and circumaural earphones

ISO 3382-1 (2009) Acoustics. Measurement of room parameters. Part 1: performance spaces

ISO 11904-1 (2002) Acoustics determination of sound immission from sound sources placed close to the ear—Part 1: technique using a microphone in a real ear (MIRE technique)

Lee J (2009) MEMS microphone. Course notes MECH 207. http://mech207.engr.scu.edu/SensorPresentations/Lee%20-%20MEMS_Microphone%20Combined.pdf

Lewis J (2012a) Microphone specifications explained. Analog devices. Application note AN-1112. http://www.analog.com/static/imported-files/application_notes/AN-1112.PDF

Lewis J (2012b) Microphone array beamforming. Analog devices, Application note AN-1140. http://www.analog.com/static/imported-files/application_notes/AN-1140.pdf

Miyara F (2007) Introducción a la Electroacústica. Monografía Cátedra de Fundamentos de Audio UNR. http://www.fceia.unr.edu.ar/acustica/audio/electroac.pdf. Accessed 26 Feb 2016

Miyara F (2013) Vibración de membranas. Monografía Cátedra de Fundamentos de Audio UNR. Publicación interna

Richter U (Ed) (2003) Characteristic data of different kinds of earphones used in extended high frequency range for pure-tone audiometry. PTB report PTB-MA-72, Braunschweig, February 2003. http://cds.cern.ch/record/1113557/files/PTB-MA-72.pdf. Accessed 17 Oct 2016

Sennheiser Electronic Corporation (no date) HDA 200. Audiometric headphone. http://en-us.sennheiser.com/downloads/a135337001d00197026b38d9c278f44d.pdf. Accessed 27 Feb 2016

Chapter 6
Digital Signal Processing

6.1 Discrete Signals

Once a continuous signal $x(t)$ has been sampled one gets a *discrete signal*,[1] i.e., a series of samples $x(kT_s)$, where T_s is the sampling period. Abusing notation, when it is not necessary to make explicit the chronometric time, we shall write the discrete signal as $x(k)$. Unless specifically noted, we will assume that the signal complies with the Nyquist condition for sample rate $F_s = 1/T_s$ (see Chap. 3 and Appendix J).[2]

6.2 Discrete Impulse

The *discrete impulse*, also called *unit pulse* is a signal with great conceptual importance since it plays the same role as Dirac's impulse (or Dirac's delta) in continuous-time systems. This signal may be considered as the result of sampling a Dirac's delta $\delta(t)$ that has been previously filtered by an ideal low-pass anti-aliasing filter. To see this, remember that the Fourier transform of $\delta(t)$ is 1 over the whole bilateral spectrum:

$$\Delta(\omega) = F\{\delta(t)\}(\omega) = \int_{-\infty}^{\infty} \delta(t)\, \mathrm{e}^{-j\omega t} \mathrm{d}t = 1, \tag{6.1}$$

[1]Observe that a discrete signal is not necessarily digital. There are examples of analog discrete signals, such as those in switched-capacitor filters (Miyara 2004, Ghausi et al 1981).

[2]This chapter is not intended to provide a complete treatment of digital signal processing but only a brief introduction. The interested reader may be willing to consult specific texts such as the one by Proakis and Manolakis (Proakis et al. 1996).

F. Miyara, *Software-Based Acoustical Measurements*, Modern Acoustics and Signal Processing, DOI 10.1007/978-3-319-55871-4_6

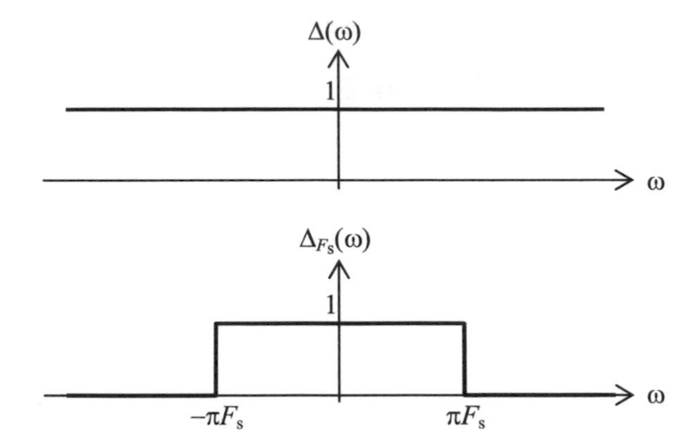

Fig. 6.1 Spectrum of a Dirac's delta $\delta(t)$ before and after being filtered by an ideal low-pass filter with cutoff frequency $F_s/2$

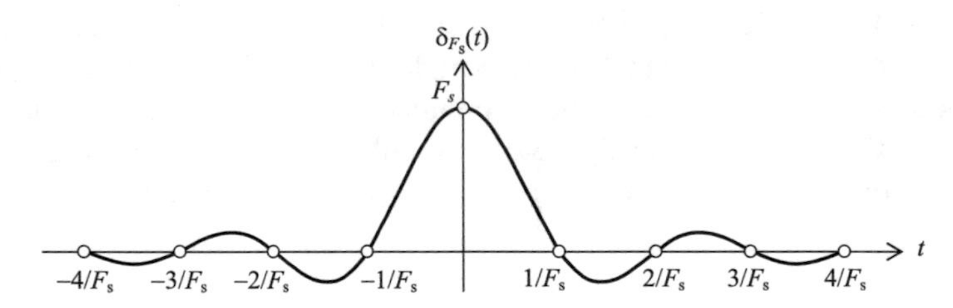

Fig. 6.2 A Dirac's impulse after being filtered by an ideal low-pass filter with cutoff frequency $F_s/2$

so when filtered by an ideal low-pass filter with cutoff frequency $F_s/2$, its spectrum will be limited to the $\pm F_s/2$ range, where it is equal to 1, as shown in Fig. 6.1.

We can retrieve the time-domain filtered signal by means of the inverse Fourier transform:

$$\delta_{F_s/2}(t) = \frac{1}{2\pi} \int_{-\pi F_s}^{-\pi F_s} e^{j\omega t} d\omega = F_s \frac{\sin \pi F_s t}{\pi F_s t}. \tag{6.2}$$

This function is plotted in Fig. 6.2. We note that this signal is non-causal (since it starts in $-\infty$, i.e., before the input departs from 0). This situation is characteristic of the response of any ideal filter.[3] However, in this very particular case, if we sample the signal at a sample rate F_s, all samples, except the one at $t = 0$, are zero.

[3]There is a theorem by Pahley and Wiener one of whose corollaries is that a bounded-spectrum signal (as one having gone through an ideal filter) cannot be identically 0 in any interval.

The filtered impulse has a value at the origin equal to F_s, which has frequency units. This is because the integral of the impulse is 1, but in the case of the discrete impulse signal it is more convenient to normalize it to 1. Therefore, it is defined as

$$u_o(k) = \begin{cases} 1 & k = 0 \\ 0 & k \neq 0 \end{cases} \quad (6.3)$$

6.3 A Signal as Its Convolution with a Discrete Impulse

Given a discrete signal $x(k)$, we can consider it as the sum of a series of time-shifted and weighted discrete impulses:

$$x(k) = \sum_{m=-\infty}^{\infty} x(m)\, u_o(k-m). \quad (6.4)$$

This formula will be quite useful when working with discrete systems. The second member is the *convolution product* between x and u_o.

6.4 Discrete Systems

Generally speaking, a system is any device or set of components, generally interdependent, with one or more inputs and one or more outputs. In our case we are interested in discrete *dynamic systems* that somehow process input signals (for instance, filtering them).

Consider the case of a *linear time-invariant system* (LTI) with one input and one output. Linearity means that given two input signals $x_1(k)$ and $x_2(k)$ that produce output signals $y_1(k)$ and $y_2(k)$ and two constants a and b, then the combined input $ax_1(k) + bx_1(k)$ will produce an output $ay_1(k) + by_2(k)$. Time invariance means that if $y(k)$ is the output to input $x(k)$, then $y(k-n)$ will be the output to the delayed input $x(k-n)$.

This type of system can be characterized by its *impulse response*, which we shall designate $h(k)$. If we apply linearity and time invariance to the expression of $x(k)$ given in (6.4), the response can be expressed as

$$y(k) = \sum_{m=-\infty}^{\infty} x(m)\, h(k-m). \quad (6.5)$$

This is the superposition of a series of time-shifted impulse responses weighted by the successive values of the samples. In other words, the response may be expressed as the *convolution product* between the input signal and the impulse response of the system, which is notated as

$$y(k) = x * h(k). \tag{6.6}$$

With a simple variable change in (6.5) it is easy to see that the convolution product is commutative.

$$y(k) = h * x(k) = \sum_{m=-\infty}^{\infty} h(m)\, x(k-m). \tag{6.7}$$

As an example, suppose the following impulse response:

$$h(k) = \begin{cases} 0 & k < 0 \\ Ka^{-k} & k \geq 0 \end{cases} \tag{6.8}$$

Then the response to any signal $x(k)$ will be

$$y(k) = \sum_{m=0}^{\infty} Ka^{-m} x(k-m). \tag{6.9}$$

This expression gives a closed form solution in terms of an infinite series, but it has the problem that in order to get the output in any single instant we have to know the impulse response and input signal over the whole time axis. This is not convenient since we seldom know the complete time history of the signal (except in the case of theoretical expressions). In particular, it would be completely useless in the case of a real-time system with a continuous input signal stream.[4]

6.5 Finite- and Infinite-Impulse-Response Systems

Discrete systems may have an infinitely long impulse response or, on the contrary, the impulse response may have a finite-length $M + 1$. In the first case they are called *infinite-impulse-response systems,* IIR, and in the second one, *finite-impulse-response systems,* FIR. In the latter case, Eq. (6.7) can be written as

[4]It might be applied to a finite-length signal that has been recorded, but it would have a high computational cost.

$$y(k) = \sum_{m=0}^{M} h(m)\, x(k-m). \tag{6.10}$$

Unlike the general case, here it would be possible to compute the response since it involves only a finite number of terms.

6.6 Difference Equation of a Discrete System

With a little trick it is possible to simplify the calculation of the response of an IIR system reducing it to a recursive expression. It consists in calculating the response at discrete time k from the response at $k - 1$. Consider the example of Eq. (6.9):

$$y(k-1) = \sum_{m=0}^{\infty} Ka^{-m} x(k-1-m) = a \sum_{m=0}^{\infty} Ka^{-(m+1)} x(k-(m+1)). \tag{6.11}$$

Calling $n = m + 1$, we have

$$y(k-1) = a \sum_{n=1}^{\infty} Ka^{-n} x(k-n) = a\left(\sum_{n=0}^{\infty} Ka^{-n} x(k-n) - Kx(k)\right), \tag{6.12}$$

which can be rewritten as

$$y(k-1) = a(y(k) - Kx(k)),$$

from which

$$y(k) = \frac{1}{a} y(k-1) + Kx(k). \tag{6.13}$$

This equation allows to compute the successive values of $y(k)$ from the previous value and the current value of the input. Starting from an initial value, for instance $y(0) = 0$, it is possible to calculate one by one all output values without the need to know more than a single input value at a time. In particular, given a discrete signal that can be described analytically, we can get an explicit solution. For instance, if $x(k)$ is a unit step in 0, we have

$$\begin{aligned}
y(1) &= K \\
y(2) &= \frac{K}{a} + K = K(a^{-1} + 1) \\
y(3) &= \frac{K(a^{-1} + 1)}{a} + K = K(a^{-2} + a^{-1} + 1) \\
&\vdots \\
y(k) &= K \sum_{n=0}^{k-1} a^{-n} = K \frac{1 - a^{-k}}{1 - a^{-1}} = \frac{Ka}{a - 1}\left(1 - a^{-k}\right)
\end{aligned} \tag{6.14}$$

i.e., a constant minus a decreasing exponential (for $a > 1$).

Equation (6.13) is called the *difference equation* of the discrete system, and it is analog to the differential equation of a continuous-time system. In fact we can see that it may be considered as an approximation of the differential equation. To this end bear in mind that for small increments the derivative can be estimated as the incremental quotient. If in (6.13) we make the time explicit, we can get an incremental quotient subtracting $y(k-1)$ and dividing by the sampling period T_s:

$$\frac{y(kT_s) - y((k-1)T_s)}{T_s} = \frac{1}{T_s}\left(\frac{1}{a} - 1\right) y(k-1) + \frac{K}{T_s} x(k). \tag{6.15}$$

This equation is an approximation of this one:

$$y'(t) + \frac{a-1}{T_s a}\, y(t) = \frac{K}{T_s} x(t) \tag{6.16}$$

It is a first-order differential equation whose impulse response is

$$h(t) = \frac{K}{T_s}\, e^{-\frac{a-1}{aT_s}t}. \tag{6.17}$$

Even if this result does not resemble the original impulse response, $h(t) = K\,a^{-k}$, when the slope is small, i.e., $a \cong 1$, it turns out that

$$\begin{aligned}
e^{-\frac{a-1}{a}} &\cong 1 - \frac{a-1}{a} = \frac{1}{a} \\
\frac{t}{T_s} &\cong k.
\end{aligned}$$

As regards the factor $1/T_s$, not present in the original impulse response, it is precisely the one we removed when normalizing to 1 the amplitude of the unit impulse.

Anyway, the approximation of the derivative used in (6.15) is not good for rapidly varying signals. We shall see later that we can get a better approximation for the original differential equations if instead of approximating the derivative we express the equation in its integral form and then approximate the integral.

We can generalize Eq. (6.13) considering an equation in which the output depends on M past input values and N past output values:

$$\begin{aligned} y(k) &= b_0 x(k) + b_1 x(k-1) + \cdots + b_M x(k-M) \\ &\quad - a_1 y(k-1) - \cdots - a_N y(k-N) \end{aligned} \tag{6.18}$$

This system is said to be of order N since the present output depends on N past values of the output. As in the first-order case, it is possible to write a recursive algorithm to compute the output signal. However, there are more efficient methods to be described later.

If in (6.18) we assume that $a_n = 0$ for all $n = 1, \ldots, N$, i.e.,

$$y(k) = b_0 x(k) + b_1 x(k-1) + \cdots + b_M x(k-M), \tag{6.19}$$

where the present output $y(k)$ does not depend any longer on past output samples but only on M past input samples, we are in the presence of a *FIR system*, i.e., one whose impulse response has a finite duration. If, for instance, $b_0 = 1$, $b_1 = 0.5$, $b_2 = 0.25$ and the remaining coefficients are zero, $b_m = 0$, then the impulse response will be:

$$\begin{aligned} h(0) &= b_0 1 + b_1 0 + b_2 0 = 1 \\ h(1) &= b_0 0 + b_1 1 + b_2 0 = 0.5 \\ h(2) &= b_0 0 + b_1 0 + b_2 1 = 0.25 \\ h(k) &= b_0 0 + b_1 0 + b_2 0 = 0 \qquad k \neq 0, 1, 2 \end{aligned}$$

It is apparent that, in general, the impulse response of a system such as (6.19) is given by the b_k coefficients:

$$h(k) = \begin{cases} b_k & k = 0, 1, \ldots, M \\ 0 & k > M \end{cases} \tag{6.20}$$

6.7 Frequency Response of a Discrete System

Consider a discrete system with impulse response $h(k)$. We will use the following test signal:

$$x(k) = e^{jk\Omega}, \tag{6.21}$$

where Ω is the normalized angular frequency:

$$\Omega = 2\pi \frac{f}{F_s}. \tag{6.22}$$

This variable is such that $\Omega = \pi$ corresponds to the Nyquist frequency, i.e., $F_s/2$.[5] The complex exponential is equivalent to a sinusoidal signal, but it is more convenient because it is analytically easier to work with. The system response is the convolution with the impulse response $h(k)$:

$$y(k) = h * e^{jk\Omega} = \sum_{m=-\infty}^{\infty} h(m)\, e^{j(k-m)\Omega} = e^{jk\Omega} \sum_{m=-\infty}^{\infty} h(m)\, e^{-jm\Omega},$$

i.e.,

$$y(k) = x(k) \underbrace{\sum_{m=-\infty}^{\infty} h(m)\, e^{-jm\Omega}}_{\text{Frequency response}}. \tag{6.23}$$

The sum turns to be the frequency response of the system, since for each frequency it gives the quotient between the output and input amplitudes. As can be seen, it is equivalent to the Fourier transform of the discrete impulse response. Note that the Fourier transform of a discrete signal is a function of a continuous frequency variable (since Ω varies continuously) and it is periodic with period 2π. However, any frequency above π (equivalent to $F_s/2$) is representing aliased components.

As an example, let us calculate the frequency response of the system of Eq. (6.8),

$$\sum_{m=-\infty}^{\infty} h(m) e^{-jm\Omega} = \sum_{m=0}^{\infty} K a^{-m}\, e^{-jm\Omega} = \frac{K}{1 - a^{-1}\, e^{-j\Omega}}. \tag{6.24}$$

Figure 6.3 illustrates this frequency response for $a = 1.25$ and $K = 0.2$ (K was chosen to have a DC response equal to 1).

In Fig. 6.4 the frequency response has been plotted in dB and logarithmic frequency scale. Also plotted for comparison is the frequency response of a continuous-time low-pass filter with the same cutoff frequency, as obtained by comparison with the equivalent differential Eq. (6.16):

$$f_c = \frac{1}{2\pi\tau} = \frac{a-1}{2\pi a} F_s. \tag{6.25}$$

[5]Some authors use ω for the normalized frequency. Here we use Ω the avoid notation inconsistencies.

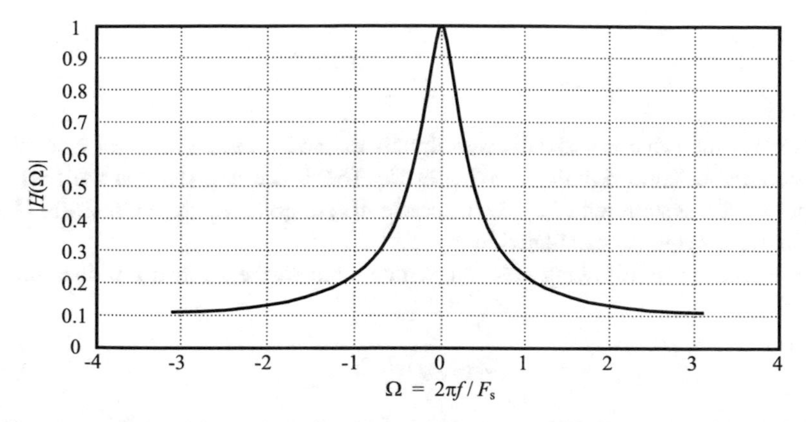

Fig. 6.3 Bilateral frequency response of the system given in Eq. (6.24)

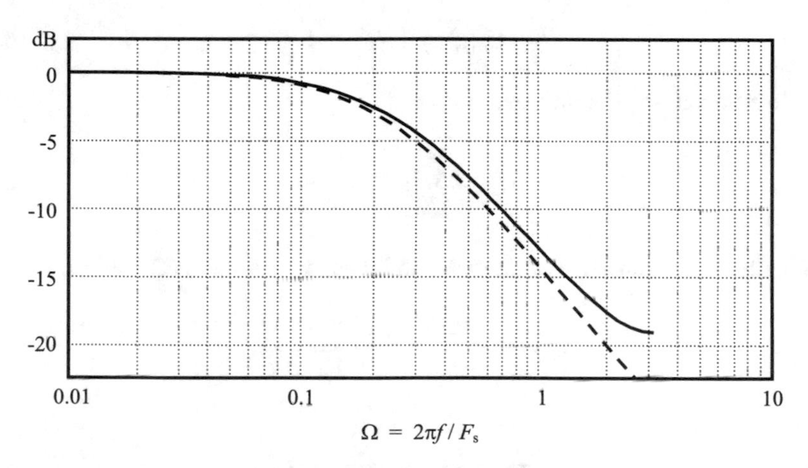

Fig. 6.4 Frequency response in logarithmic frequency scale. In *dashed line*, the frequency response of an equivalent continuous-time system

We can see that at low frequency, where the derivative is correctly approximated by the incremental quotient and hence the difference equation approximates the differential equation, both responses are similar. At high frequency, particularly near Nyquist's frequency $F_s/2$, the frequency responses differ considerably.

6.8 *z* Transform of a Discrete Signal

The *z transform* of a discrete signal $x(k)$ is defined as (Proakis et al. 1996)

$$X(z) = \sum_{k=-\infty}^{\infty} x(k)\, z^{-k}. \tag{6.26}$$

It plays the same role for discrete signals as the Laplace and Fourier transforms for continuous-time signals. Looking at Eq. (6.23) we can see that the frequency response of a system with impulse response $h(k)$ is given by the z transform $H(z)$ of the impulse response calculated in $z = \mathrm{e}^{\,j\Omega}$.

Sometimes the following notation, which stresses the z transform operator Z, is useful:

$$Z\{x(k)\} = X(z). \tag{6.27}$$

Let us see some examples. The z transform of the unit impulse is, considering that except $u_{\mathrm{o}}(0)$ all the other values are 0,

$$U_{\mathrm{o}}(z) = 1\, z^{-0} = 1. \tag{6.28}$$

The z transform of a unit step, $u_1(\mathrm{k})$, is

$$U_1(z) = \sum_{k=0}^{\infty} z^{-k} = \frac{1}{1 - z^{-1}} \tag{6.29}$$

Finally, the z transform of an exponential for $k > 0$, i.e., $g(k) = b^k u_{\mathrm{o}}(k)$ is

$$G(z) = \sum_{k=0}^{\infty} b^k z^{-k} = \frac{1}{1 - b\, z^{-1}}. \tag{6.30}$$

6.9 *z* Transform of a Difference Equation

Let us see now how to solve a difference equation in the spectral domain by means of the z transform.

As a first example we shall apply the z transform to both sides of the difference Eq. (6.13), but we need first to know how to transform a delayed signal $y(k-1)$.

$$\begin{aligned}
Z\{y(k-1)\}(z) &= \sum_{k=-\infty}^{\infty} y(k-1)\, z^{-k} \\
&= \sum_{k=-\infty}^{\infty} y(k-1)\, z^{-(k-1+1)} \\
&= z^{-1} \sum_{k=-\infty}^{\infty} y(k-1)\, z^{-(k-1)} = z^{-1}\, Y(z)
\end{aligned} \tag{6.31}$$

In words, a one-sample delay translates into a multiplication of the original z transform by z^{-1}. The transform of Eq. (6.13) yields, then

$$Y(z) = \frac{1}{a} z^{-1} Y(z) + KX(z). \tag{6.32}$$

We can get the z *transfer function* as

$$\frac{Y(z)}{X(z)} = \frac{K}{1 - \frac{z^{-1}}{a}}. \tag{6.33}$$

If we compare with Eq. (6.24) that gave the frequency response of the same system, we see that it is equivalent to replace $z = e^{j\Omega}$. This is analogous to what happens in a continuous-time system, where replacing $s = j\omega$ in the Laplace transfer function we also get the frequency response.

We can generalize the preceding result applying the z transform to the general difference Eq. (6.18). It yields

$$\begin{aligned} Y(z) = {} & \left(b_0 + b_1 z^{-1} + \cdots + b_M z^{-M}\right) X(z) \\ & - \left(a_1 z^{-1} + \cdots + a_N z^{-N}\right) Y(z) \end{aligned} \tag{6.34}$$

hence,

$$Y(z) = \frac{b_0 + b_1 z^{-1} + \cdots + b_M z^{-M}}{1 + a_1 z^{-1} + \cdots + a_N z^{-N}} X(z), \tag{6.35}$$

from where we get the *transfer function*

$$\frac{Y(z)}{X(z)} = \frac{b_0 + b_1 z^{-1} + \cdots + b_M z^{-M}}{1 + a_1 z^{-1} + \cdots + a_N z^{-N}}. \tag{6.36}$$

A particular case of Eq. (6.36) is when the input is a discrete impulse, in which case the output is the impulse response. In that case $X(z) = 1$ and $Y(z) = H(z)$, from where it turns out that the transfer function is the z transform of the impulse response:

$$H(z) = \frac{b_0 + b_1 z^{-1} + \cdots + b_M z^{-M}}{1 + a_1 z^{-1} + \cdots + a_N z^{-N}}. \tag{6.37}$$

From Eq. (6.35) it results, therefore, that the z transform of the output is the product of the z transforms of the input and the impulse response:

$$Y(z) = H(z)\, X(z). \tag{6.38}$$

6.10 z Transform of a Convolution

Equation (6.5) expresses the time-domain output of a discrete system as the convolution between the input and the impulse response, and Eq. (6.38) expresses a similar relationship in the transformed domain. This is a general property: the z transform of a convolution $u*v$ is the product of the z transforms of u and v.[6] To see this, let us apply the definition:

$$Z\{u * v\} = \sum_{n=-\infty}^{\infty} \left(\sum_{m=-\infty}^{\infty} u(m)v(n-m) \right) z^{-n}.$$

Now we introduce z^{-n} inside the second sum and reverse the summation order

$$Z\{u * v\} = \sum_{m=-\infty}^{\infty} \left(\sum_{n=-\infty}^{\infty} u(m)v(n-m)\, z^{-n} \right).$$

Now we add and subtract m in the exponent of z and operate:

$$Z\{u * v\} = \sum_{m=-\infty}^{\infty} \left(u(m) \sum_{n=-\infty}^{\infty} v(n-m)\, z^{-(n-m)} z^{-m} \right) \tag{6.39}$$

Reordering and changing variables,

$$\begin{aligned} Z\{u * v\} &= \sum_{m=-\infty}^{\infty} \left(u(m)z^{-m} \sum_{n=-\infty}^{\infty} v(n-m)\, z^{-(n-m)} \right) \\ &= \sum_{m=-\infty}^{\infty} \left(u(m)z^{-m} Z\{v\} \right) \\ &= \left(\sum_{m=-\infty}^{\infty} u(m)z^{-m} \right) Z\{v\} = Z\{u\}\, Z\{v\} \end{aligned} \tag{6.40}$$

6.11 z Transform and Frequency Response

The sum in Eq. (6.23) is the same as the z transform of the impulse response calculated in $e^{j\Omega}$. This implies that the frequency response can be calculated as $H(e^{j\Omega})$, where the interval $-\pi < \Omega < \pi$ corresponds to $-F_s/2 < f < F_s/2$.

[6] This property also holds for the Fourier and Laplace transforms.

6.12 Solution of a Difference Equation

The resolution of a differential equation follows the same guidelines for continuous-time differential equations. The problem is separated into getting a particular solution, y_{P}, and getting a solution to the associated homogeneous equation (i.e., that one in which the input is 0), y_{H}. Because of linearity, the general solution can be expressed as:

$$y(k) = y_{\mathrm{P}}(k) + y_{\mathrm{H}}(k). \tag{6.41}$$

To get the solution to the homogeneous equation

$$y(k) + a_1 y(k-1) + \cdots + a_N y(k-N) = 0, \tag{6.42}$$

we will assume it is of the form

$$y_{\mathrm{H}}(k) = \lambda^k. \tag{6.43}$$

Substituting,

$$\lambda^k + a_1 \lambda^{k-1} + \cdots + a_N \lambda^{k-N} = 0,$$

which can be rewritten, taking out λ^{k-N} as a common factor,

$$\left(\lambda^N + a_1 \lambda^{N-1} + \cdots + a_N\right) \lambda^{k-N} = 0.$$

This equation is satisfied for all k if and only if

$$\lambda^N + a_1 \lambda^{N-1} + \cdots + a_N = 0. \tag{6.44}$$

This algebraic equation admits N complex solutions, the roots λ_n of the polynomial in the first member, called *characteristic polynomial* of the system. If the coefficients are real, as is generally the case, the roots are either real or are grouped into conjugate pairs. If there are no multiple roots, the general solution to the homogeneous equation is a linear combination of the solutions of the form (6.43):

$$y_H(k) = c_1 \lambda_1^k + c_2 \lambda_2^k + \cdots + c_N \lambda_N^k. \tag{6.45}$$

As regards the particular solution, if the input presents some simple analytic forms, the method is to assume that the response has a specific form related to the input form and try to solve for the associated parameters. In Table 6.1 a few cases are given.

Table 6.1 Form of the particular solution of a difference equation for some typical analytic input signals

Input		Particular solution
Type	Expression	
Constant	A	B
Exponential	Aa^k	Ba^k
Power	Ak^b	$B_0k^b + B_1k^{b-1} + \cdots + B_b$
Sinusoid	$A\sin(\Omega k + \varphi)$	$B\sin(\Omega k + \psi)$

As an example, let us get the particular solution to the difference Eq. (6.13) when excited with a ramp input

$$x(k) = k.$$

According to Table 6.1, the particular solution has the general form

$$y_P(k) = B_0 k + B_1.$$

Replacing in (6.13) and reordering

$$((a-1)k+1)\,B_0 + (a-1)\,B_1 = aKk$$

This is valid for any k so we choose $k = 0$ and $k = 1$. We get the following system of equations:

$$\begin{cases} B_0 + (a-1)B_1 = 0 \\ aB_0 + (a-1)B_1 = aK \end{cases}$$

Solving it we finally get,

$$y_P(k) = \frac{aK}{a-1}k - \frac{a^2K}{a-1}. \tag{6.46}$$

It is necessary to add to this solution the general solution of the homogeneous equation, which is obtained by solving the characteristic equation

$$\lambda - \frac{1}{a} = 0.$$

The single root is $\lambda = a^{-1}$, so the homogeneous solution turns out to be

$$y_H(k) = c\,a^{-k},$$

which, as expected, has the form of the impulse response.

6.13 Poles and Stability of a Discrete System

A system is said to be *stable* if, starting from rest, the response to a finite-duration signal returns asymptotically to rest. In order for that to happen it suffices that the homogeneous solutions asymptotically approach zero as time tends to ∞. The homogeneuos solutions are of the form (6.43), where λ is any of the roots λ_k of the characteristic equation (6.44), which are also the denominator roots of the z transfer function (6.37), i.e., its *poles* (values of z where the function tends to ∞).

We can see that in order to get a stable system, all roots λ_k must satisfy

$$|\lambda_k| < 1. \tag{6.47}$$

This is the case if the poles of the transfer function fall inside the unit radius circle centered in 0, as shown in Fig. 6.5.

6.14 Inversion of a Rational z Transform

An alternative method to find the response of a system to an input signal consists in the application of Eq. (6.38), multiplying the z transform of the input by the transfer function (that, as was already noted, is the same as the z transform of the impulse response). This yields the z transform of the output signal, which can subsequently be inverted to get the time-domain response. While there is a theoretical inversion formula to get the inverse z transform, it is not applied except in theoretical studies. Instead, alternative methods are often used in special cases. The most relevant case for our purposes is when the z transform to invert is rational, i.e., of the form $P(z)$ / $Q(z)$, where P and Q are polynomials.

In this case it is possible to expand the function into partial fractions. In the most frequent case in which there are no multiple roots, the expansion is

$$\frac{P(z)}{Q(z)} = \sum_{m=1}^{N} \frac{A_m z}{z - p_m} \tag{6.48}$$

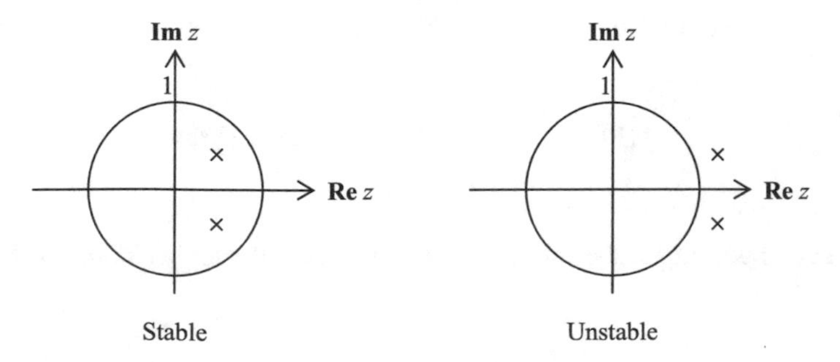

Fig. 6.5 Location of a pair of conjugate poles of the z transfer function of a stable and an unstable discrete system

where p_m are the complex poles (the zeros of the denominator) $Q(z)$. The additive terms in (6.48) are particular in that their inverse transform can be readily obtained from Eq. (6.30), where we see that the form is very similar to the z transform of a discrete exponential. Rewriting,

$$\frac{P(z)}{Q(z)} = \sum_{m=1}^{N} \frac{A_m}{1 - p_m z^{-1}} \tag{6.49}$$

so its inverse z transform is

$$Z\left\{\frac{P(z)}{Q(z)}\right\}(k) = \sum_{m=1}^{N} A_m p_m^k \, u_\mathrm{o}(k). \tag{6.50}$$

In many situations, however, the main interest in working with z transforms is related to the spectral information of the output signal or to the frequency response of the system, rather than to the possibility of getting an explicit form of the response.

On the other hand, when real-life signals are involved, it is impossible to find an analytic representation that allows computing the z transform of the input. All we can hope is to get an empiric representation of the long-term average spectrum of the signal, which allows us to derive interesting information on the long-term spectral behavior of the response.

6.15 Continuous- to Discrete-System Bilinear Conversion

Suppose, we have a continuous-time system with Laplace transfer function $H(s)$, and we want to get an equivalent discrete-time system. Although posed in these terms the problem has no general solution, under certain conditions there is a way to transform the Laplace variable s into the z variable that provides reasonably good results.

We saw earlier that the approximation of the derivative by the incremental quotient may be too coarse at high frequency (near the Nyquist limit). We can think, instead that a signal is the integral of its derivative. If $u(t) = y'(t)$, then

$$y(kT_\mathrm{s}) = \int_{(k-1)T_\mathrm{s}}^{kT_\mathrm{s}} u(t)\,\mathrm{d}t + y((k-1)T_\mathrm{s}) \tag{6.51}$$

We can approximate the integral in a short interval such as T_s using the trapezoidal rule:

$$\int_{(k-1)T_s}^{kT_s} u(t)\,\mathrm{d}t \cong \frac{u(kT_s)+u((k-1)T_s)}{2}T_s, \tag{6.52}$$

From which, replacing the associated discrete signals,

$$y(k) \cong \frac{u(k)+u(k-1)}{2}T_s + y(k-1).$$

Now we apply the z transform and rearrange:

$$Y(z)\,(1-z^{-1}) \cong \frac{1+z^{-1}}{2}T_s\,U(z).$$

Finally

$$U(z) \cong \frac{2}{T_s}\frac{1-z^{-1}}{1+z^{-1}}\,Y(z). \tag{6.53}$$

Since U(z) is the z transform of the derivative of y, the multiplicative operator between z transforms

$$\frac{2}{T_s}\frac{1-z^{-1}}{1+z^{-1}}$$

is equivalent to the differentiation operator in the continuous-time domain. This suggests that the variable change

$$s \rightarrow \frac{2}{T_s}\frac{1-z^{-1}}{1+z^{-1}} = \frac{2}{T_s}\frac{z-1}{z+1} \tag{6.54}$$

allows us to convert Laplace transfer functions in z transfer functions between a system's input and output signals.

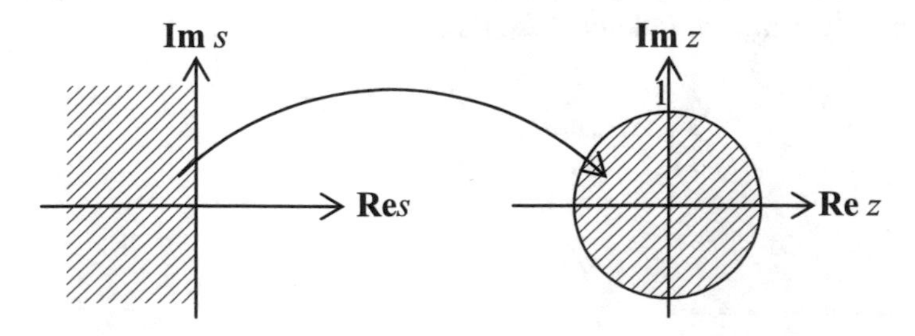

Fig. 6.6 Transformation of the real negative half-plane into the interior of the unit circle by means of the bilinear transformation

The variable change (6.54) is the *bilinear transformation*. It can be shown that it transforms the negative real half-plane (where the stable poles of a continuous-time system are located) into the interior of the unit circle, the stable region for a discrete-time system as shown in Fig. 6.6 (Proakis et al. 1996). This means that a stable continuous-time system is correctly transformed into a stable discrete-time system.[7]

As an example, consider the first-order filter with time constant given by the following transfer function

$$H(s) = \frac{1}{1 + \tau s}. \tag{6.55}$$

Replacing s as in (6.54) we get (allowing for the abuse of notation)

$$H(z) = \frac{1}{1 + \tau \frac{2}{T_s} \frac{z-1}{z+1}} = \frac{z+1}{z\left(1 + \frac{2\tau}{T_s}\right) + \left(1 - \frac{2\tau}{T_s}\right)}. \tag{6.56}$$

This z transfer function provides a better match for the first-order frequency response.

Problems

(1) (a) Create a non-aliased discrete impulse centered at a time instant halfway between two sampling instants. How many nonzero oscillations are necessary if the error caused by truncation is required to be less than 1 LSB for 16 bit? (b) Plot its FFT spectrum using the function **`fft`**. (c) Simulate an aliased discrete impulse. To this end, create a non-aliased discrete impulse at a sample rate M times the target one, and then keep every Mth sample. Try with different values of M. (d) Repeat part (b). (e) Repeat parts (a) through (d) moving the impulse to different fractional positions $0 < \alpha < 1$ between two consecutive samples. Notice the spectrum dependency on α in the case of the aliased discrete impulse.

(2) Often it is necessary to get a periodic signal with a given spectral envelope. This can be accomplished applying a filter with the proper frequency response to a sum of N cosine harmonics of equal amplitude, where N is the maximum integer not exceeding $F_s/2$. (a) Show that this sum can be computed more efficiently using the following formula:

$$\sum_{k=1}^{N} \cos\, kx = \cos\left(\frac{N}{2}x\right) \frac{\sin\left(\frac{(N+1)}{2}x\right)}{\sin\left(\frac{x}{2}\right)} - 1$$

[7] Note, however that this theoretical result may prove false in practice when digitization comes in, due to cumulative arithmetic precision errors. Unfortunately, this is often the case for high-order filters whose original poles are too close to the imaginary axis.

Hint: Express each cosine as the real part of a complex exponential. Use the summation formula of N successive powers and rearrange the result to make the trigonometric functions explicit. (b) Show that the result is equivalent to the superposition of infinite equally spaced non-aliased impulses (i.e., a *pulse train*). (c) Write a Scilab script that implements this formula. Hint: Note that when $\sin(x/2)$ is 0 the formula yields NaN's (NaN stands for "not-a-number," a result obtained in mathematical software when attempting to compute 0/0 expressions). In such case, the quotient should be replaced by its limit. Note also that when $\sin(x/2)$ is not 0 but very small, arithmetic errors will add significant distortion. In this case, the quotient should also be replaced by its limit. (d) Apply the algorithm to the case in which the period of the pulse train is an exact multiple of the sampling period. Plot examples in which the pulses fall exactly on sampling instants and in which they do not.

(3) A system has an impulse response given by

$$h(k) = 3 \times 2^{-k} - 2 \times 3^{-k}$$

(a) Find its difference equation. (b) Check that the first few samples of the impulse response match the expected values. Hint: Since there are two decay modes, the difference equation is of the form $y(k) + a\ y(k-1) + b\ y(k-2) = c\ x(k) + d\ x(k-1)$. Replace the convolutional expression of $y(k)$ and apply the discrete impulse $u_o(k)$ as an input. Taking $k = 1, 2, 3, 4$will provide a system of four equations with the four unknowns a, b, c and d.

(4) Find the difference equation of a system whose impulse response is [1, 0.3, 0.09, 0.027, 0, 0, 0, …].

(5) Assuming $F_s = 44\ 100$ Hz, find and plot the frequency response of the following: (a) The system from problem (3). (b) The system from problem (3) with the impulse response truncated to the first ten samples. (c) The system from problem 4.

(6) (a) Find a closed formula for the z transform of the impulse response of the systems from problems (3) and (4). (b) Find the corresponding difference equations directly from the z transform and compare with the results from problems (3) and (4). (c) Find again the frequency response of the systems from problems (3) and (4) using the z transform.

(7) Find the response of the systems from problems (3) and (4) when excited with a unit step.

(8) (a) Find the difference equation of a system whose impulse response is

$$h(k) = 1.125^{3-k} \sin(30k)$$

Hint: Use Euler's identity. (b) Find its z transform. (c) Find and plot its frequency response. (d) Find its response to a sine wave $\sin(25k)$ starting at $k = 0$. (e) Find its response to a unit step function.

(9) (a) Search and download from the Internet an impulse response of a hall and some anechoic recordings of speech or music (there are many repositories available for free). If the recordings are longer than about 10 s, select a smaller portion using Audacity or any other editor. Then load that portion into Scilab as variable **x** using the function **wavread**. Load also the impulse response as variable **h** and use the function **convol** to get the response of the hall when excited with the anechoic signal. (b) Repeat with a quasi-anechoic recording of your own voice uttering a short phrase. You can make the recording using a digital recorder or a computer with a microphone connected to its audio input. Try to speak close to the microphone, but avoid blowing into the microphone to prevent popping.

(10) (a) Use the bilinear transformation to convert the third-order continuous-time low-pass Butterworth filter whose Laplace transfer function is

$$H(s) = \frac{1}{1 + 2s + 2s^2 + s^3}$$

Hint: s is in rad/s. Choose a proper sample rate. (b) Plot the frequency response of the continuous-time and the discrete-time versions of the filter.

(11) (a) Write a Scilab script that directly implements an nth order filter given the coefficients of its difference equation and the signal to be filtered. (b) Apply it to a first-order filter with a sine wave input. Include in a single figure both the input and the output plots using the function **subplot**.

(12) (a) Use the function **syslin** to create a transfer function **H** from the numerator and denominator polynomials (which can be created from their coefficients or their roots using the function **poly**). (b) Use the function **flts** to apply the transfer function to an input **x** and compare with the result obtained in 11. (c) Use the function **frmag** to compute the frequency response of **H** and plot it.

References

Ghausi MS, Laker KR (1981) Modern Filter Design, Active RC and Switched Capacitor. Prentice-Hall, Inc., Englewood Cliffs, New Jersey, USA

Miyara F (2004) Filtros Activos (in Spanish). Internal publication Universidad Nacional de Rosario, Argentina. Accessed 10 Apr 2017. http://www.fceia.unr.edu.ar/enica3/filtros-t.pdf

Proakis JG, Manolakis DG (1996) Digital signal processing. Principles, algorithms and applications (third edition). Prentice-Hall International, Upper Saddle River, New Jersey, USA

Chapter 7
Basic Algorithms for Acoustical Measurements

7.1 Introduction

In this chapter, we shall discuss several of the basic algorithms to be applied to acoustical measurements. We shall use a kind of mathematical software with native handling of vectors, matrices, and arrays, in which vector and matrix operations with real and complex numbers are invoked as if they were simple numbers. This type of software is very interesting for digital signal processing since many calculations require the application of the same process to massive amounts of data structurally organized as vector or matrices.

Another interesting feature is the possibility of working symbolically, which allows us to work with symbolic expressions, simplifying them, solving equations for explicit solutions, differentiating and integrating functions.

There are several programs of this kind, such as MATLAB, Scilab, Octave, and Euler. The most widely used is MATLAB. While it is the most developed one, it is a very expensive commercial software package. The other three alternatives are compatible with the General Public License, GPL promoted by the Free Software Foundation, FSF. In the case of Scilab, although the license is not GPL but CeCILL, the FSF itself has declared it to be GPL-compatible. In all cases, the source code in C++ is available for download, study, modification, and even distribution, as far as any derivative version is distributed under the same license.

We could use any of these programs but we have chosen Scilab since it has a highly developed interface, very similar to that of MATLAB, so the programming work will be facilitated for readers that are used to MATLAB. In Appendix M, a brief introduction to Scilab's features and programming is given. Reference information can be obtained in the Scilab manual (Scilab no date).

As in the case of the digital audio editor Audacity, the decision to use free software is based in the philosophy of spreading applied knowledge as far as possible, and to achieve, through cooperative work, permanent feedback and improvement. One of the factors that make this approach viable is the ease and

F. Miyara, *Software-Based Acoustical Measurements*, Modern Acoustics and Signal Processing, DOI 10.1007/978-3-319-55871-4_7

economy of the distribution process through Internet-based data networks or through low-cost copies on CD or DVD, as well as the possibility to resort to a network of users and developers to get technical support and suggest new features and improvements.

7.2 Opening a .wav File

The first step prior to any data processing is to acquire the signal from a .wav file by means of the function **wavread**. Suppose the file path and name is c:/sounds/sound.wav. Then

```
[x, Fs, Nbits] = wavread('c:/sounds/sound.wav');
N = length(x);
```

copies the signal to a variable **x**. This is a row vector in the case of a monophonic file or a two-row matrix in the case of a stereophonic file. The signal is automatically converted from the original 16 bit signed version to a normalized version between -1 and 1. With this syntax,[1] the number of bits per sample and the sample rate are also retrieved, and copied into the variables **Nbits** and **Fs**.

In case the file is too large, in order to prevent memory saturation issues it may be necessary to load only part of the file. For instance,

```
N1 = 100000;
N2 = 1000000;
[x, Fs, Nbits] = wavread('c:/sounds/sound.wav', [N1,N2]);
```

will read from sample **N1** to sample **N2**.

7.3 Energy Average and Equivalent Level

The most basic process is to find the energy average or *effective value* (also called *RMS value*). As seen in Chap. 1, the effective value of a continuous-time signal $x(t)$ is defined as

$$X_{\mathrm{ef}} = \sqrt{\frac{1}{T}\int_0^T x^2(t)\,\mathrm{d}t}. \tag{7.1}$$

[1]There are more alternatives that can be investigated by writing "**help wavread**".

In the case of a discrete signal $x(n)$ with a length of N samples, the preceding integration must be computed numerically as

$$X_{\mathrm{ef}} = \sqrt{\frac{1}{NT_{\mathrm{s}}}\sum_{n=1}^{N} x^2(n)T_{\mathrm{s}}} = \sqrt{\frac{1}{N}\sum_{n=1}^{N} x^2(n)} \tag{7.2}$$

We can observe that although the numerical approximation is initially written using the sampling period T_{s} as the increment interval, it finally cancels out. The calculation can be performed in Scilab by means of the code

```
Xef = sqrt(sum(x.^2)/N);
```

Note how the component-wise power operation, **x.^2**, and the sum of the components of a vector allowed a concise and efficient code.

7.4 Calibration

Calibration consists in finding the relationship between the arbitrary value of the digitized output of a transducer and its original value, so that from the digitized samples it will be possible to retrieve the original values of the signal samples. In our case, as has been remarked earlier, there are several proportionality constants involved: The microphone sensitivity, S_{m}, the gain of the instrument's signal conditioning, G_{c}, the gain of the analog/digital converter's preamplifier, G_{d}, the converter's constant, $K_{\mathrm{A/D}} = 2^{N\mathrm{bits}-1}/V_{\mathrm{ref}}$, and the normalization constant $K_{\mathrm{n}} = 2^{1-N\mathrm{bits}}$ used by Scilab. In consequence, the dimensionless value $x(n)$ with which the software represents the sound pressure level at instant t_n is

$$x(n) = S_{\mathrm{m}}\, G_{\mathrm{c}}\, G_{\mathrm{d}}\, K_{\mathrm{A/D}} K_{\mathrm{n}}\, p(t_n) = K\, p(t_n). \tag{7.3}$$

As already commented in Sect. 4.2.9, in practice it is not possible to control each of the constants that form K.

The simplest way to obtain K is by means of a calibration signal whose nominal level, L_{cal}, is known (typically, 94 or 114 dB). The normalized signal x_{cal} corresponding to an interval where the calibration tone has been recorded is retrieved following the procedure of Sect. 7.2. Its effective value is computed as:

$$X_{\mathrm{cal}} = \sqrt{\frac{1}{N_{\mathrm{cal}}}\sum_{n=1}^{N_{\mathrm{cal}}} x_{\mathrm{cal}}^2(n)}. \tag{7.4}$$

The sound pressure corresponding to L_{cal} is

$$P_{cal} = P_{ref}\ 10^{L_{cal}/20}, \tag{7.5}$$

where P_{ref} = 20 μPa. Then we can compute K as

$$K = \frac{X_{cal}}{P_{ref} 10^{L_{cal}/20}}. \tag{7.6}$$

If, as is often the case, we intend to work with sound pressure levels instead of plain sound pressure, then

$$L_p = 20\ \log \frac{X_{ef}/K}{P_{ref}} = 20\ \log\ X_{ef} - 20\ \log\ X_{cal} + L_{cal}. \tag{7.7}$$

In order to implement this calculation there are two options. The first one is to have a separate file containing the calibration tone recording, load it by means of the function **wavread**, and finally obtain X_{cal} using the previous method. The second option is to have the calibration tone recorded in the same file that contains the signal. In this case, using a digital audio editor as Audacity (see Chap. 5), we locate by simple inspection the calibration tone and determine its initial and final samples[2] **Ncal1** and **Ncal2**.

The same is done with the portion of signal of interest, getting its initial and final samples **N1** and **N2**. Assuming the calibration tone level is 94 dB, the following code

```
Lcal = 94;
Ncal = Ncal2 - Ncal1 + 1;
N = N2 - N1 + 1;
Xcal = sqrt(sum(x(Ncal1:Ncal2).^2)/Ncal);
Xef = sqrt(sum(x(N1:N2).^2)/N);
Lp = 20*log10(Xef/Xcal) + Lcal;
```

computes the sound pressure level. Note the use of the syntax **x(N1:N2)** to denote the subsequence of **x** encompassed between indices **N1** and **N2**.

7.5 Energy Envelope

The energy envelope (see Fig. 7.2, Chap. 1) can be obtained applying a low-pass filter with a given time constant τ to the squared signal. The Laplace transfer function is

[2]It may be convenient, as explained in Chap. 5, to adjust these samples to the nearest zero crossings.

$$H(s) = \frac{1}{1 + \tau s}. \tag{7.8}$$

We need an equivalent discrete-time digital filter so we must convert the Laplace transfer function into the z transfer function. Since the standard time constants ($\tau_F = 0.125$ s for F response and $\tau_S = 1$ s for S response) correspond to cutoff frequencies much lower than the Nyquist frequency, it is valid to approximate the derivative by the incremental quotient over successive samples. The preceding filter is equivalent to the differential equation

$$\tau\, y'(t) + y(t) = x(t),$$

which can be approximated as the following difference equation:

$$\tau \frac{y(k) - y(k-1)}{T_s} + y(k) = x(k)$$

$$(1 + F_s\tau)y(k) - F_s\tau\, y(k-1) = x(k)$$

Applying the z transform and rearranging we have

$$H(z) = \frac{z}{(1 + F_s\tau)z - F_s\tau} \tag{7.9}$$

This expression is the division of two polynomials in z, $B(z)/A(z)$, and it is suitable for the application of filtering algorithms.

In order to implement this filter in Scilab, we need first to declare this transfer function as a linear system by means of the function **syslin**. In the following code, **p** and **q** are declared as the numerator and denominator, i.e., the polynomials whose roots are 0 and $F_s\tau$ / (1 + $F_s\tau$).

```
Fs = 44100;
tau = 0.125;
p = poly(0,'z');
q = (1 + Fs*tau) * poly(Fs*tau/(1 + Fs*tau), 'z');
H = syslin('d', p, q);
```

In this example, we have chosen $\tau = 0.125$ s, which corresponds to fast (F) response. For slow (S) response we would choose $\tau = 1$ s.

An alternative way would be to use the **iir** function to synthesize a low-pass Butterworth filter with cutoff frequency $1/(2\pi\tau)$:

```
H = iir(1,'lp','butt',[1/(2*%pi*tau)/Fs, 0],[0,0]);
```

The next step is to apply this transfer function to the squared signal **x.^2**,[3] which can be done by means of the function **flts**, and then we apply the square root:

```
y = sqrt(flts(x.^2, H));
```

Finally, the signal **x** and its energy envelope **y** can be plotted. The following code sets window 1 as the default one and plots both signals versus time by means of the command **plot2d**:

```
t = [0:length(x)-1] / Fs;
xset('window', 1);
plot2d(t, [x', y'])
```

Note the use of syntax **x'** to represent the transpose of **x**, since to plot two signals corresponding to a single time vector they must be arranged as column vectors. An alternative syntax would be

```
plot2d(t, x, t, y)
```

In some cases we may have an overflow for long signals (more than 15 000 points). Then we may perform a graphic decimation taking a sample every n, where n is conveniently chosen:[4]

```
sub = 1:n:$;
plot2d(t(sub), [x(sub)', y(sub)'])
```

This is actually equivalent to downsampling (decimating) the signal, creating very likely some aliasing, since in fact the sample rate is divided by n. In the case of the energy envelope, it is safe to downsample, since the low-pass filter acts indeed as an effective anti-aliasing filter because of its extremely low cutoff frequency (in the order of 1 Hz).

The decimation can also be carried out by means of the function **intdec**, to be studied in Sect. 8.3. It applies an anti-aliasing filter prior to the decimation.

7.6 A Weighting

Many acoustical measurements require the use of weighting filters, for instance, A weighting or C weighting. According to the model of sound level meter, the audio output may be before or after filtering. However, for greater versatility in the

---

[3]Note that in the preceding discussion, we freely used x and y as a generic input to the filter and its response, as is customary. In this special case the input to the filter is actually the square signal.

[4]Scilab has the wildcard character $ to indicate the last component of a vector or a matrix dimension.

postprocessing it is better to record without any weighting filter (flat weighting). While in theory it is possible to remove the A weighting with an inverse A filter, in practice we face the fact that the analog electronics of any recording system adds low frequency noise ($1/f$ noise). Since the inverse A filter has a high gain at low frequency, it could be amplifying the noise, which would introduce a large systematic error.

It is thus necessary to have an algorithm to apply the A weighting. This can be attained implementing a discrete filter with the desired frequency response. The transfer function of an A filter, derived from the International Standard IEC 61672-1 (2002), Proakis and Manolakis (1996) is

$$A(s) = \frac{1.2589049 \times 76617.2^2\, s^4}{(s+129.43)^2\,(s+676.7)\,(s+4636.36)\,(s+76617.2)^2} \tag{7.10}$$

There are two approaches for implementing this filter. The first one is a FIR (finite impulse response) structure; the second one, an IIR (infinite impulse response). The former uses a linear combination of P past input samples, where P must be sufficiently large as to allow a transient response of comparable duration to that of the A filter. Since two of its poles are of very low frequency, with a time constant $\tau = 1/129.43 = 7.7$ ms (the transient response is of the form $t\,e^{-t/\tau}$), the required value of P is high, causing the algorithm to be slow.

The second approach is conceptually more complicated, since it requires not only the past values of the input but also of the output (it is a recursive filter) and might present stability issues. However, the algorithm is faster, since the required order is much smaller, so much less operations are necessary. It is possible to get a first version of the filter by means of the bilinear transformation (see Eq. 6.53)

$$s = 2F_s\frac{1-1/z}{1+1/z} = 2F_s\frac{z-1}{z+1}. \tag{7.11}$$

In order to design an A filter, we first note that if p_k is a pole of a polynomial in the s domain, then

$$z_k = \frac{2F_s + p_k}{2F_s - p_k} \tag{7.12}$$

is the corresponding zero of the rational function we get by applying the bilinear transformation (7.11) to the polynomial. At the same time, the denominator will be $(z + 1)^m$ where m is the degree of the polynomial. We can apply this idea to both the numerator and denominator of the transfer function (7.10).

The following lines of code implement this method. We first compute the poles after the bilinear transformation and adjust the constant that multiplies the whole transfer function. Then we compute the numerator and the denominator and we finally create a linear system structure (using the function **syslin**) which implements the digital A filter.

```
Fs = 44100;
p1 = -2*%pi*20.60;
p2 = -2*%pi*107.7;
p3 = -2*%pi*737.9;
p4 = -2*%pi*12194;

K = 1.2589049*p4^2;
K1 = 2*Fs

p1z = (K1 + p1)/(K1 - p1);
p2z = (K1 + p2)/(K1 - p2);
p3z = (K1 + p3)/(K1 - p3);
p4z = (K1 + p4)/(K1 - p4);

Kz = K * K1^4/(K1 - p1)^2/(K1 - p2)/(K1 - p3)/(K1 - p4)^2;

z = poly(0,'z');

Aznum = Kz * (z - 1)^4 * (z + 1)^2;
Azden = (z - pz1)^2*(z - pz2)*(z - pz3)*(z - pz4)^2;


Az = syslin('d', numAz, denAz)
```

The resulting function is (use **format** to get more decimals)

```
                                  2              3              4              5              6
0.255739 - 0.511479z - 0.255739z + 1.022957z - 0.255739z - 0.511478z + 0.255739z
------------------------------------------------------------------------------------
                                         2              3                4                 5   6
  0.0043522 - 0.1418439z + 1.4208986z - 4.453265z + 6.1894402z - 4.0195821z + z
```

The bilinear function works very well in the low and middle frequency region, but introduces errors in the response at high frequencies. At the 20 kHz end, the frequency response is about 24 dB lower than that of the continuous transfer function (7.10). However, the tolerance allowed for class 1 (precision) instruments by the Standard IEC 61672-1 (2002) at 20 kHz is $-\infty$. It can be verified that the whole frequency response is within the required tolerance at all frequencies for class 1.

7.7 Statistical Analysis

Statistical analysis is used to assess the variability of noise, as well as for the application of some criteria and for estimating the background noise level. It consists in the determination of the statistical distribution of the values of the energy envelope, and it is expressed as the *exceedance levels* L_N, where L_N is the level that is exceeded an N % of the time. Statistically, L_N is the same as the $100 - N$ percentile of the envelope values expressed logarithmically in dB. The function **perctl** allows to compute the percentiles of the components of a vector. In a single function call we can get several percentiles, as indicated by a second argument vector.

If we assume that vector y contains the energy envelope samples of x, then we compute the exceedance levels $N = 0, \ldots, 100$ with the following code:

```
N = [1:99];
yN = perctl(y, 100 - N);
yN = [max(y), yN(:,1)', min(y);
LN = 20*log10(yN/Xcal) + Lcal;
```

where **Lcal** and **Xcal** are the values defined in Sect. 7.4. In the case of very large files, the memory requirements may be too stringent since the function **perctl** involves the computationally costly sorting function **gsort**. It is thus convenient to downsample **y**. This can be safely done without aliasing issues because the filter used to take the energy envelope acts as an anti-aliasing filter as well. The second code line is, hence, replaced by

```
yN = perctl(y(1:100:$), 100 - N);
```

It is also convenient to increase the *memory stack size* by means of **stacksize**, for instance:

```
stacksize(10000000);
```

This variable can be permanently changed by editing the corresponding entry in the scilab.star file, which is located in the Scilab directory:

```
newstacksize = 10000000;
```

The preceding method has the drawback that increasing the memory stack size may not be enough. It may also happen that a given file is so large that it cannot even be loaded in memory as a variable. In that case it is convenient to split the problem by processing smaller data blocks. However, it is not possible to get the global percentiles from the percentiles of several blocks. The alternative procedure will be to work with histograms, since partial histograms can be combined to get the global histogram. Although Scilab's function **histc** computes the histogram, it will be illustrative to generate a specific code to get the histogram of a block.

First, a range of relevant values of variable **y** is selected. The choice depends on the particular problem. For instance, in the case of urban noise this range could span from 20 dB to 120 dB. Once referred to the digital reference, we will assume these values translate into −100 dB and 0 dB, respectively (a reasonable assumption when using 16 bit wav files):

```
mini = -100;
maxi = 0;
```

Then we divide that range into **K** bins, where **K** depends on the number of available data points. The larger the amount of data, the larger the possible number of bins and the narrower each bin. It is convenient to take a number of bins sufficiently large to represent the most important features of the statistical distribution, but not so large as to represent irrelevant details. For instance, one minute of audio corresponds to 26 460 decimated samples, allowing to split the data into 1000 bins:

```
K = 1000;
```

Next, the values of **y** are discretized so that **mini** corresponds to **1** and **maxi** to **K + 1**:

```
Ybin = floor(K * (y - mini) / (maxi - mini)) + 1;
```

Finally, the successive values of **Ybin** are explored and used as an index to the bins. For instance, if **Ybin(i)** is equal to *k*, then the *k*-th bin is increased by 1:

```
histo = zeros(1, K + 1);
for i = 1:length(y)
  histo(Ybin(i)) = histo(Ybin(i)) + 1;
end
```

The first line is meant to preallocate memory space for the variables.[5] Figure 7.1 illustrates this process for bin *k*.

Once the histogram of two or more consecutive blocks (about 60 s each) has been computed, we can get the combined histogram by adding the histograms component by component (i.e., adding together the numbers of samples in

[5]Preallocation of memory space is very convenient when the size of the data is previously known. This improves efficiency since it prevents the risk that the program have to move the variable elsewhere when the variable has grown so much that there is no contiguous space available.

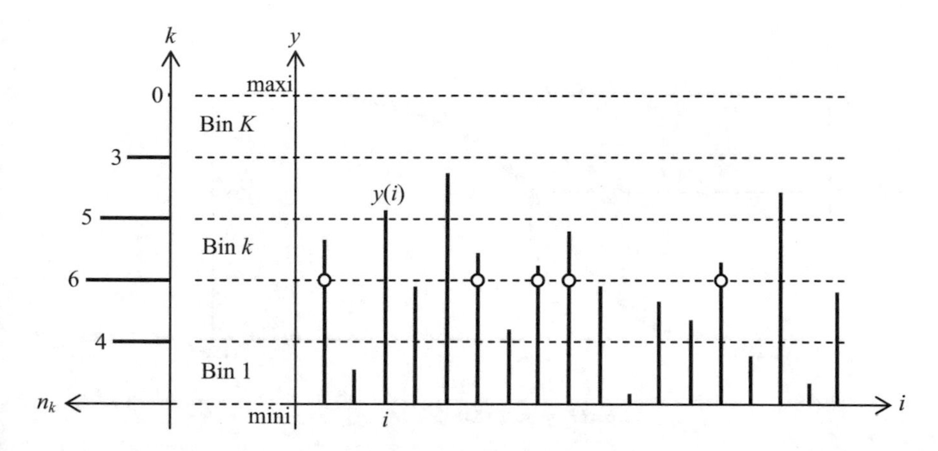

Fig. 7.1 Determination of the histogram of a variable y distributed into K bins. The *small circles* indicate the instances of y within bin k, after discretization

homologous bins). In fact, we can have a vector variable that contains the cumulative histogram to which the new histograms can be added.

After this process, we will have the histogram of a large amount of audio (for instance, 30 min). The next step is to estimate the cumulative probability function. The function **cumsum** computes the cumulative frequency function. If we normalize it to 1 dividing by the sum of all frequencies (which is equal to the last component of the cumulative frequency distribution), we get the cumulative distribution function:

```
f = cumsum(histo);
cpf_y = f/f($);
```

After preallocating the variable that will contain the percentiles,

```
perc_y = zeros(1, 101);
```

follows the code that computes the percentiles. Basically, we need to find the inverse function of the cumulative probability function **cpf_y**. There are several possible approaches, for instance to use linear or spline interpolation. The method implemented here simply consists in detecting for what values of the index **k** it turns out that **cpf_y(k) < p/100**. The amount of indices that comply with this inequality is equivalent to the index **ko** such that **cpf_y(ko)** is as closest as possible to **p/100**. This index must be finally converted into the corresponding value that the index represents, i.e., the values of **y**:

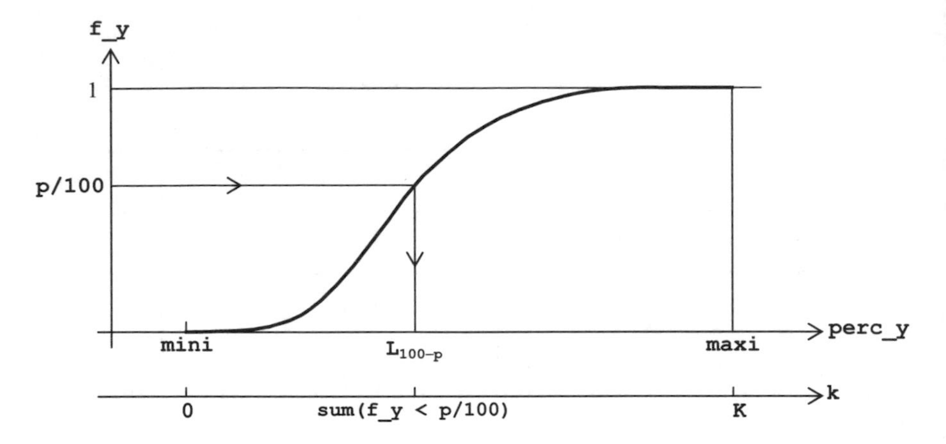

Fig. 7.2 Determination of the percentile **p** by means of the cumulative probability function. The axis parallel to the abscissas is the index axis. Each index corresponds to an increment **delta/K** of the variable

```
for p = 0:100
  perc_y(p + 1) = mini + delta * sum(cpf_y < p/100);
end
```

Figure 7.2 illustrates schematically this procedure. Note that the exceedance level N corresponds to percentile $p = 100 - N$. This is due to the fact that the exceedance level is defined as the level that is *exceeded* an N % of the time, while the p-th percentile is the level that *exceeds* a p % of the variable's values.

Problems

(1) (a) With a digital recorder or a computer record some voices and noises during about one minute. Then open the file in Audacity, place the cursor at the beginning and, using *Generate/Tone...* insert a 1 kHz sine wave with an amplitude of 0.25 and a duration of 3 s. We will assume it corresponds to a calibration tone of 94.0 dB. Export the file as a wave file. Hint: Most probably the original recording is in stereo format. You may prefer keeping things simple at the beginning, so you can choose to convert to mono using *Tracks/Stereo track to mono*. (b) Open the file in Scilab using **wavread** with the keyword **'size'** to get the size in samples, the sample rate and the bits per sample. From the sample rate calculate the first and last samples of the calibration tone. (c) Open the file using the sample interval corresponding to the calibration tone and find its digital equivalent level (referred to the maximum digital value, i.e., 1). (d) Open the file using the sample interval corresponding to the recorded signal and find its digital equivalent level. (e) Compute the acoustic equivalent level using the assumption from part (a). Hint: To prevent memory issues, it

may be necessary to increase the stack by using the function **stacksize**. Do experiment!

(2) Repeat problem (1) using the (unweighted) signal from a sound level meter as the recorder audio input. In this case, record the calibration signal from an acoustic calibrator (or the electric calibration signal, if the sound level meter has this feature). Take note of the instrument's reading. Note: If the microphone is a free-field one, as in most sound level meters, the reading is 0.15 dB below the actual calibrator level. Hint: In this case, while still in Audacity, select most of the calibration tone, use Z to detect zero crossings, and copy the start and end samples that can be seen in the selection toolbar at the bottom. In order to show sample numbers you can click on the small triangle at the right of any of the selection limits to open the drop-down menu to change the selection format. Repeat for the signal portion for which you want to measure the equivalent level.

(3) When the signal is very long (several minutes to hours) memory will not be enough to host all the signal. Write a script that loads about 10 s at a time and computes the equivalent level. Hint: Find a simple recursive formula that updates the equivalent level each time a new 10 s interval is added.

(4) Find the energy envelope applying the bilinear transformation to Eq. (7.8) instead of approximating the differential equation. Compare the results of both approaches in the case of the signal of problem (1).

(5) Compute the A-equivalent level in the cases of problems (1) and (2).

(6) (a) Find the exceedance levels L_5, L_{10}, L_{50}, L_{90}, L_{95} in the cases of problems (1) and (2). (b) Write a script that reads a large wav file by sections and computes its complete set of long-term exceedance levels L_1, …, L_{99}.

References

IEC 61672-1. "Electroacoustics—sound level meters—part 1: specifications". International Electrotehcnical Commission. 2002

Proakis JG, Manolakis DG (1996) Digital signal processing. Principles, algorithms and applications (third edition). Prentice-Hall International, Upper Saddle River, New Jersey, USA

Scilab (no date) Scilab manual. http://www.scilab.org/download/5.1.1/manual_scilab-5.1.1_en_US.pdf. Accessed 10 Apr 2017

Chapter 8
Spectrum Analysis

8.1 Introduction

Nowadays there are many commercially available measurement instruments that can measure octave and one-third octave band sound spectrum according to the International Standard IEC 61260. Even if the cost of highly reliable electronics keeps falling, these instruments are still very expensive. The modern trend is to manufacture instruments with an embedded programmable processor[1] for which there are software modules that allow not only to extend the instrument capabilities but also to update the overall design without changing the hardware. While this makes it possible to amortize the investment in hardware design and cut costs down, unlike what can be imagined, prices have indeed risen.

A viable alternative is to use an inexpensive instrument with no special capabilities except providing a calibrated analog output to be digitally recorded by means of an external recorder in PCM (uncompressed) format. The signal will be later analyzed by means of a computer algorithm (or a low-cost autonomous DSP system). This allows almost unlimited types of analysis.

8.2 Spectrum Analysis Paradigms

Originally, the spectral analysis was performed by means of an analog bandpass filter of adjustable frequency (constant bandwidth analyzers) or a bank of bandpass filters with center frequencies uniformly distributed along a logarithmic frequency

[1]An embedded system is a computing system with some complexity integrated to a specific-purpose piece of equipment. However, as technology evolves, the border between a general-purpose computer and an embedded system is less clear.

F. Miyara, *Software-Based Acoustical Measurements*, Modern Acoustics and Signal Processing, DOI 10.1007/978-3-319-55871-4_8

axis (constant percentage analyzers). The first case applies to pure tone signals and the second to wideband or mixed signals.[2]

At the beginning, these filters were implemented as passive electric circuits (inductor and capacitor networks) and later as active circuits (operational-amplifier-based circuits). As they were large-sized, heavy, and expensive, they were difficult to carry around, which severely limited field work. Therefore, they were mostly used at the laboratory. Later on, the size was progressively reduced hand in hand with the miniaturization of electronics. This boosted generalized use with good impact on research of frequency-dependent acoustic phenomena.

High-quality tape recorders were also developed, which made it possible to record calibrated signals to be analyzed at the laboratory with the described equipment.

With the arrival of digital computing systems, it was possible to implement numerical algorithms for spectrum analysis. There are three possible approaches for spectrum analysis of discrete signals. The first one consists in using digital band filters that emulate traditional analog band filters as a means for separating the different spectral components of the signal to analyze. The second one consists in implementing the spectral analysis by means of the Fourier transform, converting the signal from the time domain to the frequency domain. The third one is a mixed approach that uses direct and inverse Fourier transforms in order to implement digital filters that allow an almost ideal frequency response, and hence, a very detailed spectral analysis.

8.3 Digital Filters

The first paradigm uses algorithms based on difference equations such as

$$y(k) = -\sum_{n=1}^{N} a_n y(k-n) + \sum_{m=0}^{M} b_m x(k-m) \tag{8.1}$$

which expresses the present value of the output $y(k)$ as a linear combination of N past values of the output and M past values of the input. This equation corresponds to a differential equation in the continuous-time domain. If $a_n = 0$ for $n = 1, \ldots, N$, we have a finite impulse response (FIR) filter. If, on the contrary, $a_n \neq 0$ for at least one n, we have an infinite frequency response (IIR) filter. The z transfer function is

$$H(z) = \frac{b_0 z^M + b_1 z^{M-1} + \cdots + b_{M-1} z + b_M}{z^N + a_1 z^{N-1} + \cdots + a_{N-1} z + a_N} z^{N-M} \tag{8.2}$$

[2]It is recommended to refer to Sect. 1.7 for general concepts on spectrum analysis.

In the filters obtained from the Laplace transfer function by means of the bilinear transformation (6.53), it turns out that $M = N$, so

$$H(z) = \frac{b_0 z^N + b_1 z^{N-1} + \cdots + b_{N-1} z + b_N}{z^N + a_1 z^{N-1} + \cdots + a_{N-1} z + a_N} \tag{8.3}$$

For spectral analysis purposes, these filters are designed to emulate the analog prototypes, i.e., bandpass filters whose lower and upper cutoff frequencies are given by

$$f_l = f_o \, 2^{-\frac{1}{2b}}, \quad f_u = f_o \, 2^{\frac{1}{2b}}, \tag{8.4}$$

where f_o is the center frequency of the pass band and b is the denominator of the octave fraction (b = 1 for octave band and b = 3 for one-third octave band).

In order to design an IIR band filter we can use the Scilab function **iir**. This function applies the bilinear transformation (see Sect. 6.15) to the corresponding continuous-time domain prototype. It accepts two possible syntaxes

```
hz = iir(n, ftype, fdesign, frq, delta)
[p, z, g] = iir(n, ftype, fdesign, frq, delta)
```

where:

hz is a discrete *transfer function structure*. This structure has four components:

hz(1) contains the description of the structure: **!r num den dt !**
hz(2) contains the numerator polynomial
hz(3) contains the denominator polynomial
hz(3) indicates whether it is a discrete filter (“c”: continuous; “d”: discrete)

p is a column vector that contains the poles of the discrete transfer function
z is a column vector that contains the zeros of the discrete transfer function
g is the constant required to have unity gain in the pass band(s)
n is the filter order;

ftype is the type of filter

“**lp**”:	Lowpass	“**hp**”:	Highpass
“**bp**”:	Bandpass	“**sp**”:	Bandstop

fdesign is the approximation type

“**butt**”: Butterworth (maximally flat)
“**cheb1**”: Type I Chebychev (constant ripple in the pass band)
“**cheb2**”: Type II Chebychev (constant ripple in the stop band)
“**ellip**”: Elliptic or Cauer (constant ripple in both bands)

Frq is a two-component vector containing the normalized lower and upper cutoff frequencies in the range $0 \leq F \leq 0.5$, where 0.5 is equivalent to $F_s/2$.
Note: This normalization is required since the discrete filter is not associated with any specific sample rate. For a bandpass filter of lower and upper frequencies f_i and f_s this argument will be **[fi/Fs, fs/Fs]**;

delta is a two-component vector with components δ_1 and δ_2, applicable to Chebychev and Cauer filters, as appropriate, where δ_1 is the maximum ripple in the pass bands and δ_2 is the maximum ripple in the stop bands. Note that they are not in decibels but in a linear scale in the range $0 < \delta < 1$

Once the filter has been obtained, the signal can be filtered by means of the function **filter**, with the syntax

```
[y, zf] = filter(num, den, x [,zi])
```

where

y is the filtered output

zf is the final state or condition, i.e., the last max{N, M} response values. It is useful when block-wise filtering is performed to prevent memory saturation issues

num is the filter numerator, expressed as coefficients in decreasing power order or else as a polynomial. In the latter form it is possible to use the variable **hz (2)** obtained with the first syntax of **iir**

den is the filter denominator expressed as coefficients in decreasing power order or else as a polynomial. In the latter form it is possible to use the variable **hz (3)** obtained with the first syntax of **iir**

x is the input signal, expressed as a row vector

zi is the initial state or condition, i.e., the max{N, M} response values before the first value. It is useful when block-wise filtering is performed to prevent memory saturation issues. In this case it is made equal to **zf** from the previous filtered block

It is important to bear in mind that for frequencies much smaller than the Nyquist limit, F_s, IIR filters may become unstable. The reason is that since IIR filters are recursive, truncation errors tend to accumulate, even if the representation has 16 decimal digits. Filters with $f_o \ll F_s$ have their poles very close to the unit circumference. For example, a one-third octave band Butterworth filter centered at 100 Hz with a sample rate of 44 100 Hz has the following poles:

$$p_{1,2} = 0.9994220 \pm 0.0129037j$$
$$p_{3,4} = 0.9986432 \pm 0.0135627j$$
$$p_{5,6} = 0.9992838 \pm 0.0155332j$$
$$p_{5,6} = 0.9985264 \pm 0.0146502j$$

whose moduli are, respectively, 0.9995053, 0.9987353, 0.9994045, and 0.9986338. While these poles are theoretically stable because they lie inside the unit circle (the stability region for discrete filters), error accumulation leads to a divergent response.

The solution to this problem consists in downsampling the input signal in such a way that the poles are not any longer so close to the stability limit. This can be achieved with the function **intdec**. Remember that before downsampling it is necessary to apply an anti-aliasing filter. This function carries out such operation by means of an efficient anti-aliasing algorithm. The syntax is as follows:

```
xsub = intdec (x,lom)
```

where

xsub is the downsampled version of **x**
x is the vector containing the signal to be downsampled
lom is the factor by which the original sample rate is multiplied to get the new sample rate, i.e., F_{s2}/F_{s1}

A criterion to prevent instability is that the sample rate F_s be at most 7 times the center frequency f_o of the filter. Note that to achieve this, the signal has to be generally downsampled, so, to prevent aliasing, frequencies larger than 3.5 f_o must be removed. While it might not seem important, it could be relevant in some particular cases, since the output of a continuous-time bandpass filter has a small but nonzero output even at very high frequencies. This output will be completely absent after the downsampling process. From the point of view of the applicable Standard (IEC 61260) this is indeed not important, since outside the pass band the lower tolerance limit is $-\infty$ dB.[3] But if for any reason we were interested in assessing the residual signal outside the pass band, we would be in trouble.

The following code performs the downsampling process, but only if it is necessary. Note that the downsampling factor is chosen to be a power of 2, since in this case the algorithm is more efficient.

[3]International Standard IEC 61260 specifies bandpass filters features, in particular their frequency response tolerance.

```
if fo < Fs/7
  p = 1;
  q = 2^ceil(log2(Fs/fo/7));
  xx= intdec(x, p/q);
end
Fs2 = Fs/q;
```

The following code lines define the lower and upper cutoff frequencies and design the corresponding filter as a third-order Butterworth approximation:

```
Finf = 2^(-1/6) * fo/Fs2);
Fsup  =  2^( 1/6) * fo/Fs2);
hz  =  iir(n, 'bp', 'butt', [Finf Fsup], []);
```

Next the filter is applied to the downsampled signal. In this case, the optional arguments **zi** and **zf** are not necessary

```
yy  =  filter(num, den, xx);
```

The effective value can be computed from **yy** using the procedure described in Sects. 7.3 and 7.4, notwithstanding the sample rate. However, if for any reason it were important to have the filtered signal itself (for instance, to listen to it), it is possible to oversample back to the original sample rate

```
if fo > Fs/7
  y  =  intdec(x, q/p);
end
```

Note that the final length of the filtered signal after oversampling might differ slightly from the original length. If this is important it is possible to add zeros (*zero-padding*) or to remove some final samples according to the result.

If the only aim is to get the level of the one-third octave band, it is neither necessary nor convenient to oversample, since it implies a waste of computing time. It is possible to apply the procedures from Sects. 7.3 and 7.4 to the downsampled signal.

The iterative application of this procedure to all bands allows to get the complete band spectrum of the signal. Once the filtered signals have been computed, it is possible to split each one into short temporal segments or blocks and analyze them. This yields the temporal evolution of the band spectrum.

It is also possible to design bandpass FIR filters by means of the function **wfir** and proceed as in the case of IIR filters. However, particularly at low frequency the length of the response may be considerable, making it computationally inefficient. A viable alternative will be to use FFT filters, as will be studied later on.

8.4 Discrete Fourier Transform (DFT)

Consider a periodic signal $x(t)$. We can represent it by means of a trigonometric series or a *complex Fourier series*

$$x(t) = \sum_{n=-\infty}^{\infty} c_n \, e^{jn\omega t}, \tag{8.5}$$

where $\omega = 2\pi/T$. The *Fourier coefficients*, c_n, are given by

$$c_n = \frac{1}{T}\int_0^T x(t)\, e^{-jn\omega t} dt. \tag{8.6}$$

If the signal is band limited, i.e., it has no spectral content above a given frequency value f_{max}, then (8.5) may be written with a finite number N of terms

$$x(t) = \sum_{n=-N/2}^{N/2-1} c_n \, e^{jn\omega t}, \tag{8.7}$$

where $N = 2f_{max}T$.[4] It can also be shown that the latter can be expressed as

$$x(t) = \sum_{n=0}^{N-1} c_n \, e^{jn\omega t}, \tag{8.8}$$

where, for $= N/2, \ldots, N-1$,

$$c_n = c_{n-N}. \tag{8.9}$$

This is a case of time aliasing, where, because of the periodicity, of the complex exponential, different spectra represent the same temporal signal.

Now suppose that we have a signal sampled at instants kT/N, where $k = 0, \ldots, N-1$. Since it is band limited, the signal can be represented by means of Eq. (8.8) at instants kT/N

$$x\left(k\frac{T}{N}\right) = \sum_{n=0}^{N-1} c_n e^{jn\frac{2\pi}{T}k\frac{T}{N}} = \sum_{n=0}^{N-1} c_n e^{j2\pi\frac{nk}{N}} \tag{8.10}$$

This holds for $k = 0, \ldots, N-1$, so this is a system of N equations with N unknowns (the Fourier coefficients $c_1, \ldots, c_N$). To solve it we multiply by a similar exponential

[4]Note that if $x(t)$ is periodic of period T, the maximum frequency must be an exact multiple of $1/T$.

$$x\left(k\frac{T}{N}\right)\,\mathrm{e}^{-j2\pi\frac{mk}{N}}=\sum_{n=0}^{N-1}c_n\mathrm{e}^{j2\pi\frac{(n-m)k}{N}},$$

and add with respect to k

$$\begin{aligned}\sum_{k=0}^{N-1}x\left(k\frac{T}{N}\right)\,\mathrm{e}^{-j2\pi\frac{mk}{N}}&=\sum_{k=0}^{N-1}\sum_{n=0}^{N-1}c_n\mathrm{e}^{j2\pi\frac{(n-m)k}{N}}\\&=\sum_{n=0}^{N-1}c_n\sum_{k=0}^{N-1}\mathrm{e}^{j2\pi\frac{(n-m)k}{N}}\end{aligned}\tag{8.11}$$

If $n = m$ the exponentials are all 1, therefore the internal sum is equal to N. If $n \neq m$, the internal sum has the form Σa^n, whose value, for $a \neq 1$, is

$$\sum_{k=0}^{N-1}a^n=\frac{1-a^N}{1-a}=\frac{1-\mathrm{e}^{j2\pi(n-m)k}}{1-\mathrm{e}^{j2\pi\frac{(n-m)}{N}}}=0$$

Note that in the exponential of the numerator, the N that was dividing the exponent has been canceled, so the exponential is 1 and the sum is 0.

Therefore, of all the terms of (8.11) the only one different from 0 is that one for which $n = m$

$$\sum_{k=0}^{N-1}x\left(k\frac{T}{N}\right)\,\mathrm{e}^{-j2\pi\frac{mk}{N}}=Nc_m,\tag{8.12}$$

i.e.,

$$c_m=\frac{1}{N}\sum_{k=0}^{N-1}x\left(k\frac{T}{N}\right)\,\mathrm{e}^{-j2\pi\frac{mk}{N}},\tag{8.13}$$

Observe that rearranging Eq. (8.13) it may be seen as a numerical approximation of the integral in (8.6)

$$c_m=\frac{1}{T}\sum_{k=0}^{N-1}x\left(k\frac{T}{N}\right)\,\mathrm{e}^{-j2\pi\frac{mk}{N}}\frac{T}{N},\tag{8.14}$$

where T/N represents the width of a short interval Δt. Interestingly, and, to some extent, surprisingly, this "approximation" is exact.

Since in (8.13) the temporal scale (given by T/N) is not relevant, we can omit it, preserving only the value of the successive samples.

We can call

$$X(m) = \sum_{k=0}^{N-1} x(k)\, \mathrm{e}^{-j2\pi\frac{mk}{N}}. \tag{8.15}$$

the *discrete Fourier transform* (DFT) of $x(0), \ldots, x(N-1)$.

The original signal is, thus, the *inverse discrete Fourier transform* (IDFT) of $X(0), \ldots, X(N-1)$, defined as

$$x(k) = \frac{1}{N} \sum_{m=0}^{N-1} X(m)\, \mathrm{e}^{j2\pi\frac{mk}{N}}. \tag{8.16}$$

Note that the Fourier coefficients c_n are not the same as the discrete Fourier transform. If we need those coefficients, it is necessary to divide by N

$$c_m = \frac{1}{N} X(m), \tag{8.17}$$

Some authors define the discrete Fourier transform including the factor $1/N$ (Randall et al. 1978). Although this seems to make more sense, since the so-defined transform directly provides the Fourier coefficients and it is easier to interpret, we will adhere here to the definition given in (8.15) since mathematical software such as Octave, Scilab, or MATLAB use definition (8.15) for the direct transform and (8.16) for its inverse.

8.5 Fast Fourier Transform (FFT)

The discrete Fourier transform turns out to be a very useful tool, not only for spectral analysis but for many other applications. However, especially for large N, it is computationally prohibitive. Indeed, the computation of each component of the DFT requires N products and N sums, hence the computation of all its components will demand N^2 products and N^2 sums. For an N-sample transform with $N = 4096$, this implies more than 33 million floating-point operations. An attentive scrutiny of (8.15) reveals that there are many repeated calculations that could be omitted. Indeed, while the exponent of $\mathrm{e}^{-j2\pi/N}$ varies between 1 and N^2, the exponential actually takes only N different values.

In 1965, Cooley and Tuckey found a way to reorganize the computation to take advantage of this circumstance when the number of samples to be transformed is a power of 2, i.e., $N = 2^n$ (Cooley et al. 1965). Their algorithm, called *fast Fourier transform, FFT*, requires only $^3/_2 N \log_2 N$ floating-point operations, which in the previous example implies a reduction in a factor of 455 the operation count. A brief description of this algorithm is given in Appendix N.

The introduction of the FFT has had a huge impact on the development of many digital signal processing and analysis techniques. It must be stressed, however, that they are actually applications of the DFT made possible thanks to the computational efficiency of the FFT. In what follows we will apply it to spectrum analysis.

In Scilab, the FFT is performed with the following code:

```
y = fft(x, -1);
```

where **x** is a previously loaded variable. The argument −1 indicates the sign of the exponent of the exponentials that appear in the definition of the DFT. The inverse transform is similarly computed, changing −1 by 1 (it includes the division by N)

```
x = fft(y, 1);
```

Both functions are computed more efficiently when the number of samples is a power of 2. If it is not, we can take a portion of the signal of length **N** in which **N** is a power of 2

```
y = fft(x(k:k + N-1), -1);
```

This code is an example where the samples are not taken starting from the beginning but from sample **k**.

It is also possible, and often very useful, to compute several transforms at the same time (actually, on a single function call), for instance when we have several signal blocks arranged either as the columns of a matrix **x**:

```
y = fft(x, -1, 1);
```

or else as its rows

```
y = fft(x, -1, 2);
```

This function has several options that the reader may be willing to investigate opening the help page (**help fft**).

8.6 Spectrum Analysis with the FFT

Before introducing specific techniques based on the FFT, we will discuss some of its limitations. First, when we decide to use the DFT (or FFT) for performing the spectrum analysis of a signal portion, we are implicitly assuming a periodic model, with period N, for the signal, since from Eq. (8.16) it is clear that $x(k)$ is periodic. However, the original signal could be non-periodic, as shown in Fig. 8.1, or it could be periodic with a period different from N. It could also happen that the original

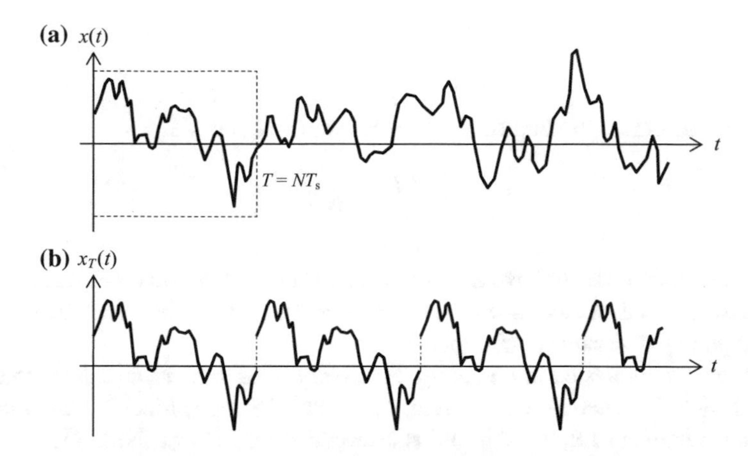

Fig. 8.1 **a** Temporal evolution of a signal. **b** Periodic extension of the highlighted portion in (**a**)

continuous-time signal is periodic but with a period that is not commensurable with the sampling period, so the discrete version would not be periodic.

Whatever the case, we are assuming a harmonic structure for $x(k)$ that is not consistent with the real one. This is an example of time-domain aliasing.

Another problem is created by any possible discontinuity arising from the very process of periodic extension. Periodic signals representable as a finite Fourier series such as (8.8) have a bounded spectrum, so they are continuous (as a matter of fact they have infinite continuous derivatives). Any discontinuity introduces frequencies above the Nyquist limit $F_s/2$, causing aliasing. This problem, called spectral leakage, shall be controlled by the use of windowing, as will be shown later on.

8.6.1 *Line Spectrum Analysis*

The spectral components presented by the FFT according to Eq. (8.16) can be interpreted in terms of the physical time. Indeed, we first multiply and divide the exponent by T_s and then, considering that $F_s = 1/\,T_s$, we have

$$x(k) = \frac{1}{N}\sum_{m=0}^{N-1} X(m)\, e^{j2\pi \frac{mk}{N}\frac{T_s}{T_s}} = \frac{1}{N}\sum_{m=0}^{N-1} X(m)\, e^{j2\pi m \frac{F_s}{N} k T_s}. \tag{8.18}$$

As we can see, this is equivalent to a series of harmonics of frequencies mF_s/N, where the fundamental frequency is F_s/N. This value coincides with the separation between the frequencies of the harmonics, Δf, or *spectral resolution*

$$\Delta f = \frac{F_s}{N}. \tag{8.19}$$

The period corresponding to the fundamental frequency is

$$T = \frac{N}{F_s} = NT_s \tag{8.20}$$

We see that in order to have a better resolution (i.e., to reduce Δf) it is necessary to increase the analysis window (i.e., to increase T). This is one more instance of the time-frequency uncertainty principle.

Now we are interested in obtaining the energy of each harmonic. Note that for a physical signal in which $x(k)$ is real, as shown in Appendix O, the following conjugate symmetry relationship holds between the values of the DFT:

$$X(N-n) = \overline{X(n)}. \tag{8.21}$$

This implies that the sum of the corresponding terms is

$$2|X(n)|\cos\left(2\pi mk/N + \arg(X(n))\right).$$

The amplitudes of the real spectrum are, hence

$$X_r(n) = \begin{cases} \frac{X(0)}{N} & n = 0 \\ \frac{2\,|X(n)|}{N} & n = 1, \ldots, N/2 - 1 \\ \frac{X(N/2)}{N} & n = N/2 \end{cases} \tag{8.22}$$

The first and the last components have RMS values equal to their amplitude (the first one because it is the DC component and the latter because it is a signal with only two opposite values). The $N/2 - 1$ remaining components are sinusoidal tones whose RMS values are $\sqrt{2}$ times smaller than their amplitudes.

It is important to note, once more, that these are not the actual components present in the original signal but those corresponding to the periodic model that has been adopted for the signal.

8.6.2 The Problem of Frequencies Close to $F_s/2$

When a signal has spectral content at frequencies very close to the Nyquist frequency $F_s/2$, the number of samples per cycle is quite small. This implies that if the sampling time is too short those high-frequency components may not be correctly represented. In Fig. 8.2 an example is shown. As can be observed, if only the shown samples were used to compute the RMS value, it would be grossly underrated.

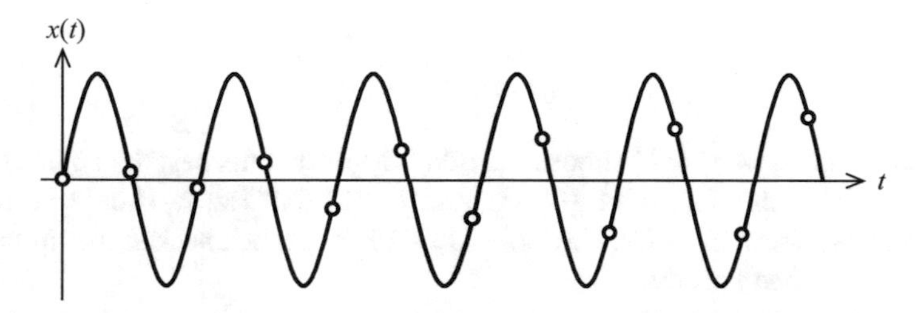

Fig. 8.2 Pure tone of a frequency slightly smaller than the Nyquist frequency $F_s/2$

However, the signal satisfies the condition of the sampling theorem so, in the long run, the signal could be perfectly reconstructed. We can pose a criterion that guarantees that the sampling process is not only theoretically correct but also practically. To that end we need a large enough number of samples so that some of them reach values close to the peak value. Moreover, to ensure that the effect of taking a non-integer number of half cycles is negligible, the peak should be reached several times (see Fig. 1.3, Chap. 1).

Observing the plot in Fig. 8.2 it is easy to see that if we keep taking samples we will eventually find the intended values close to the peak. This will happen at a discrete time k_o such that

$$\sin\left(2\pi f\frac{k_o}{F_s}\right) \cong 1 \tag{8.23}$$

Note, however, that it is not sufficient to ask that the sine argument is close to $\pi/2$, since even for $k_o = 1$ the argument is considerably larger than $\pi/2$ (it is approximately equal to π if $f \cong F_s/2$). The required approximation will hold for some integer values k_o and m such that

$$2\pi f\frac{k_o}{F_s} \cong 2\pi m + \frac{\pi}{2},$$

i.e.,

$$k_o\frac{f}{F_s/2} \cong 2m + \frac{1}{2}.$$

The idea is to split the first member into an integer part equal to 2 m and a fractional part close to 1/2. Since $f \cong F_s/2$ we can write

$$k_o - k_o\left(1 - \frac{f}{F_s/2}\right) \cong 2m + \frac{1}{2} = 2m + 1 - \frac{1}{2}.$$

It will suffice to adopt

$$\begin{cases} k_{\mathrm{o}} \cong \frac{1}{2}\,\frac{F_{\mathrm{s}}}{F_{\mathrm{s}}-2f} \\ 2m+1 = k_{\mathrm{o}} \end{cases} \tag{8.24}$$

In other words, we should choose the odd integer that fits best the approximation above. For instance, if F_{s} = 44 100 Hz and f = 22 049 Hz, k_{o} should be the odd integer that is closest to 11 025, i.e., k_{o} = 11 025. It can be checked that in this case the peak is reached exactly.

Except if we can somehow guarantee that we are covering exactly a one-fourth period, as needed to compute the RMS value exactly, it will be convenient to take

$$k \gg k_{\mathrm{o}}. \tag{8.25}$$

In practice, there is no such guarantee, since in the first place we do not know the tone frequency.

An alternative way to solve this problem is to prevent that the signal contains frequencies too close to $F_{\mathrm{s}}/2$. From (8.24), if we want that

$$k < k_{\text{máx}} \tag{8.26}$$

we can adopt

$$k_{\mathrm{o}} = k_{\text{máx}}/10$$

so the frequency must comply with

$$f < \frac{F_{\mathrm{s}}}{2}\left(1 - \frac{1}{2k_{\mathrm{o}}}\right) = \frac{F_{\mathrm{s}}}{2}\left(1 - \frac{5}{k_{\text{máx}}}\right) \tag{8.27}$$

For instance, if we intend that the analysis frame is at most 4096 samples long, the frequency should be less than 22 023 Hz. As we can see, it is sufficiently close to $F_{\mathrm{s}}/2$ to be almost completely removed by any anti-aliasing filter.

On the other hand, in most practical situations the high-frequency content of typical noises is attenuated by several dissipative phenomena, so this will very seldom constitute a serious limitation.

8.6.3 The Problem of Subharmonic Frequencies

A similar problem to the already analyzed one appears at the low frequency end, particularly at subharmonic frequencies, i.e., those that are below the fundamental frequency corresponding to the analysis window. Indeed, if the window is too short, we may be sampling portions of the signal that are not representative of the true contribution to the long-term spectrum.

The same kind of criterion of the last section can be applied here. It is assumed that the minimum frequency for which a reasonable uncertainty can be guaranteed is $10F_s/N$, which corresponds to ten cycles in the analysis window. For a sample rate of 44 100 Hz and an analysis window of 4096 samples, this limit turns to be 107 Hz, too high a frequency to simply ignore the problem. However, if an error of 0.5 dB is admissible, Eq. (1.6) shows that two cycles are enough, so the minimum frequency can be reduced to $2F_s/N$, i.e., 22 Hz.

8.6.4 Precise Detection of Pure Tones

When a signal contains pure tones, it is often of interest to be able to find the corresponding frequencies and amplitudes. If a given tone did coincide with some spectral line of the discrete Fourier transform, i.e., if its frequency were a multiple of F_s/N, then except for any other spectral content, the FFT spectrum would contain a single and neat line associated with that frequency.

If, instead, the frequency were different from all spectral lines, we would have multiple associated lines, a phenomenon known as *spectral leakage*. It is possible, however, to estimate with better precision the corresponding frequency and amplitude. The method consists in selecting the three highest lines associated with a spectral peak and applying a parabolic interpolation, as shown in Fig. 8.3.

Spectral peaks are identified by the presence of a clear-cut relative maximum, in this case, Y_2. This is not to be confused with the maximum of the approximation, Y_{max}, which is a priori unknown.

We pose the following system of equations whose parameters are the frequencies f_k and the absolute values Y_k of the FFT, and whose unknowns are the parameters of the parabola. We have chosen an analytic form of the parabola for which f_o is a parameter (in fact, the only one we are interested in)

$$\begin{cases} Y_{max} + K(f_1 - f_o)^2 = Y_1 \\ Y_{max} + K(f_2 - f_o)^2 = Y_2 \\ Y_{max} + K(f_3 - f_o)^2 = Y_3 \end{cases} \tag{8.28}$$

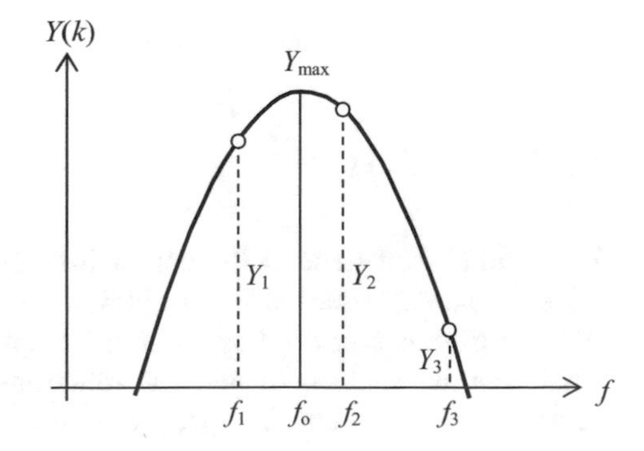

Fig. 8.3 Parabolic approximation of the distribution of spectral lines corresponding to a pure tone

Subtracting in pairs we cancel Y_{max}, and dividing the resulting equations we cancel K

$$\frac{K\left((f_2-f_o)^2-(f_1-f_o)^2\right)}{K\left((f_3-f_o)^2-(f_1-f_o)^2\right)}=\frac{Y_2-Y_1}{Y_3-Y_1}$$

Although this seems to be a second degree equation in f_o, actually thc squares cancel each other out

$$\frac{f_2^2-2f_2f_o+f_o^2-(f_1^2-2f_1f_o+f_o^2)}{f_3^2-2f_3f_o+f_o^2-(f_1^2-2f_1f_o+f_o^2)}=\frac{Y_2-Y_1}{Y_3-Y_1}$$

$$\frac{f_2^2-2f_2f_o-f_1^2+2f_1f_o}{f_3^2-2f_3f_o-f_1^2+2f_1f_o}=\frac{Y_2-Y_1}{Y_3-Y_1}$$

$$\frac{f_2^2-f_1^2-2(f_2-f_1)f_o}{f_3^2-f_1^2-2(f_3-f_1)f_o}=\frac{Y_2-Y_1}{Y_3-Y_1}$$

Clearing denominators

$$\begin{aligned}&(f_2^2-f_1^2)(Y_3-Y_1)-(f_3^2-f_1^2)(Y_2-Y_1)\\&\quad=2((f_2-f_1)(Y_3-Y_1)-(f_3-f_1)(Y_2-Y_1))\,f_o\end{aligned}$$

Finally

$$f_o=\frac{(f_2^2-f_1^2)(Y_3-Y_1)-(f_3^2-f_1^2)(Y_2-Y_1)}{2((f_2-f_1)(Y_3-Y_1)-(f_3-f_1)(Y_2-Y_1))} \tag{8.29}$$

As regards the amplitude, it can be estimated by the superposition of the energy of the three highest lines

$$Y_o\cong\sqrt{Y_1^2+Y_2^2+Y_3^2} \tag{8.30}$$

Note that the amplitude is not Y_{max}.

8.6.5 Windows

A very important issue in the use of the FFT is that because of border discontinuities (different values at the beginning and the end of the analysis interval), the FFT spectrum is corrupted by an undue emphasis at high frequency. Indeed, in the classic Fourier analysis of periodic functions, it is shown that the Fourier coefficients (which represent the spectral content at the successive harmonics) tend to zero faster as more derivatives of the function are continuous. If the function itself presents discontinuities, such as in the case of a square wave, the coefficients will tend to 0 as $1/n$ for $n \to \infty$. If the function is continuous but the derivative is not, as in a triangle wave, they tend to 0 as $1/n^2$. In general, if the $(m-1)$th derivative is continuous but the m-th is not, the coefficients tend to 0 as $1/n^{m+1}$. When we select a portion of signal and extend it periodically, we may be creating a discontinuity that will alter the high-frequency range of the spectrum (see Fig. 8.1)

The solution is to multiply the signal by one of several possible *window functions*, $w(n)$, in the time domain

$$y(n) = w(n)\,x(n). \tag{8.31}$$

These are functions that are 0 or very close to 0 at both ends of the analysis interval, and reach a maximum of 1 at the center. For example, the Hann window[5] is defined as

$$w_{\text{hann}}(n) = \frac{1}{2}\left(1 - \cos\frac{2\pi n}{N-1}\right); \tag{8.32}$$

the Blackman window, as

$$w_{\text{blackman}}(n) = 0.42 - 0.5\cos\frac{2\pi n}{N-1} + 0.08\cos\frac{4\pi n}{N-1}; \tag{8.33}$$

and the Blackman–Harris window, as

$$w_{\text{blackman-harris}}(n) = 1 - 1.36\cos\frac{2\pi n}{N-1} + 0.39\cos\frac{4\pi n}{N-1} - 0.032\cos\frac{6\pi n}{N-1}. \tag{8.34}$$

Unlike the preceding ones, the Hamming window, defined as

$$w_{\text{hamming}}(n) = 0.54 - 0.46\cos\frac{2\pi n}{N-1}, \tag{8.35}$$

is the only one (except for the square or *boxcar* window, which is the same as using no window) that is nonzero at the ends.

Figure 8.4 plots the temporal waveform presented by each one of these windows.

The use of these windows has two consequences. The first one is that the signal energy is altered, since the windows are 1 only at the center of the interval.

[5]Sometimes it is incorrectly called *Hanning window* when his author is really called Julius von Hann. This is most likely due to the fact that there is another window, the Hamming window, named after its author, Richard Wesley Hamming. The spelling is similar and can easily lead to confusion, so it is important to avoid the incorrect name.

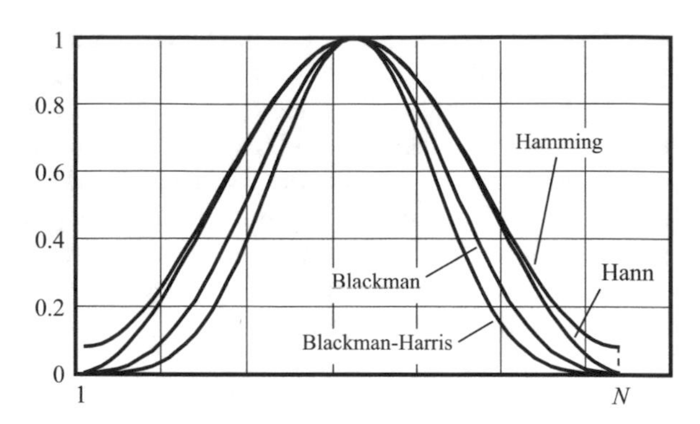

Fig. 8.4 Temporal waveform of several window functions to be used in FFT spectrum analysis

A correction will therefore be necessary, and it may depend on the type of signal. The second consequence is that each sinusoidal component of the original signal will be replaced by several components, so the energy of a harmonic will be spread over several neighboring frequencies. This imposes a limit on the capability to separate frequencies very close to each other, so the analysis resolution gets worse with respect to the theoretical value F_s/N. However, except in very particular cases, we are mostly interested in continuous spectra where pure tones that need to be separated are not very close from each other. It is the case of most environmental noises.

The Hann window presents a good frequency discrimination at the expense of considerable high-frequency noise corruption of the signal. The Blackman window, offers an excellent tradeoff between resolution loss and residual noise at high frequency. The Blackman–Harris window, has somewhat worse frequency resolution but it has extremely low high-frequency residual noise. Finally, the Hamming window, has a good spectral resolution and, at the same time, it spreads the high-frequency noise in a rather uniform fashion.

The Hann window can be implemented in Scilab by means of the function **window**, which allows several window types such as rectangular, triangular, Hann, Hamming and Kaiser. The FFT including the pointwise product by the window is

```
y = fft(x(k:k + N-1) .* window('hn',N), -1);
```

The Blackman window, on the contrary, is not implemented in Scilab 5.5.2 (the current version at the time of this writing), so it is necessary to compute it directly or to add it to the function **window**. In the first case we compute a vector **black** with the Window values computed according to (8.33)

```
black = 0.42 + 0.5*cos(2*%pi*[0:N]/(N-1))    ...
      + 0.08* cos(4*%pi*[0:N]/(N-1));
y = fft(x(k:k + N-1) .* black, -1);
```

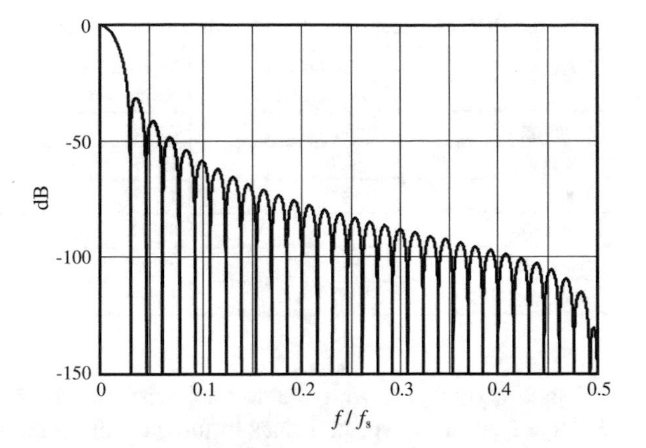

Fig. 8.5 Unilateral spectrum of the Hann window for $N = 64$, where the scale of the frequency axis is relative to the sample rate

The spectrum of the windows (which is analyzed in more detail in Appendix P) has great importance, since the application of the window implies that the signal spectrum is convolved with the window spectrum. The window that would produce the least effect on the spectrum is one whose spectrum was a Dirac delta, since it is the only function that convolved with another one does not change it. Hence, the more the window spectrum resembles the Dirac delta, the less will be its negative effects on the signal spectrum.

Therefore, it is desirable to have a very slim low frequency lobe (main lobe) and much smaller lateral lobes. As an example, Fig. 8.5 shows the unilateral spectrum of the Hann window, with a relatively slim main lobe and a first lateral lobe −33 dB below the main lobe. From then on, there is further reduction of the amplitudes of the subsequent lateral lobes towards the high-frequency end.

The main lobe bandwidth determines the minimum separation between two spectral lines (or two pure tones) in order that they can be discriminated by the analyzer In the case of a rectangular window the main lobe has a width equal to

$$\Delta f = \frac{2}{T} = \frac{2F_s}{N}. \tag{8.36}$$

where T is the window duration. In the case of the Hann window this width is doubled and in the case of the Blackman window it is tripled.

In the choice of a window there is a compromise between three factors: (1) The discrimination power, given by the width of the main lobe; (2) the degree of spectral leakage, given by the difference in decibels between the main lobe and the first lateral lobe; and (3) the degree of time localization, given by the analysis window duration.

For instance, the Blackman window has a good performance as regards spectral leakage, but in order to achieve a good frequency discrimination it requires, according to (8.36), to increase N, losing time discrimination (time-frequency uncertainty principle). Table 8.1 gives the amplitude of the first lateral lobe and the main lobe bandwidth.

Table 8.1 Relevant properties of some windows used for spectral analysis

Window	Lateral lobe peak [dB]	Main lobe bandwidth [Hz]	Maximum error for white noise [dB]
Rectangular	−13	$2\ f_s/N$	–
Triangular	−27	$4\ f_s/N$	2.5
Hann	−32	$4\ f_s/N$	3
Hamming	−43	$4\ f_s/N$	3.4
Blackman	−58	$6\ f_s/N$	2
Blackman–Harris	−92	$8\ f_s/N$	1.5

Another problem with the use of windows is the accumulation of spectral leakage from many spectral lines in the case of spectrally complex signals such as white noise. This happens because each spectral line adds some residual noise due to spectral leakage and the superposition of all of them in fact increases the total noise, causing an error. The maximum error is also given in Table 8.1.

Another potential risk affects non-stationary signals or signals with rapid transients with a short duration relative to the window duration. Indeed, a very short pulse will be affected by quite a different weighting if it is located at the center of the window or close to one of its ends. This problem can be dealt with by overlapping techniques. The Hann window, for instance, allows a 50 % overlapping with a perfect reconstruction, since a series of overlapped windows yield a sum identically equal to 1.

8.6.6 FFT Band Spectrum Analysis

The second type of analysis is the constant percentage band spectrum, i.e., one where each band is wider than the preceding one by a specified percentage. In the case of octave band filters, for instance, each band is 100 % larger than the preceding one. This type of analysis can be performed by an energy superposition of the spectral lines of an FFT transform that falls in each band of interest. Since the size of the bands grows exponentially with the band index, the lowest frequency band contains very few spectral lines and the highest frequency one, many lines. According to the size N of the analysis window, it might happen that in the lowest band there were no spectral line. Table 8.2 gives the frequency limits, the bandwidth and the number of FFT spectral lines for several values of N corresponding to each octave band, when $F_s = 44\ 100$ Hz. As can be seen, it is convenient to adopt $N = 8192$ in order to guarantee that the lowest band contains at least four lines. The larger the number of spectral lines in a band, the more precise will be the result, although, because of the uncertainty principle, also the lesser will be the time localization.

In the case of one-third octave bands the window length must be even larger since the first bands are very narrow. For instance, the one-third octave band

Table 8.2 Frequency limits, bandwidth and number of spectral lines of an FFT according to the size N of the analysis window (F_s = 44 100 Hz) for octave bands

f_o [Hz]	f_l [Hz]	f_u [Hz]	B [Hz]	N = 4096	N = 8192	N = 16 384	N = 32 768
31.5	22.4	4.50	22.6	2	4	8	17
63	45.0	90.0	45.0	4	8	17	33
125	90.0	180	90.0	8	17	33	67
250	180	355	175	16	33	65	130
500	355	710	355	33	66	132	264
1000	710	1400	690	64	128	256	513
2000	1400	2800	1400	130	260	520	1040
4000	2800	5600	2800	260	520	1040	2081
8000	5600	11 200	5600	520	1040	2081	4161
16 000	11 200	22 400	11 200	1040	2081	4161	8322

centered on 20 Hz has a bandwidth of 4.63 Hz, while the *octave* band centered on 31.5 Hz has a bandwidth of 22.3 Hz. This implies that the separation between consecutive lines F_s/N must be very small, thus requiring a very long FFT window.

The number m_1 of lines in the first band of fraction b of an octave, f_{o1} can be computed as

$$m_1 = N\frac{f_{o1}}{F_s}\left(2^{\frac{1}{2b}} - 2^{-\frac{1}{2b}}\right). \tag{8.37}$$

In order to represent the sound energy contained in that band it is necessary, in principle, to have at least one spectral line in that band. However, to have an acceptable precision in spite of the spectral leakage for a Hann window, which corresponds to four spectral lines, we need that $m_1 \geq 4$, which is attained if

$$N \geq 4\frac{F_s}{f_{o1}}\frac{1}{2^{\frac{1}{2b}} - 2^{-\frac{1}{2b}}} \tag{8.38}$$

For octave bands (b = 1) it results $N \geq$ 7983, and for one-third octave bands (b = 3), $N \geq$ 38 696, from where, adopting the closest powers of 2 we should take, respectively, N = 8192 and N = 65 536 (in both cases, for F_s = 44 100 Hz).

The effective or RMS value of the hth band can be computed as

$$X_{\text{oct},\,h} = \sqrt{\sum_{k\,=\,[2^{-1/(2b)}Nf_{oh}/F_s]\,+1}^{[2^{1/(2b)}Nf_{oh}/F_s]} 2\left|\frac{X(k)}{N}\right|^2} \tag{8.39}$$

where $X(k)$ is the FFT of $x(n)$, f_{oh} is the center frequency of band h and [ ] represents the integer part. If a time window $w(n)$ has been previously applied, its attenuation may be corrected dividing by

$$C_w = \sqrt{\frac{1}{N}\sum_{n=1}^{N} w^2(n)}. \tag{8.40}$$

since this will be the factor by which the frequency components will be affected. This correction assumes that the signal is stationary in a temporal scale corresponding to the analysis window. If it is not, it is necessary to take overlapping windows to prevent cancelation of short events close to the ends of the window. The most favorable case is when the overlapping windows sum is identically equal to 1, as in the case of the Hann window with a 50 % overlapping. In this case no correction is needed.

The algorithm described in what follows assumes that the window size is 8192. The first step is to apply the Blackman window

```
N = 8192;
black = 0.42 + 0.5*cos(2*%pi*[0:N]/(N-1))   ...
      + 0.08* cos(4*%pi*[0:N]/(N-1));
xb = x .* black;
```

Note that we are using here the component-wise product (i.e., **.***) between the values of vectors **x** and **black**. Next the module of the line spectrum is computed, considering the correction factor mentioned before in (8.22), equal to **1/N** for the first line and **2/N** for the rest

```
e = 2/N * abs(fft(xb, N));
e(1) = e(1)/2;
```

This spectrum contains 8192 lines, from which only the first 4096 are physically significant.

The code that follows computes, for each band, the minimum and maximum indices of the lines contained in that band. To that end, starting from the 1000 Hz band we compute the exact center frequencies multiplying and dividing by 2 repeatedly until the lower and upper audio limits. Although they are not exactly equal to the standard frequencies for the bands of 31.5 Hz and 63 Hz, the difference is very small and the code is greatly simplified. Dividing and multiplying by $\sqrt{2}$ we can compute the limits of each band.

Finally, since the spectral lines are uniformly distributed at a constant distance of **Fs/N** from each other, dividing by this value we get the approximate limit indices. These values are converted to integers by mans of functions **ceil** (minimum greater integer) and **floor** (maximum smaller integer)

```
for h = 1:10
  imin(h) = ceil(1000 * 2^(h - 6.5) * N / Fs);
  imax(h) = floor(1000 * 2^(h - 5.5) * N / Fs);
```

```
  if imax(h) > N/2 - 1
    imax(h) = N/2 - 1;
  end
end
```

The `if` clause prevents the existence of lines beyond the Nyquist frequency, which could be the case in the upper band, since by definition it ranges up to approximately 22 400 Hz but it must be limited to 22 050 Hz.

Next the energy content of each band is computed summing the square of the spectral lines of the band (energy superposition) and taking the square root

```
for h = 1:10
  y(h) = sqrt(sum(e(imin(h):imax(h)) .^2));
end
```

Although for the sake of clarity this step has been isolated from the previous loop, it could be included inside it.

Finally, it is necessary to correct the values previously computed for the energy reduction because of the Blackman window. To this end, we compute a factor **Q** equal to the inverse of the RMS value of the window

```
Q = 1 / sqrt(mean(black.^2));
Yef = y * Q / sqrt(2);
```

The division by $\sqrt{2}$ is just to get the RMS value of the bands, to which the calibration Eq. (7.7) can be applied.

8.6.7 Spectral Density and Spectrum Averaging

There are many cases in which it is interesting to measure the spectral density, for instance for the characterization of a relatively stationary wideband noise, for the application of dynamic models based on transfer functions, for source identification and even for failure prediction. One problem with the discrete Fourier transform is that since it is a short time transform it cannot correctly represent a long-term average spectrum.

We can think of taking an FFT with a very large size, but here there are two issues. First, the number of operations and particularly the large amount of required memory makes it impractical to compute an FFT of tens of millions samples. Second, while the spectral resolution increases in proportion to the number of samples, no spectral line happens to stabilize, since each one is only computed once.

This problem can be solved if we take many FFT spectra corresponding to successive frames and average the homologous lines (i.e., those corresponding to the same frequency). In this case, we will indeed be computing many times a quantity with the same expected value, so the mean value tends to stabilize, as intended (see Sect. 1.7.4).

On the other hand, from the point of view of computational efficiency, computing several short FFT's requires less operations than a single one covering the same total time. Indeed, if we call N the size of each small FFT and M the number of analysis windows, the number of operations for a single N-sample FFT is $KN \log_2 N$, where K is a constant, so the number of operations to compute M FFTs will be $MKN \log_2 N$. Computing a giant MN-sample FFT requires, instead, $KMN \log_2(MN)$ operations,

$$MKN \log_2(MN) = MKN \log_2 N + MKN \log_2 M \tag{8.41}$$

As we can see, in this case we need to perform an excess of $MKN \log_2 M$ operations. For instance, analyzing one minute of audio signal at a sample rate of 44 100 Hz, i.e., a total signal size of 2 646 000 samples, using a single FFT would require 56 million operations. Using, instead, FFTs of $N = 4096$ samples, would require $M = 646$ frames and the total number of operations will be, hence, 32 million. We are saving almost 50 % of computational load. The saving improves as the total signal length increases.

8.7 FFT Filters

IIR filters allow transfer functions with a low order (number of delay taps), making them extremely efficient and particularly well-suited for real-time processing. However, they can become easily unstable because of the accumulation of truncating errors. Indeed, because of their recursive nature, the value of each output sample appears in all subsequent samples. Besides, it is impossible to get a linear-phase response, a desirable feature in order to improve the transient response.

FIR filters, on the other hand, are inherently stable, as each sample affects only a fixed number of future samples, and allow linear-phase designs (Proakis et al. 1996). They present the problem that they require a rather high order to achieve a given frequency response. This is because selective filters exhibit long impulse responses, which in turn requires a large number of terms. The computational time is, thus, considerable.

An interesting possibility is to carry out the filtering process in the frequency domain, which reduces to a simple multiplication of the Fourier transform $X(k)$ of the signal by the filtering window

$$Y(k) = X(k)\, W(k), \tag{8.42}$$

where $W(k)$ represents the frequency response for frequency kF_s/N. The time-domain filtered signal is retrieved by means of the inverse Fourier transform.

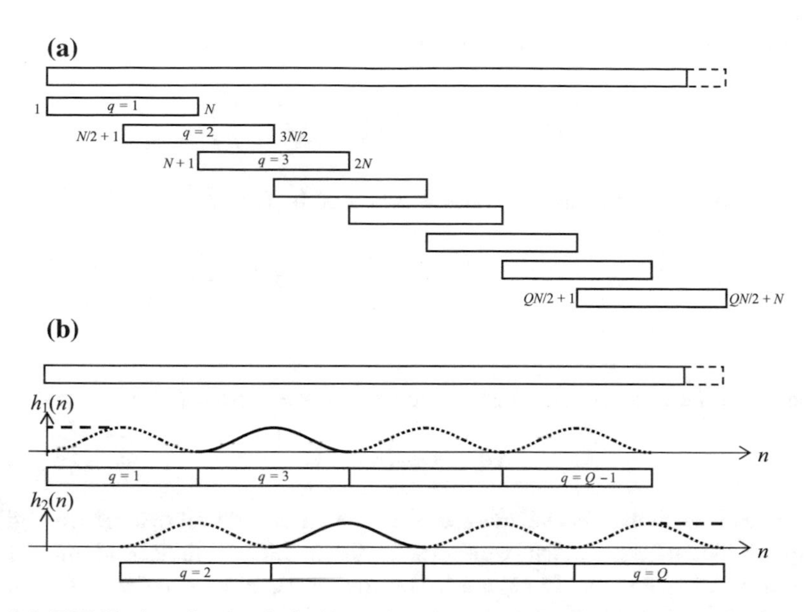

Fig. 8.6 FFT filtering of a signal of arbitrary length. **a** Decomposition in frames of length N and overlapping $N/2$. **b** Hann windows for odd and even frames. Two individual frames have been highlighted. The *dashed horizontal lines* show the modifications in both initial and final frames

This method, implemented by means of the FFT and IFFT allows a dramatic reduction of the computing time. Indeed, computing N samples using a FIR filter of order N would require N^2 sums and products, while with the procedure just described the number of operations is of an order of $N \log_2 N + N$. For $N = 4096$ this means a reduction by a factor of 315.

As it turns out, the FFT filtering introduced above is limited to a single analysis window of length N. Even if it is possible to increase N to fit the available signal length, on the one side it is progressively less efficient and on the other side it results in an unnecessary increase in spectral resolution. In order to extend this method to arbitrarily long signals, the signal is split into N-sample frames with an overlapping of 50 % (i.e., $N/2$). The FFT is computed for each frame, it is multiplied by the intended filtering window and finally the IFFT is computed. So far we have the filtered signal for each frame. The overlapped regions contain signal portions that are very similar to each other except at the ends, because of transient behavior caused by the Gibbs phenomenon.[6] These differences can be greatly reduced by means of a weighted sum of both overlapped parts so that close to the borders the weight tends to 0 and in the central part it tends to 1. The Hann window is ideal since in the overlapped region the sum of both windows is identically equal to 1. Figure 8.6 depicts this method's structure.

[6]The Gibbs phenomenon consists in the presence of relatively large oscillations in a band limited Fourier approximation (such as a Fourier series that has been truncated to a finite number of terms) in the neighborhood of a discontinuity (such as the flanks of a square wave).

If the window corresponding to the first frame is, for $1 < n < N$,

$$h_1(n) = 0.5\left(1 - \cos\left(\frac{2\pi(n-1)}{N}\right)\right), \tag{8.43}$$

and the window corresponding to the second one is, for $N/2 + 1 < n < 3\,N/2$,

$$h_2(n) = 0.5\left(1 - \cos\left(\frac{2\pi(n - N/2 - 1)}{N}\right)\right) = 0.5\left(1 + \cos\left(\frac{2\pi(n-1)}{N}\right)\right), \tag{8.44}$$

it can be verified immediately that in the overlapped region, $N/2 + 1 < n < N$,

$$h_1(n) + h_2(n) = 1. \tag{8.45}$$

It is easily seen that this behavior can be generalized for all overlapped regions, except for the first and last half windows. To fix this, for the initial and final frames the windows are modified to have a flat weighting equal to 1 in the half windows where there is no overlapping.

This type of method is known as *overlap-add* and was introduced by Crochiere (Crochiere 1980). As regards the process described above, in order to have an integer number of frames we must previously extend the original signal with an adequate number of zeros (a process known as *zero-padding*).

FFT filters constitute an excellent solution to the problem of filtering with an arbitrary frequency response leaving practically no side effects. They have even zero phase, i.e., a special case of linear-phase response with frequency. When a filter has linear phase, all spectral components that go through the filter have the same delay. Indeed, the time delay for a spectral component of frequency f and period T caused by a phase response $\varphi(f)$ is

$$\Delta T(f) = \frac{\varphi(f)}{2\pi}T = \frac{Kf}{2\pi}\frac{1}{f} = \frac{K}{2\pi} \tag{8.46}$$

In the case of the FFT filter, the constant K is zero. However, this implies that the filter is not causal, since for a signal starting at some intermediate instant of the first window the response may start before the signal, i.e., there is a pre-transient, though the effect is in general not significant. This phenomenon is sometimes called *pre-echo*, and it owes to the very nature of the DFT, since it is representing a signal over a window with components (complex harmonics) whose amplitudes are constant throughout the window. A signal starting with some null samples will be correctly represented by N components whose amplitudes are the solution of a system of equations (see Eq. 8.10), but if because of the filtering action some of those components are removed or altered, there is no guarantee that the sum of the remaining terms will still be zero at the beginning of the frame.

As an extreme case that illustrates the problem, consider a unit impulse located at the middle of the window. The absolute value of the discrete transform of such a pulse is constant with frequency. If we apply a filter that removes all frequency components but a single one, there will remain only two conjugate lines whose sum is a sine wave with constant amplitude throughout the whole window. This phenomenon is similar to the spectral leakage but in the time domain. It is thus, a sort of *temporal leakage*.[7]

However, noncausality extends at most for a period equal to a window length, so if the window is not too long, for most metrological purposes it has no relevant consequences.

Another drawback of the FFT filtering is the presence of a slight amplitude modulation of frequency $2/T$, where T is the duration of the analysis window. This modulation can be reduced by increasing the overlapping to 75 %, at the expense of longer computation time. In this case it is also possible to use a Blackman window, since it can be split into two Hann windows of different amplitudes and frequencies

$$w_{\text{blackman}}(n) = 0.5\left(1 - \cos\frac{2\pi n}{N-1}\right) + 0.08\left(1 - \cos\frac{4\pi n}{N-1}\right) \tag{8.47}$$

The FFT filters can be used to implement filters with a practically ideal frequency response, such as lowpass, highpass, bandpass, and bandstop. They might be used to remove a parasitic tone that is contaminating the signal with minimal effect on the spectrum. In this case it is advisable to use a high resolution (large N) and select a window that allows a few lines around the frequency to remove.

When designing a filtering window for an FFT filter, it is necessary to take into account the symmetry conditions of the FFT of a real signal given by (8.21), to prevent nonphysical imaginary components. Equation (8.21) indicates what components must appear with the same amplitude in the filtering window. However, since in Scilab and similar mathematical environments the indices start at 1 instead of 0, the FFT is implemented as

$$X(n) = \sum_{k=1}^{N} x(k)\, \mathrm{e}^{-j2\pi\frac{(n-1)(k-1)}{N}}, \tag{8.48}$$

so the symmetry relationship is, for $n = 1, \ldots, N$,

$$X(N-n+2) = \overline{X(n)}. \tag{8.49}$$

Therefore, the filtering window must satisfy

$$W(N-n+2) = \overline{W(n)}. \tag{8.50}$$

[7]There is distinct duality between spectral and temporal phenomena due to the similarity of the direct and inverse Fourier transforms (they differ essentially in the exponent signs).

Note that while in most cases the window will be real, a window with a nonzero imaginary part allows to introduce also a phase response.

Normally the window is designed for $n = 1, \ldots, N/2$ and then (8.50) is applied. The following piece of code implements this from a window **w** defined as a column vector for frequencies smaller than $F_s/2$:

```
w = [w; zeros(N/2-length(w), 1)];
w = [w; 0; conj(w($:-1:2))];
```

The FFT computation may take advantage of the possibility that offers the function **fft** of computing several FFT's using a single code line. Suppose we have stored in an $N \times m$ matrix (where N is the size of the window) the overlapped frames of a signal as column vectors. Then

```
XX = fft(xx, -1, 1);
```

computes m transforms, one for each column. The last argument indicates that the transforms are computed along dimension 1, i.e., the rows. This means that the m signals to be transformed are the m columns of the matrix, each one of which contains N rows.

FFT filters can be used to implement octave band and one-third octave band filters. In that case, the window must be designed for frequencies smaller than $F_s/2$ according to the following definition:

$$W(n) = \begin{cases} 0 & \text{if } 1 \le n < \frac{f_o}{F_s} 2^{-1/2b} \\ 1 & \text{if } \frac{f_o}{F_s} 2^{-1/2b} \le n < \frac{f_o}{F_s} 2^{1/2b} \\ 0 & \text{if } \frac{f_o}{F_s} 2^{1/2b} \le n \le N/2 \end{cases}. \tag{8.51}$$

where $b = 1$ for octave bands and $b = 3$ for one-third octave bands. The following piece of code implements this design:

```
finf = fo / 2^(1/b/2);
fsup = fo * 2^(1/b/2);
w = zeros(1, floor(finf/Fs*N));
w = [w, ones(1, floor(fsup/Fs*N) - floor(finf/Fs*N))];
w = [w, zeros(1, N/2 - floor(fsup/Fs*N))];
```

In http://www.fceia.unr.edu.ar/acustica/measurements there is a complete implementation of the FFT filter and its application to band filtering.

It is interesting to note that if one needs to split the signal into a complete set of bands, the algorithm can be improved as regards computational efficiency if instead of computing the same FFT for each band it is computed only once and after applying the filtering windows, an IFFT is computed for each band. This reduces the computational time by a factor of almost 2.

It is also important to bear in mind that unlike IIR filters, where in order to compute the RMS value it is necessary to have the whole temporal response of the filter, in the case of using FFT techniques this is unnecessary. Indeed, once the spectral representation has been obtained, the energy in each band can be computed as the sum of the energies of all spectral lines in the band, as shown in Eq. (8.39). Conceptually, we are applying Parseval's identity, according to which the energy may be obtained either in the time domain or in the spectral domain.

For this reason, the use of an FFT filter makes sense only in the cases where it is interesting to have the filtered signal.

8.8 Some Applications of the FFT

8.8.1 Determination of Tonality

In some cases, such as in urban noise assessment or in some psychoacoustic studies, it is necessary to determine the degree of tonality of a signal. This attribute refers to the presence of spectral components that cause an average listener to perceive a tonal or musical quality in the signal. The presence of perceivable pure tones usually worsens the negative effects of noise. Several criteria on annoyance as well as on hearing impairment penalize tonal noises adding some constant to the evaluation level.[8]

The International Standard ISO 1996-2 provides a procedure to assess the audibility of tones. It is based on psychoacoustic criteria on masking.

Suppose we have a digital recording at a sample rate of 44 100 Hz of a noise that contains a 500 Hz tone. We want to find out if it will be audible using the ISO 1996-2 method.

At 500 Hz, the critical bandwidth is $B_{\mathrm{crit}} = 100$ Hz (see ISO 1996-2 and Sect. 8.8.4). The effective bandwidth of the narrow band analysis method must be at most a 5 % of the critical bandwidth, i.e.,

$$B_{\mathrm{ef}} < 0.05 B_{\mathrm{crit}} = 5\,\mathrm{Hz}. \tag{8.52}$$

If we carry out the analysis by means of an N-sample FFT with the Hann window, the effective bandwidth of the method is (Appendix P)

$$B_{\mathrm{ef}} = 1.5\,\frac{F_{\mathrm{s}}}{N}, \tag{8.53}$$

[8]In noise assessment, the *evaluation level* is a parameter computed from an objective measurement adding corrections that take into account a variety of contextual aspects or the type of noise or source.

from where

$$N = \frac{1.5\ F_s}{B_{ef}} > \frac{1.5 \times 44100\ \text{Hz}}{5\ \text{Hz}} = 13\ 230.$$

We choose the next power of 2, i.e., $N = 16\ 384$. The effective bandwidth is, thus, 4.04 Hz, which complies with the requisite. In the critical band of 500 Hz, i.e., the interval [450 Hz; 550 Hz], there is a number M of spectral lines given by

$$M = \frac{B_{crit}}{F_s/N} \cong 37.$$

Suppose that the tone has a sound pressure level L_{pt} = 65 dB, defined as the energy superposition of all neighboring lines that reach at least a level 6 dB below the peak that characterizes the tone

$$L_{pt} = 10\ \log\left(\sum_{L_{p,k} > L_{peak} - 6\ \text{dB}} 10^{L_{p,k}/10}\right) \tag{8.54}$$

Suppose also that the masking noise has a mean level within the critical band $L_{pn,mean}$ = 44 dB, defined as

$$L_{pn,mean} = 10\ \log\left(\frac{1}{H} \sum_{L_{p,k} < L_{peak} - 6\ \text{dB}} 10^{L_{p,k}/10}\right) \tag{8.55}$$

where H is the number of spectral lines within the critical band that satisfy the condition $L_{p,k} < L_{peak} - 6$ dB. Figure 8.7 shows an example.

According to the Standard, the total level of the masking noise is obtained with the following expression:

$$L_{pn} = L_{pn,mean} + 10\ \log \frac{B_{crit}}{B_{ef}} = 44\ \text{dB} + 14\ \text{dB} = 58\ \text{dB} \tag{8.56}$$

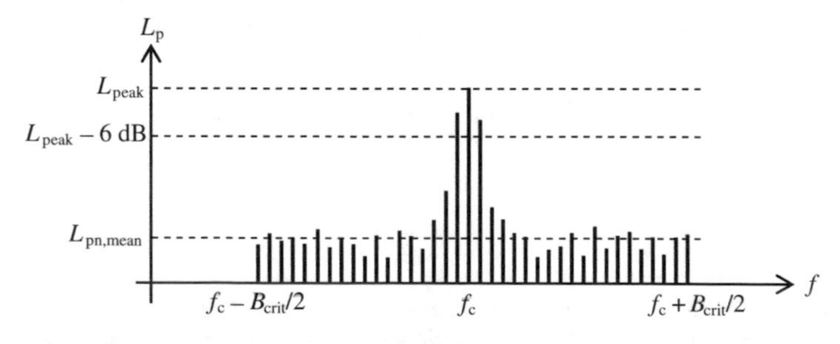

Fig. 8.7 FFT spectrum of a noise with a tonal component in the critical band centered at f_c, represented by several spectral lines

The fact that the correction is given by 10 log $(B_{\text{crit}}/B_{\text{ef}})$ instead of 10 log$(B_{\text{crit}}/(F_s/N))$, where F_s/N is the spectral line separation is because a wideband noise is overestimated by the FFT, because of spectral leakage (see Appendix P).

The difference between the levels of the tone and the masking noise is

$$Y = L_{\text{pt}} - L_{\text{pn}} = 65\text{dB} - 58\text{dB} = 7\text{dB}$$

The *tonal audibility* is defined as

$$\Delta L_{\text{ta}} = L_{\text{pt}} - L_{\text{pn}} - MT$$

where *MT* is the *masking threshold*, a psychoacoustic parameter that may be computed by means of the following simplified model:

$$MT = \log \frac{1}{1 + (f_c/502\text{ Hz})^{2,5}} - 2\text{ dB} \tag{8.57}$$

It results

$$\Delta L_{\text{ta}} = 65\text{ dB} - 58\text{ dB} - (-2.3\text{ dB}) = 9.3\text{ dB}$$

This means that the tone is 9.3 dB above the masking threshold.

The tonality adjustment K_t is the value to be added to the measured (or computed) A-weighted equivalent level. It is given by

$$K_t = \begin{cases} 6\text{ dB} & 10\text{ dB} < \Delta L_{\text{ta}} \\ \Delta L_{\text{ta}} - 4\text{ dB} & 4\text{ dB} \le \Delta L_{\text{ta}} \le 10\text{ dB}\,. \\ 0\text{ dB} & \Delta L_{\text{ta}} < 4\text{ dB} \end{cases} \tag{8.58}$$

Our example corresponds to the second case, thus,

$$K_t = 9.3\text{ dB} - 4\text{ dB} = 5.3\text{dB}$$

8.8.2 FFT Convolution

The convolution with an impulse response is an operation that allows the implementation of filters and other types of signal processing in the time domain. The main difficulty is that a long-term convolution requires a prohibitively large number of operations.

It is possible to compute the convolution if we remember that the transform of a convolution is the product of the transforms of the signals to be convolved. However, in the case of short duration or periodic signals representable by a DFT, we must use the *circular convolution*, defined as

$$x * y(k) = \sum_{h=0}^{N-1} x(h)\, y((k-h) \mod N) \tag{8.59}$$

where a mod b (a modulus b) is the rest of the division of a by b. We can write

$$x * y(k) = \sum_{h=0}^{N-1} \left(\frac{1}{N} \sum_{m=0}^{N-1} X(m) e^{j\frac{2\pi mh}{N}} \right) \left(\frac{1}{N} \sum_{n=0}^{N-1} Y(n) e^{j\frac{2\pi(k-h)n}{N}} \right).$$

Note that the periodicity of the complex exponential makes it possible to remove the modulus operator. Operating,

$$\begin{aligned} x * y(k) &= \frac{1}{N^2} \sum_{h=0}^{N-1} \sum_{m=0}^{N-1} \sum_{n=0}^{N-1} X(m) Y(n) e^{j\frac{2\pi mh}{N}} e^{j\frac{2\pi(k-h)n}{N}} \\ &= \frac{1}{N^2} \sum_{h=0}^{N-1} \sum_{m=0}^{N-1} \sum_{n=0}^{N-1} X(m) Y(n) e^{j\frac{2\pi nk}{N}} e^{j\frac{2\pi(m-n)h}{N}} \end{aligned}$$

Reversing the summation order and taking common factors,

$$x * y(k) = \frac{1}{N^2} \sum_{n=0}^{N-1} \sum_{m=0}^{N-1} X(m) Y(n)\, e^{j\frac{2\pi nk}{N}} \sum_{h=0}^{N-1} e^{j\frac{2\pi(m-n)h}{N}}$$

The internal sum is N for $m = n$ and 0 for $m \neq n$. Finally,

$$x * y(k) = \frac{1}{N} \sum_{n=0}^{N-1} X(n) Y(n)\, e^{j\frac{2\pi nk}{N}} \tag{8.60}$$

i.e., the circular convolution is the inverse transform of the product of the transforms.

Suppose that we want to compute the convolution between an arbitrary signal $y(k)$ and a short signal $x(k)$ (for instance, a finite impulse response or an infinite impulse response approximated by a finite response). In Fig. 8.8 it is shown how to compute the value corresponding to the discrete instant k. When k increases, the backward version of x slides to the right.

In the case of a circular convolution, the sliding takes place along a circular path. The way to use the circular convolution (which would allow to apply the FFT and the IFFT) is to split the long signal into N-sample frames, and to complete each frame with N zeros (zero padding). Figure 8.9 shows an example of a convolution of $N = 4$ samples. Two zero-padded signals expanded to $2N$ samples replace the original frames (the zeros are not shown for clarity). In this case the sliding is circular. The positions for the calculation at two discrete instants are shown.

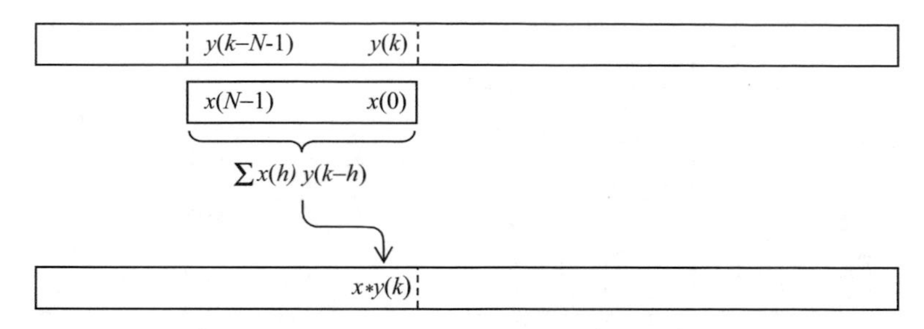

Fig. 8.8 Convolution of an arbitrarily long signal $y(k)$ and a short signal $x(k)$. The computation of the convolution at k is shown

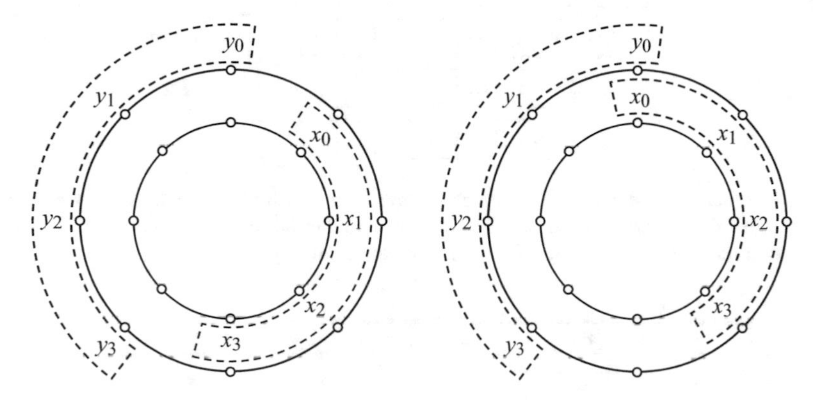

Fig. 8.9 Circular convolution of two signals y_k and x_k of length $N = 4$ that have been padded with N zeros (not shown). Two positions of the sliding backward signal x_k are shown. The calculation requires to multiply the internal and external values and to sum

Figure 8.10 presents the same convolution in linear fashion for all the possible positions of the sliding signal. This convolution is performed applying the FFT to both signals, multiplying and applying the inverse FFT.

Once we have this method to convolve two signals using the FFT, it is applied to M successive frames of N samples each taken from an arbitrarily long signal, each padded with zeros. Each convolution has a length $2N$, but they are overlapped in N samples. Summing the results of all convolutions yields the total convolution. This process is depicted in Fig. 8.11.

The computational load is reduced from MN^2, as would be with the direct calculation, to $2MN \log_2 2N = 2MN\,(1 + \log_2 N)$. The reduction is greater when N is large, as is the case for long impulse responses.

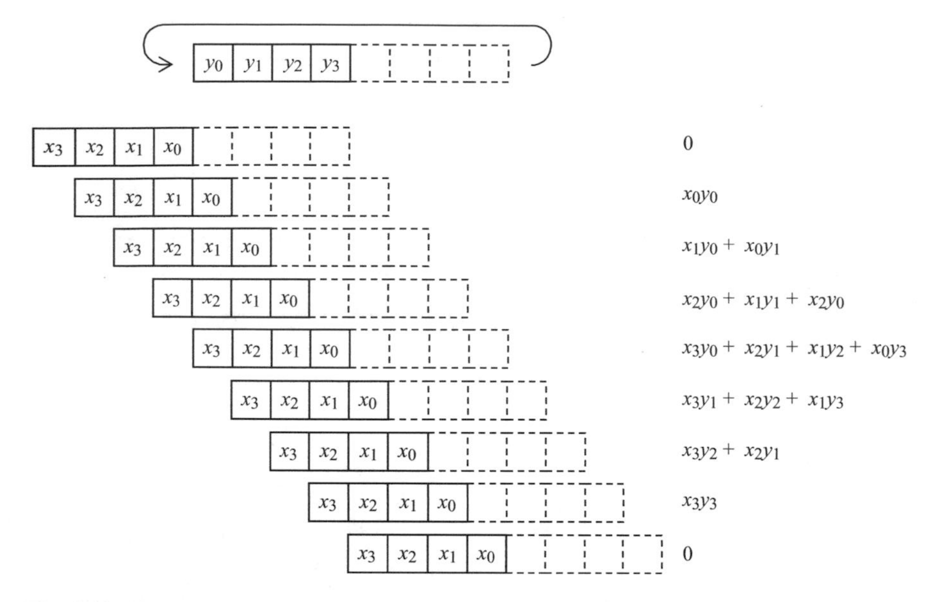

Fig. 8.10 Linear convolution of two signals y_k and x_k of length $N = 4$, computed by means of a circular convolution after padding with N zeros (indicated as *empty bins*). The resulting convolution has actually period $4N$, but only $2N$ values are kept

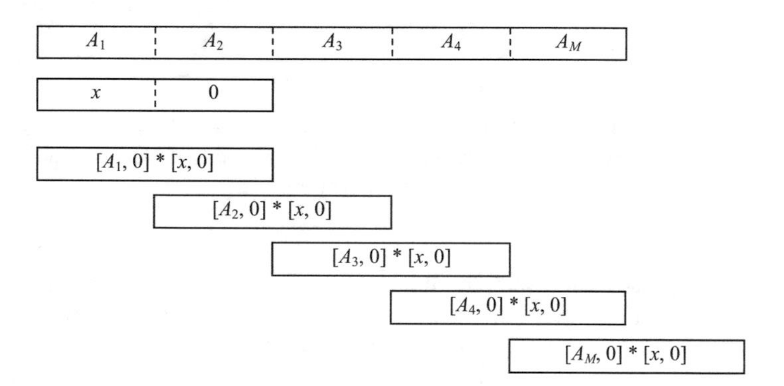

Fig. 8.11 Linear convolution of a signal split into frames of length N designated as $A_1, \ldots, A_5$ and a signal x of length N. Partial convolutions are computed after padding with N zeros. The resulting convolutions, 50 % overlapped, are added together

8.8.3 FFT Correlation

The *correlation* between two signals, also known as *cross correlation*, is an interesting operation since it reveals common features between two signals. For long time it is defined as

$$\varphi_{xy}(k) = \sum_{h=-\infty}^{\infty} x(h)\, y(h-k) \tag{8.61}$$

For short time it is

$$\varphi_{xy,N}(n_o,\ k) = \sum_{h=n_o-N}^{n_o+N-1} x(h)\, y(h-k) \tag{8.62}$$

For an efficient computation we can use again the FFT. Note that the general form is very similar to the convolution, except that the argument of signal y has the opposite sign. We can assimilate this expression to (8.59) by means of an adequate variable change

$$\varphi_{xy,N}(n_o,\ k) = \sum_{h=-N}^{N-1} x(n_o+h)\, y(n_o+h-k) \tag{8.63}$$

Calling

$$\begin{aligned} x_o(h) &= x(n_o+h) \\ y_o(h) &= y(n_o-h) \end{aligned} \tag{8.64}$$

we have

$$\phi_{xy,N}(n_o,\ k) = \sum_{h=-N}^{N-1} x_o(h)\, y_o(k-h) = x_o * y_o(k). \tag{8.65}$$

Now it is computable using the method of the previous section.

8.8.4 Critical Band Filters

In psychoacoustic experiments it is often necessary to work with critical bands. The band filters introduced in Sect. 8.7 can also be applied to implement non-uniform bands, such as the critical band series. To this end it is sufficient to replace in Eqs. (8.51) the limits of the critical bands as given by the Zwicker approximations (Zwicker et al. 1999).[9] Given the frequency f_o (in Hz) it is possible to estimate the critical bandwidth as

---

[9]There are also other approximations, such as the Buus approximation (Buus 1997), which presents some error with respect to the Zwicker approximation, but it has the advantage that the inverse is readily computed. The International Standard ISO1996-2 provides a simplified approximation according to which critical bandwidth is 100 Hz up to 500 Hz and then on they are 1.2 times the center frequency.

$$\Delta f_{\mathrm{CB}} = 25 + 75\left(1 + 1.4\left(\frac{f_{\mathrm{o}}}{1000}\right)^{2}\right)^{0.69}, \tag{8.66}$$

Here there are two possibilities. The first one is to assume that the band is distributed a half to each side of f_{o}, in which case the window is defined as

$$W(n) = \begin{cases} 0 & \text{si} \quad 1 \le n < \frac{f_{\mathrm{o}} - \Delta f_{\mathrm{CB}}/2}{F_{\mathrm{s}}} \\ 1 & \text{si} \quad \frac{f_{\mathrm{o}} - \Delta f_{\mathrm{CB}}/2}{F_{\mathrm{s}}} \le n < \frac{f_{\mathrm{o}} + \Delta f_{\mathrm{CB}}/2}{F_{\mathrm{s}}} \\ 0 & \text{si} \quad \frac{f_{\mathrm{o}} + \Delta f_{\mathrm{CB}}/2}{F_{\mathrm{s}}} \le n \le N/2 \end{cases} \tag{8.67}$$

This method does not guarantee that the critical band is actually centered at f_{o}. Another possibility is to apply Zwicker's formula for the critical band rate z_{o} (in bark),

$$z_{\mathrm{o}} = 13 \text{ arctg } \frac{f_{\mathrm{o}}}{1316} + 3.5 \text{ arctg}\left(\left(\frac{f_{\mathrm{o}}}{7500}\right)^{2}\right). \tag{8.68}$$

Once we have z_{o} corresponding to f_{o}, we get the critical band rates $z_{\mathrm{o}} \pm 0.5$ bark and then numerically invert (8.68) (using the Newton–Raphson method, for example) to find the lower and upper frequencies to use in (8.67) instead of $f_{\mathrm{o}} \pm \Delta f_{\mathrm{CB}}/2$.

8.8.5 Determination of Transfer Functions

As a last example of applications of the FFT we have the determination of the transfer function of a system. If we have two signals $x(t)$ and $y(t)$ that are, respectively, the steady state input and output of a system acquired simultaneously, then it is possible to compute the transfer function by simply dividing the FFT of both signals

$$H(n) = \frac{Y(n)}{X(n)}. \tag{8.69}$$

8.9 Contrasting Algorithms for Use in Measurements

Acoustical measurements, as is the case with any type of measurement, rely on transducers, instruments and other pieces of equipment that fulfill specific standards. This, along with the periodic calibration check, ensures that the measurement results will have a predictable uncertainty that is compatible with the intended application.

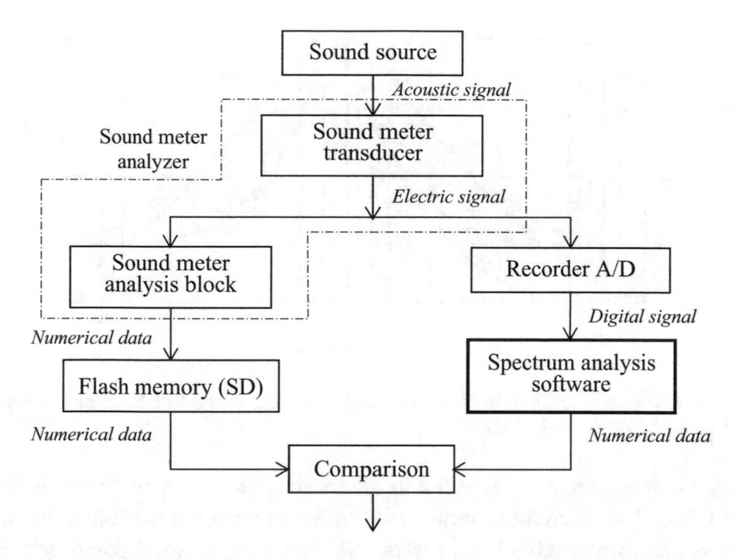

Fig. 8.12 Block diagram of the measuring process using **a** a sound meter analyzer (*left*) and **b** a measurement system formed by the transducer, the digital recorder and the analysis software (*right*)

Sometimes, instead of using an instrument that has been designed for some type of measurement (for instance, spectrum analysis), and that has undertaken a pattern or type approval process, a combination of separate parts not originally intended for this application are integrated into a measurement system. In such cases it will be essential to subject the system to a validation process. A particular instance of this arises when one of the parts is a piece of software.

In Fig. 8.12 two measurement processes are compared. At the left we have a standard instrument designed to perform spectrum analysis. At the right we have an alternative option that uses the transducer and the signal conditioning modules of the former, but includes a digital recorder and a software algorithm to compute the spectrum. The physical (acoustic and electric) signal is distinguished from the digital signal and from the abstract numerical data that represent the final measurement result.

8.9.1 International Standard IEC 61260

The fractional octave band filters intended for measurement purposes[10] are specified in the International Standard IEC 61260. As described in Sect. 1.7.5, this Standard provides, among many other items, the upper and lower tolerance limits for the frequency response of octave and one-third octave band filters. There are

[10]Other uses may include, for instance, audio system equalizers.

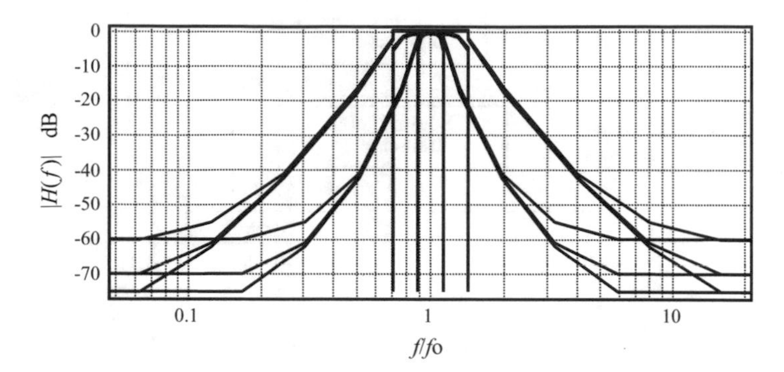

Fig. 8.13 Normalized frequency limits of tolerance for octave and one-third octave band filters according to IEC 61260. Global view

three classes, described as 0, 1, and 2 in decreasing order of stringency. Figures 1.3 to 1.6 of Chap. 1 showed those limits. Figure 8.13 compares the tolerance limits for octave and one-third octave band filters. All these limits are theoretically fulfilled by continuous-time Butterworth bandpass filters of order 3.

8.9.2 Contrasting Algorithms

The general idea is to compare the result of different spectrum analysis algorithms with the results of the same measurements performed by a commercial spectrum analyzer, such as the Brüel & Kjær 2250 (Brüel & Kjær 2016) with the BZ-7223 software module for octave band and one-third octave band analysis. This instrument has both output and input electric signals. The output is the signal from the microphone after conditioning. The instrument can be configured to analyze either the acoustic signal or the electric input signal.

Although the instrument has an electroacoustic front end and a digital signal processor, to our purposes we shall exclude the microphone and its preamplifier from the analysis. Hence, we will only study the signal path from the electric input to the numeric display. This is because the algorithms to test are intended as a substitution of the analysis block of the sound analyzer and not the complete instrument. In other words, both the analysis block and the software analyze a signal that has been already transduced, conditioned, and digitized. In the case of the external software algorithm we should include the analog/digital conversion at the recorder. Although it could certainly influence the measurement result, its analysis is not the purpose of this section.[11] Figure 8.14 shows the block diagram of the comparison process.

[11]The performance of digital recorders shall be undertaken in Chap. 9, where it is shown from tests that the non linear effects, frequency response distortion and other parameters of a good digital recorder are adequate for our purposes (see also Miyara et al. 2010).

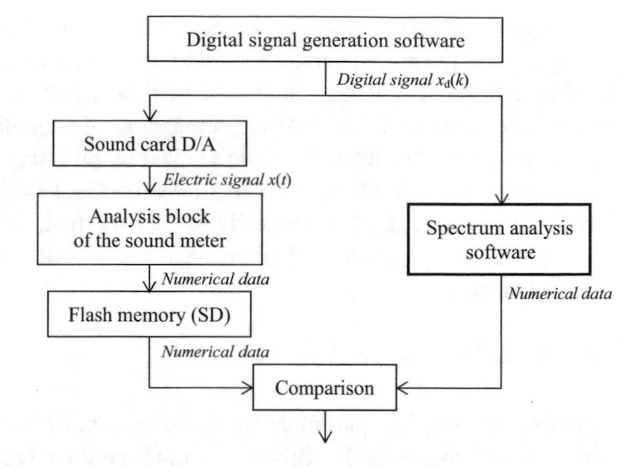

Fig. 8.14 Block diagram of the comparison process between the spectrum measurement with the sound meter analyzer and by software

8.9.3 *Procedure*

The basic procedure consists in getting the response to pure tones of several frequencies and constant amplitude. Since the test must be applicable both to an instrument that receives an electric signal and an algorithm that receives a digital signal, it is necessary to start from the same computer-generated digital signal and, in the first case, convert it to the analog domain as transparently as possible. This is attained using a computer sound card. If the sound card were ideal, it would only contribute with a conversion constant independent of frequency

$$x(kT) = K_{\mathrm{D/A,card}} x_{\mathrm{d}}(k), \tag{8.70}$$

where $x_d(k)$ is the digital signal and $K_{\mathrm{D/A,card}}$, the constant of the D/A converter. However, the sound card is not ideal so $K_{\mathrm{D/A,\ card}}$ depends on frequency and, therefore, it is necessary to find its frequency response so it can be later compensated.

8.9.4 *Sound Card Calibration*

The frequency response of the sound card is measured using the sound meter with Z weighting (zero or flat weighting). We will assume that:

(a) The sound card is essentially linear within the level range in which it will be used. This implies the need to deactivate any compression or automatic gain control feature.
(b) The frequency response of the sound card is not necessarily flat.
(c) The electric frequency response of the sound meter with *Z* weighting is essentially flat.

A digital sinusoidal signal with amplitude independent of frequency is generated and applied to the sound card. There are three possible ways to do this. The first one is to generate a .WAV file and play it with any player software, making sure that any effect, such as filter, booster, tone control, equalizer, spatializer, enhancer, compressor, etc., has been deactivated. The second way is to generate and play the signal with an audio editor, such as Audacity, Cool Edit, Sound Forge, Goldwave, etc. The third way is to generate the signal by means of a mathematical software package, such as Octave, Scilab or Matlab. In Scilab this can be done with the function **sound**

```
sound(x, Fs, Nbits)
```

Note that even if a particular program does not add any effect, effects could be added by the sound device driver and must be deactivated.

The line output of the sound card is applied (with maximum gain) to the electric input of the sound meter, and the corresponding *Z*-weighted level $L_{\mathrm{p,card}}(f_k)$ is measured for *K* frequency values covering the whole audible range. Although this value is expressed as if it were a sound pressure level, the electric reference does not have any relevant meaning for our purposes.

From the measured data we can now compute the frequency response normalized with respect to the response at $f_{\mathrm{ref}} = 1000$ Hz

$$C_{\mathrm{card}}(f_k) = L_{\mathrm{p,\,card}}(f_k) - L_{\mathrm{p,\,card}}(f_{\mathrm{ref}}). \tag{8.71}$$

This frequency response, measured over a discrete set of frequencies can be interpolated using cubic spline interpolation to get the frequency response at other frequencies that will be necessary for the following tests.

8.9.5 *Verification of the Analyzer Response*

We first set the sound meter to octave or one-third octave band spectrum analysis mode. We select a band centered at f_{on} ($n = 1, \ldots, N$) and measure its frequency response $L_{\mathrm{p}}(f_{nh})$ for several frequencies f_{nh} covering at least from four octaves below to four octaves above f_{on}. The frequency response of that band, corrected for the frequency response of the sound card, will be, therefore,

$$H_n(f_{nh}) = L_{\mathrm{p}}(f_{nh}) - C_{\mathrm{card}}(f_{nh}). \tag{8.72}$$

The frequency set $\{f_{nh}\}$ corresponding to a band is usually denser in the pass band and its vicinity in order to have a more detailed frequency response. The sound card response $C_{\mathrm{card}}(f_{nh})$ must be obtained by interpolation of the originally measured set of values

$$C_{\mathrm{card}}(f_{nh}) = \mathrm{spline}\Big((f_k,\ C_{\mathrm{card}}(f_k))_{k=1,\ldots,K}, f_{nh}\Big), \tag{8.73}$$

where $(f_k, C_{\mathrm{card}}(f_k))_{k=1,\ldots,K}$ are the K measured values.

8.9.6 Contrasting Different Algorithms

Once we have got the frequency response of a standard analyzer such as the Brüel & Kjær 2250 with the spectrum analyzer module enabled, we can proceed to test the previously described algorithms. In this case the experimental tests are carried out entirely by software. To that end, the response of each algorithm to a series of pure tones of different frequencies and constant amplitude is computed.

Three algorithms have been tested: FFT filtering (Sect. 8.7), IIR with third- and fourth-order Butterworth approximation (Sect. 8.3), and FFT band analysis (Sect. 8.6.6). FFT filterButterworth implementations are based on pre-warped bilinear approximations.[12] The following transfer functions correspond to the third-order octave band and one-third octave band filters centered at 1 kHz, with an eight-fold downsampling (F_s = 5512.5 Hz)[13]:

$$H(z) = \frac{0.0338 - 0.1013z^{-2} + 0.1013z^{-4} - 0.0338z^{-6}}{1 - 1.7009z^{-1} + 2.415z^{-2} - 1.9683z^{-3} + 1.4337z^{-4} - 0.5565z^{-5} + 0.1890z^{-6}} \tag{8.74}$$

$$H(z) = \frac{0.0018 - 0.0054z^{-2} + 0.0054z^{-4} - 0.0018z^{-6}}{1 - 2.2685z^{-1} + 4.1953z^{-2} - 4.2089z^{-3} + 3.5187z^{-4} - 1.593z^{-5} + 0.5889z^{-6}} \tag{8.75}$$

The following ones correspond to the fourth-order versions:

$$H(z) = \frac{0.011121 - 0.044485z^{-2} + 0.066728z^{-4} - 0.044485z^{-6} + 0.011121z^{-8}}{1 - 2.2767z^{-1} + 3.9032z^{-2} - 4.2818z^{-3} + 3.9331z^{-4} - 2.5225z^{-5} + 1.3424z^{-6} - 0.4396z^{-7} + 0.1135z^{-8}} \tag{8.76}$$

$$H(z) = \frac{0.00021965 - 0.0008786z^{-2} + 0.0013179z^{-4} - 0.0008786z^{-6} + 0.00021965z^{-8}}{1 - 3.0301z^{-1} + 6.7628z^{-2} - 9.3175z^{-3} + 10.2696z^{-4} - 7.8390z^{-5} + 1.7868z^{-6} - 1.8025z^{-7} + 0.5005z^{-8}} \tag{8.77}$$

[12]The bilinear transformation is not perfect. Prewarping consists in changing the band lower and upper frequencies so that at the desired frequencies the gain is exactly −3 dB as in the analog version.

[13]Downsampling is necessary to prevent numerical instability.

8.9.7 Results

Measurements have been done with the Brüel & Kjær 2250 sound level meter with the spectrum analyzer module enabled. The sound card was a Creative Audigy SE. In Fig. 8.15 is shown the frequency response of the sound card measured with the procedure described above. High-frequency fluctuations are due to the smoothing filter of the digital/analog converter. Note that different units of the same sound card model may have different frequency responses, so all subsequent tests should be done using the specific sound card that has been measured.

Figures 8.16, 8.17, 8.18 and 8.19 show the frequency response of the octave and one-third octave bandpass filters centered at 1 kHz for the B&K 2250 spectrum analyzer.

Figure 8.20 illustrates the frequency response of a 1 kHz octave band FFT filter with $N = 8192$ and an ideal filtering window (i.e., it is 1 within the pass band and 0 outside). The response is compared with that of the standard analyzer and with the tolerance limits of International Standard IEC 61260 for class 2. Figure 8.21 shows the frequency response of a 1 kHz one-third octave band FFT filter, in this case with $N = 65{,}536$. In both cases the response is very good within the band but gets worse towards the low frequency end.

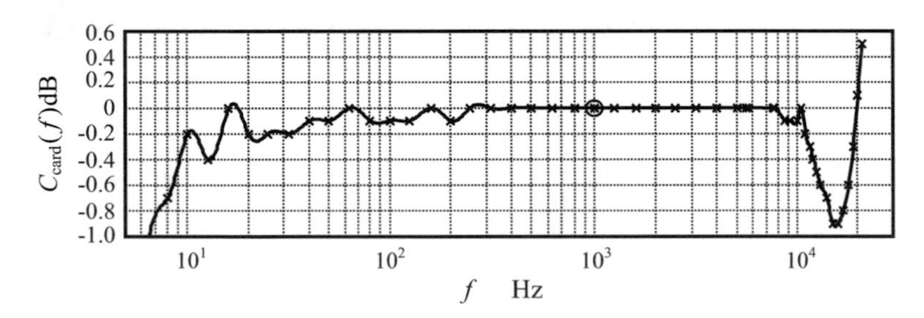

Fig. 8.15 Frequency response of a Creative Audigy SE sound card with $f_s = 96\ 000$ Hz. The crosses indicate the measured data and the curve, the frequency response with spline interpolation

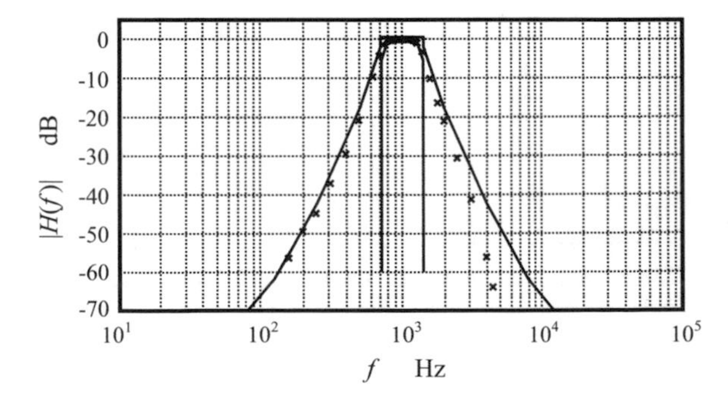

Fig. 8.16 Comparison between the measured frequency response of the B&K 2250 analyzer (×) and the tolerance limits for the Class 0 octave band centered at 1 kHz, according to IEC 61260

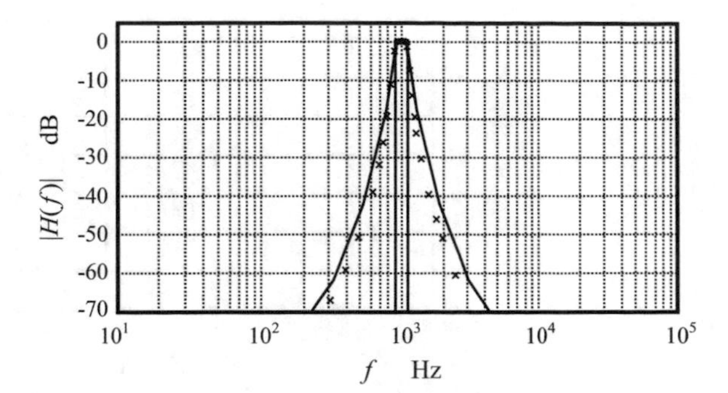

Fig. 8.17 Comparison between the measured frequency response of the B&K 2250 analyzer (×) and the tolerance limits for the Class 0 one-third octave band centered at 1 kHz, according to IEC 61260

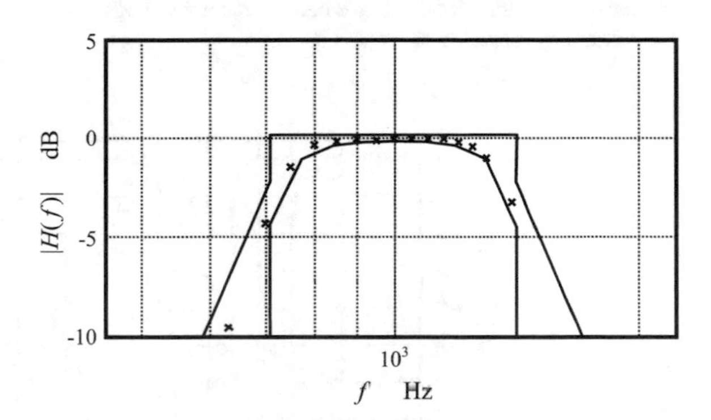

Fig. 8.18 Comparison between the frequency response of the B&K 2250 analyzer and the tolerance limits for the Class 0 octave band centered at 1 kHz, according to IEC 61260. Pass band detail

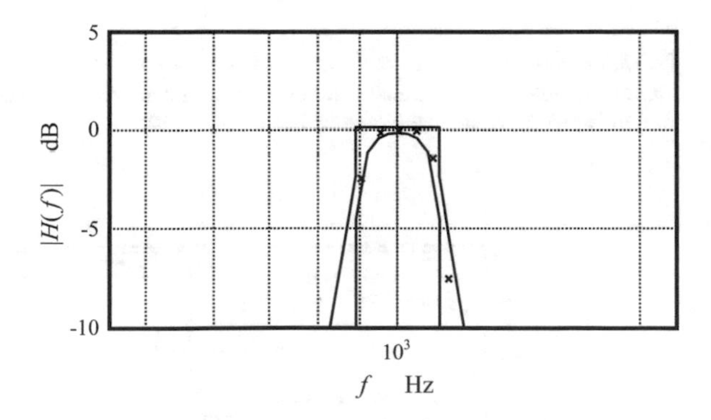

Fig. 8.19 Comparison between the frequency response of the B&K 2250 analyzer and the tolerance limits for the Class 0 one-third octave band centered at 1 kHz, according to IEC 61260. Pass band detail

In Fig. 8.22 a more detailed view is given in the proximity of one of the pass band ends. As can be noted, there is a lobular behavior as well as very steep transition into the stop band.

Figures 8.23 and 8.24 illustrate the frequency response of Butterworth IIR filters of order 3 and 4 for octave and one-third octave bands centered at 1 kHz. As can be seen, third-order filters only marginally comply with tolerance limits for Class 0 at low frequency, while fourth-order filters fully satisfy the requirements.

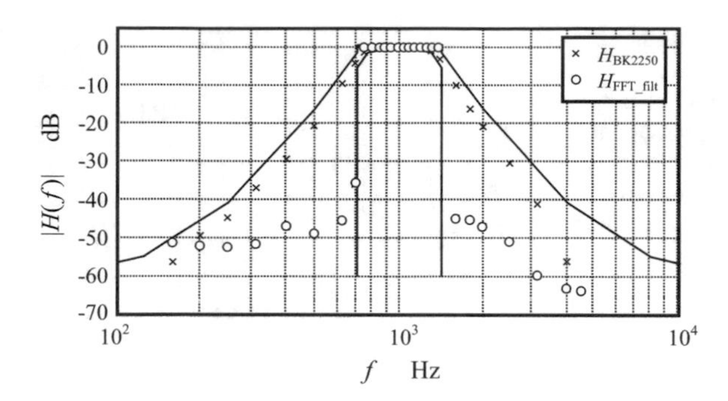

Fig. 8.20 Comparison between the frequency response of the FFT filter ($N = 8192$) and the tolerance limits for Class 2 octave band centered at 1 kHz, according to IEC 61260. Also shown is the frequency response of the B&K 2250

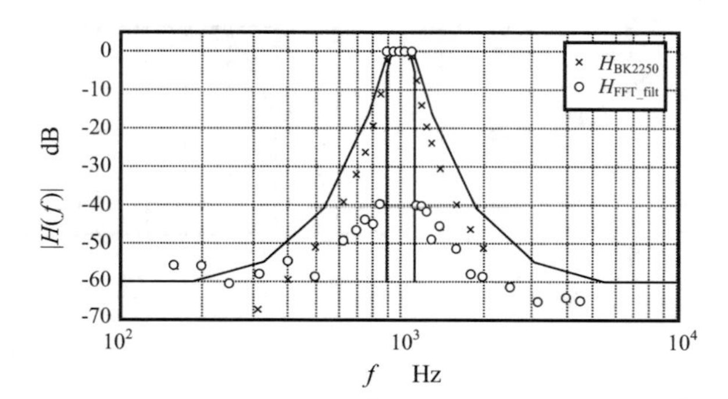

Fig. 8.21 Comparison between the frequency response of the FFT filter ($N = 65\ 536$) and the tolerance limits for Class 2 one-third octave band centered at 1 kHz, according to IEC 61260. Also shown is the frequency response of the B&K 2250

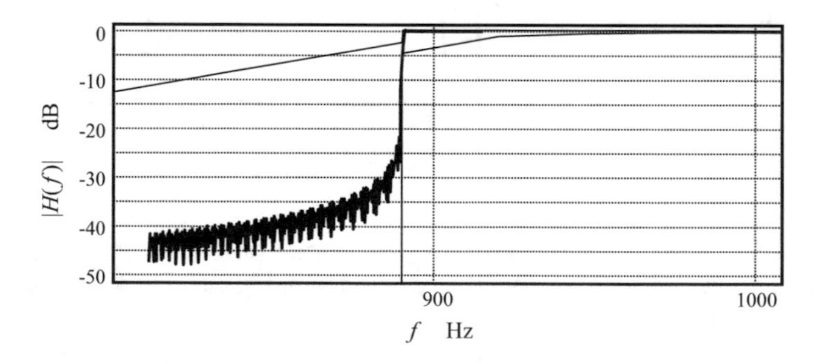

Fig. 8.22 Detailed view of the behavior of a one-third octave FFT filter in the vicinity of the lower end of the pass band

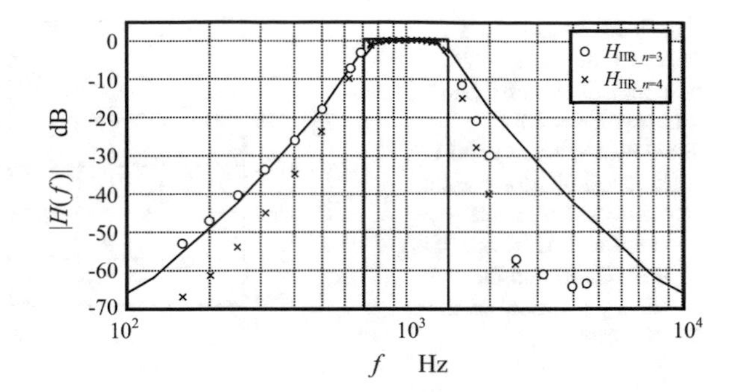

Fig. 8.23 Comparison between the frequency response of Butterworth IIR filters (order 3 and 4) and the tolerance limits for class 0 octave band centered at 1 kHz, according to IEC 61260

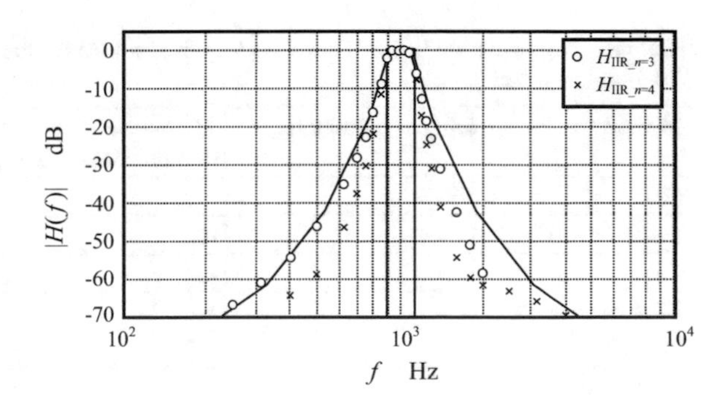

Fig. 8.24 Comparison between the frequency response of Butterworth IIR filters (order 3 and 4) and the tolerance limits for class 0 one-third octave band centered at 1 kHz, according to IEC 61260

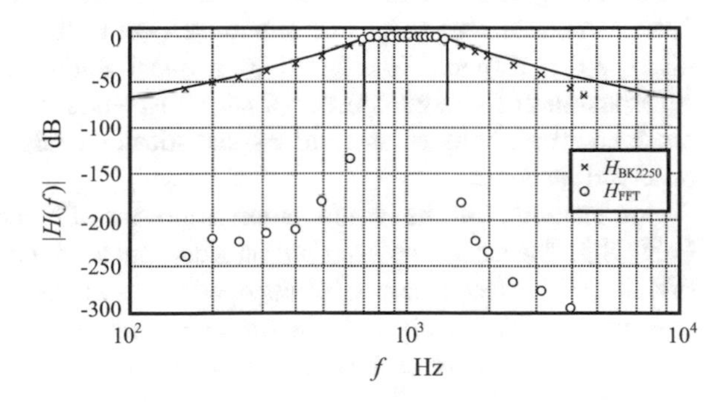

Fig. 8.25 Comparison between the frequency response of the FFT band analyzer ($N = 4096$, Blackman window) and the tolerance limits for class 0 octave band centered at 1 kHz, according to IEC 61260 and the frequency response of the B&K 2250

It is possible to implement a hybrid filter that behaves as third order at high frequency and fourth order at low frequency, which would allow a more symmetrical response, though the true advantage would be a slight improvement in computational speed.

Figures 8.25 and 8.26 plot the results of the FFT band spectrum analysis (Sect. 8.6.6) with $N = 4096$. Note the extremely fast roll-off outside the pass band. It cannot simply be explained by the attenuation of 58 dB provided by the Blackman window at the first lateral lobe. A possible explanation is that each FFT line within the band behaves as a filter with a lobular frequency response such as

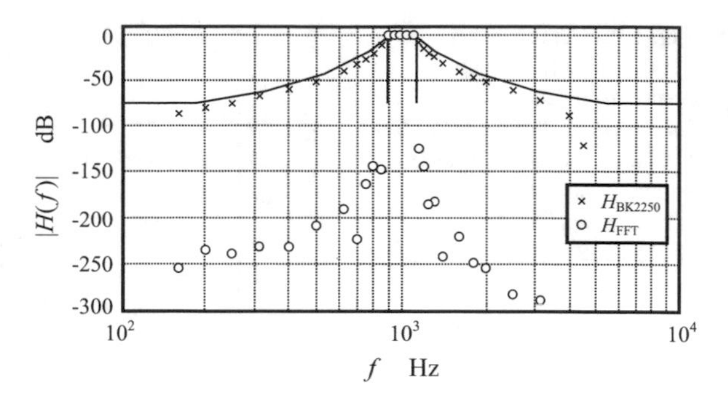

Fig. 8.26 Comparison between the frequency response of the FFT band analyzer ($N = 4096$, Blackman window) and the tolerance limits for class 0 one-third octave band centered at 1 kHz, according to IEC 61260 and the frequency response of the B&K 2250

Table 8.3 Computation times of several analysis methods for the whole sets of 1/1 octave bands and 1/3 octave bands

Method	Signal duration (s)	1/1 octave computation time (s)	1/3 octave computation time (s)
FFT	60	1.7	1.9
IIR filter $n = 3$	60	4.2	12.7
IIR filter $n = 4$	60	4.4	14.2
IIR filter $n = 10$	60	5.8	16.7
FFT filter	60	7.2	21.3

shown in Figure P.3. When an out-of-band tone is fed into the analysis algorithm, it falls at different relative positions with respect to the central lobe of each of those filters, so its response is affected by a quasi-random attenuation and phase. The superposition of a large number of such responses, all having the same frequency but random amplitudes and phases statistically tends to cancel out, yielding the observed response.

An example of the computation time for different algorithms is given in Table 8.3. The tests have been run on a computer with a processor Intel Core 2 Duo E6420, 2133 MHz clock, a 32-bit operating system and 2 GB RAM. As can be observed, the FFT analysis is the one requiring the least computation time since it takes a single FFT for each signal frame. Next we have the IIR filters, for which all bands must be filtered. Signal is downsampled for $f < F_s/7$. FFT filters are the most time consuming ones since because of overlapping they require on average two direct and two inverse transforms per frame and per band. To improve efficiency, the signal is downsampled for low frequencies. The advantage of FFT filters is a very steep roll-off when the actual filtered signals are necessary.

Computational time is influenced by many factors, such as background processes running on the computer, free memory available. We can see, however, that in all cases the spectrum analysis can be performed in a fraction of the signal duration so they could be implemented in real time. FFT-based algorithms have a latency equal to the analysis window length. IIR filters present a very small latency, allowing for DSP-based real-time analyzers.

8.10 Conclusion

From the previous discussion, we can conclude that several of the spectrum analysis algorithms described above, in combination with a sound level meter capable of providing a calibrated signal and a digital recorder (to be studied in the next chapter) allow to implement a measurement system that can replace an expensive, dedicated sound level analyzer for performing spectral analysis.[14]

Problems

(1) Obtain a formula that computes an exponentially swept sine wave starting at a frequency f_1 and ending at a frequency f_2. Hint: The argument of the sinusoidal function, i.e., its phase, is the integral of its angular frequency with respect to the time.

(2) (a) Write a script that designs a one-third octave Butterworth filter of order 3 and centered at frequency f_o and then applies it to a digital signal. Hint: Use the function `iir` to get the poles and zeros of the Z transfer function. Then use function `poly` to get the numerator and denominator (you may need to take the real part, since numerical errors may cause some small but nonzero imaginary part. Finally use the function `filter` to actually apply the filter. (b) Using $F_s = 44\ 100$ Hz, apply it to a music audio signal. (c) Apply it to an exponentially swept signal such as that of Problem (1). Plot the filtered signal and listen to the result using function `sound`. (d) Investigate the effects of lowering f_o. (e) Reorder the poles and zeros so as to isolate the quadratic bandpass stages and, instead of applying the whole filter once, apply the first stage to the input signal, then the second stage to the output of the first and then the third to the output of the second one. Investigate if an improvement can be obtained for low f_o and analyze the computing time.

(3) (a) Write a script that downsamples a signal originally sampled at $F_s = 44\ 100$ Hz in order to be able to filter it with a one-third octave Butterworth filter of order 3 centered at a frequency $f_o = 100$ Hz and finally upsamples it to get back to the original sample rate. (b) Compare the computational time of this algorithm with the filtering algorithm applied to the signal at the original sample rate. (c) Apply the algorithm to a swept sine wave and note that at frequencies beyond the intermediate Nyquist frequency there is no spectral content.

---

[14]This section is based on (Miyara et al. 2009), a paper prepared as part of a research project carried out with a grant of the National Scientific and Technological Promotion Agency (Argentina) (ANPCyT—PICT N° 38109).

(4) Write a script that computes the discrete Fourier transform directly from the definition and compare the computing time with that of the FFT for several sizes.

(5) Consider $F_s = 44\ 100$ Hz and $N = 4096$. (a) Generate a sine wave of frequency $20F_s/N$, compute its FFT and plot its absolute value from $k = 1$ to $N/2$. Note that there is a single spectral line. Explain why. (b) Generate a sine wave of frequency $(20 + \alpha)F_s/N$ with $0 < \alpha \leq 1$ and see the behavior of the amplitude for different α's. (c) Write a brief script that detects the maximum line and plots its amplitude versus α. Hint: Use the function **`max`**. (d) Repeat (c) replacing the maximum by the energy sum of the maximum and its adjacent lines. (e) Estimate the frequency using the parabolic interpolation and plot the relative error as a function of α. (f) Repeat (e) when a time window has been applied before the FFT.

(6) Consider $F_s = 44\ 100$ Hz and $N = 4096$. (a) Generate a signal that is the sum of two sine waves of frequencies f_1 and f_2. Plot the absolute value of the FFT for different combinations of amplitudes, f_1 and f_2, particularly when f_1 and f_2 get very close. (b) Analyze the possibility of discriminating both components from the FFT spectrum. Hint: When the difference is much larger than F_s/N it is trivial. Experiment to see the influence of the lateral bands of first and second order and try to find out the minimum tone separation so that the parabolic interpolation can be used reliably. The theoretical formula for the FFT of each sine wave can be obtained considering complex exponentials instead of sine waves. In this case the FFT reduces to a sum of powers Σa^n. (c) Repeat in the case that a time window has been applied, comparing several windows such as Hann, Blackman and Blackman–Harris.

(7) (a) Write a script that takes a sound (for instance a one-minute-long recording), reads it into Scilab, splits it into adjacent frames of length N, where N is a power of 2, computes the FFT and then averages the amplitudes of each frequency across the frames. Hint: The last frame may be incomplete. It can be zero-padded to complete the length N, but when averaging it must be weighted to account for its reduced duration. (b) Repeat (a) allowing for overlapping (for instance 50 % overlapping). (c) Add to the preceding algorithm the application of a window in the time domain before the FFT (for instance, Hann or Blackman). (d) Test these algorithms on a long sine wave.

(8) Generate a random Gaussian noise using function **`rand`** with key = '**`normal`**'. (a) Considering that it is actually band limited to $F_s/2$, find the theoretical spectral density. (b) Estimate the spectral density using the spectrum averaging tool from the preceding problem and compare. Explain any difference you find.

(9) Write a script that reads a very long wav file and obtains its average FFT spectrum with or without time windowing. Hint: To prevent memory issues, load a segment of less than 1 min at a time, get the average FFT spectrum and use it to update the cumulative average.

(10) (a) Write a script that computes the octave band average spectrum of an intermediately long signal (i.e., one that can be handled without memory issues) using the FFT. (b) Repeat (a) in the case of an arbitrarily long file, by reading one segment at a time, according to the idea outlined in problem (8).

(11) (a) Write a script that computes the spectral density of a long wav file (containing, for instance a background noise recording). Hint: To prevent memory issues, read the file by portions, split each portion into a number of frames, apply the FFT to each one and compute its average. (b) Assuming the recording is part of an indefinitely long stationary noise, find the uncertainty when using the result from (a) to estimate the long-term spectral density.

(12) Implement an FFT filter function that receives a signal and a filtering frequency response and returns the filtered signal. Hint: Use the overlap-add approach.

(13) Write a script that computes the convolution between a long signal and a relatively short impulse response using the FFT.

(14) In architectural acoustics it is frequently necessary to measure the impulse response of a room. There are several ways to do so, for instance to generate a very short noise such as the detonation of a firecracker or the pop of a balloon and directly record a microphone's signal. Another way is to generate a swept sine wave $x(t)$ whose frequency varies exponentially from 20 Hz to 20 kHz, record the response $y(t)$, and deconvolve it. Write a script that gets the FFT of the original signal $X(n)$ and the FFT of the room response, $Y(n)$, then computes the FFT of the impulse response as $H(n) = Y(n) / X(n)$, and finally gets the time-domain impulse response $h(k)$ as the inverse FFT.

(15) Reverberation time T_{rev} is defined as the time interval between the instant a sound source generating a reverberant sound field in a room is turned off and the instant when the sound pressure level falls 60 dB below the initial level. It is normally computed for octave or one-third octave bands. The International Standard ISO 3382-1 gives a procedure originally introduced by Schroeder (1965) to compute it from the impulse response. First perform the backward integration of the square of the impulse response filtered by the corresponding band filter centered at f_o:

$$E(t) = \int_t^\infty p_{f_o}^2(\theta)\, d\theta.$$

Then plot $10 \log(E(t)/E(0))$ versus t and fit the region between -5 and -35 dB with a straight line $y = a + bt$. The reverberation time is then computed as $T_{rev} = 60/b$. However, the formula for $E(t)$ presents two problems: we cannot integrate to ∞ and real-life measurements contain background noise. (a) Assuming an exponential energy decay, show that replacing ∞ by a finite time t_1 causes only a constant shift in the logarithmic decay curve, and if it is chosen larger than the time required for a decrease of 35 dB the described procedure can be applied. (b) If background noise is

45 dB below the maximum level, show that t_1 can be selected as the instant when the extrapolation of the linear portion of the logarithmic decay curve crosses the background noise. (c) Write a script that receives the wideband impulse response, filters it with an octave band filter and computes $T_{rev,fo}$. Hint: Use an energy envelope with $\tau = 35$ ms; estimate the -5 dB instant as the mean of the first and last instants at which -5 dB is reached and similarly for the -35 dB instant; estimate the background noise as the level of the last available sample; intersect the straight line passing through the -5 dB and -35 dB points with the background noise level in order to find t_1. Then apply Schroeder's method using the linear regression function of Scilab, **`reglin`**.

(16) Vibroacoustic correlation is a useful tool when it is necessary to identify the source of a particular noise when there are several vibrating sources or when the path between a given source and the receiver. Suppose we have a two-channel recording where channel 1 is the output of an accelerometer and channel 2 the output of a sound level meter. (a) Write a script that efficiently computes the correlation between two signals and normalizes it with respect to the product of the standard deviation of both signals. (b) Apply it to both channels of a stereo audio signal. (c) Apply it to a stereo signal generated using Audacity such that one channel is a sine wave and the other is white noise.

References

Brüel & Kjær (2016) User manual—hand held analyzers types 2250 and 2270. Brüel & Kjær BE 1713-32. Nærum, Denmark. http://www.bksv.com/doc/be1713.pdf. Accessed 26 Feb 2016

Buus, Søren (1997) Auditory Masking. In Crocker, Malcolm (ed) (1997) "Encyclopedia of Acoustics. John Wiley & Sons.

Cooley JW, Tuckey JW (1965) An algorithm for the machine computation of complex Fourier Series Math Comp 19:297–30

Crochiere R (1980) A weighted overlap-add method of short-time Fourier analysis/synthesis. IEEE Trans Acoust Speech Signal Process (ASSP) 28(2):99–102

IEC 61260:1995 Octave-Band and Fractional-Octave-Band Filters

ISO 1996-2: Acoustics—description, measurement and assessment of environmental noise—part 2: determination of environmental noise levels. Geneve, Switzerland, 2007

Miyara F, Accolti E, Pasch V, Cabanellas S, Yanitelli M, Miechi P, Marengo F, Mignini E (2010) Suitability of a consumer digital recorder for use in acoustical measurements. Internoise 2010. http://www.fceia.unr.edu.ar/acustica/biblio/consumer_recorders_for_noise_measurement_INTERNOISE_2010_993.pdf. Accessed 26 Feb 2016

Miyara F, Pasch V, Yanitelli M, Accolti E, Cabanellas S, Miechi P (2009) Contrastación de algoritmos de análisis de espectro con un instrumento normalizado. Actas de las Primeras Jornadas Regionales de Acústica AdAA 2009, Rosario, Argentina, 2009. http://www.fceia.unr.edu.ar/acustica/biblio/A032%20(Miyara)%20Contraste%20algoritmos%20analizador%20normalizado.pdf. Accessed 26 Feb 2016

Proakis JG, Manolakis DG (1996) Digital signal processing. Principles, algorithms and applications, vol 3. Prentice-Hall International, Upper Saddle River

Randall RB, Upton R (1978) Digital filters and FFT technique in real-time analysis. Brüel & Kjær. Technical Review 1978 Nº 1. http://www.bksv.com/doc/technicalreview1978-1.pdf. Accessed 26 Feb 2016
Schroeder M (1965) New method of measuring reverberation time. JASA 37:409–412
Zwicker E, Fastl H (1999) Psychoacoustics—facts and models. Springer, Berlin

Chapter 9
Testing Digital Recorders

9.1 Introduction[1]

The proposal put forward in this book involves the systematic acquisition of calibrated signals by means of a digital recorder to be later analyzed using computer software. In Chap. 8, we covered the tests designed to evaluate the software, assuming that the digital signal fed to the software algorithms accurately reflects the electric signal provided by the calibrated output of the sound meter.

There is, however, legitimate concern as regards whether the quality of the digital signal is sufficient to comply with the standards. International Standards such as ISO 1996-2 (2007) specify that in order that a piece of equipment is suitable for measurement purposes the complete measurement system including it must comply with the International Standard IEC 61672-1 (2002) on sound level meters. This means, particularly, very strict tolerance limits for the frequency response, as well as other conditions as regards self-generated noise, linearity, and transient response.

With the aim of verifying the compliance with these requisites, a series of tests were performed on three units of a widespread general purpose digital recorder, the Zoom H4, as described in the following sections. Note that in principle these tests should be performed on every unit of any brand and model we intend to use for metrological purposes, since they are not designed as part of a measurement system and it is also possible that their specifications are changed by the manufacturer over time without notice.

[1]This chapter is based on (Miyara et al. 2010), paper prepared as part of a research project carried out with a grant of the National Scientific and Technological Promotion Agency (Argentina) (ANPCyT—PICT N° 38109).

F. Miyara, *Software-Based Acoustical Measurements*, Modern Acoustics and Signal Processing, DOI 10.1007/978-3-319-55871-4_9

9.2 Specifications of the Digital Recorder

The digital recorder Zoom H4 supports three sample rates: 44.1, 48, and 96 kHz, and two resolutions: 16 bit and 24 bit. Although a sample rate of 96 kHz may improve the frequency response, in most cases the use of a resolution of 24 bit is unnecessary and it will only increase the file size, since a higher resolution is not necessarily accompanied by a real improvement in the dynamic range. The recording format may be .WAV or .MP3, but the latter is not recommended because of its psychoacoustic lossy audio compression that might alter or even suppress significant spectral components. The input signal level is specified as −10 dBm. Surprisingly, no information is provided as regards the signal/noise ratio or the distortion (Zoom Corporation, n. d.).

The recorder has two combined (balanced XLR and unbalanced 1/4″ jack) inputs. It has also two microphones mounted as an XY stereo configuration but there is no way to calibrate them. No guarantee as regards frequency response, distortion or long-term stability is given, so they are not recommended for measurement purposes. Instead, the sound meter audio output should be connected directly to one of the combined inputs using the appropriate adaptor and extension cable.

9.3 Tests

Several tests have been performed on the recorder as regards its frequency response, transient response, self-generated noise, and linearity.

9.3.1 Frequency Response

In order to study the frequency response, two types of signals were applied: constant-frequency pure tones (sine waves) and pure-tone sweeps. Pure tones are the ideal test signals, but it would require a large number of them to detect all possible anomalies such as ripples close to the Nyquist frequency (half the sample rate).

Pure-tone sweeps, instead, allow to cover all frequencies, provided that the instant frequency increases at a sufficiently low rate to prevent excessive spectral widening.

When the input frequency is close to the Nyquist limit, the discrete sequence has little more than two samples per period. The waveform ceases to be evident from the simple observation of the samples in an oscillogram, which tend to group visually in a sort of Moiré patterns (visual aliasing) such as shown in Fig. 9.1. However, since the input signal has gone through an anti-aliasing filter before being

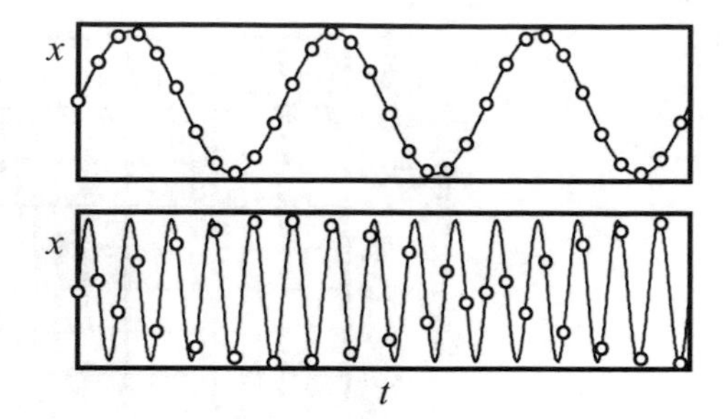

Fig. 9.1 Sample pattern of tones far below (*top*) and very close (*bottom*) to the Nyquist limit

sampled, it is theoretically possible to perfectly retrieve the original signal (through sinc interpolation—see Appendix 10).

We are interested only in the RMS (effective) value of the recorded signal. We could, for instance, oversample and apply a smoothing filter. This is, however, unnecessary since the RMS value can be directly computed from the original samples. This is true even for signals whose frequency is close to the Nyquist limit. Another approach is to detect the maximum of the samples that are within a short interval around the desired instant, which approximates the ideal sine wave peak. From this peak the RMS value maybe easily derived.

It has been verified by computer simulation that both methods are adequate. However, the approach of taking the maximum requires less samples to meet a specified tolerance. This is important in the case of frequency sweeps, since it is necessary to ensure that the RMS value detection is accomplished before the frequency changes significantly. The test has used logarithmic sweeps[2] covering the audio range with a duration of 100 s. In order to attain a good time resolution (and hence, frequency resolution), intervals of about 200 samples were used except at low frequency, where a duration of 2 complete cycles was used. In some cases, especially at high frequency, abnormally low maxima are found during a fairly long run of samples. These do not represent a real dip in the frequency response but an artifact due to the method. They are easily corrected by removal of any sudden drop that differs from the response at the neighboring intervals by a certain threshold such as 0.02 dB, and interpolation.

The best and worst responses are shown in Fig. 9.2. Both responses present a somewhat noisy appearance. This is due to a slight distortion of the signal synthesizer, electric noise, and noise inherent in the amplitude detection method. As its level is far below the main features of the response, it is altogether inconsequential.

For a given unit of the Zoom H4 the frequency response is essentially invariant over successive tests. For comparative purposes, 1/10 the IEC 61672-1 (2002) tolerance for a class 1 (precision) instrument is also shown.

When the temperature is raised from 24 ºC to 40 °C, a small difference of at most 0.2 dB can be detected in the frequency response, as shown in Fig. 9.3.

[2]Actually, in the so-called "logarithmic sweep" the frequency evolves exponentially with time.

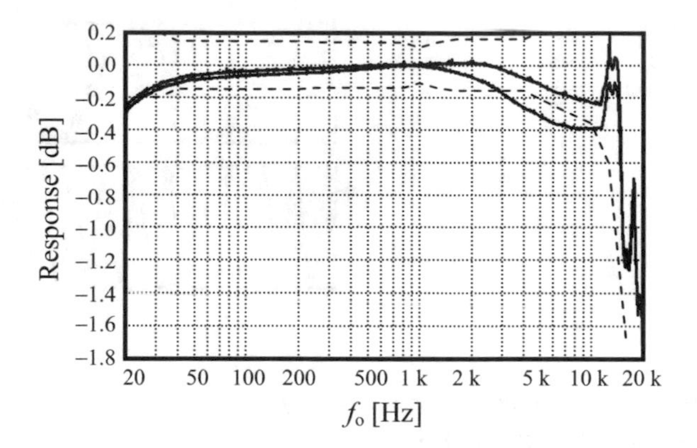

Fig. 9.2 Frequency response of the best and the worst ones of a group of three units of the digital recorder Zoom H4 obtained with a synthesized tone sweep from signal generator SRS DS 345 (Stanford Research Systems n.d.) at T = 24 °C. The sample rate was 44.1 kHz. The dashed lines show 1/10 of the upper and lower tolerance limits specified in the International Standard IEC 61672-1 (2002) for the frequency response of a class 1 sound level meter

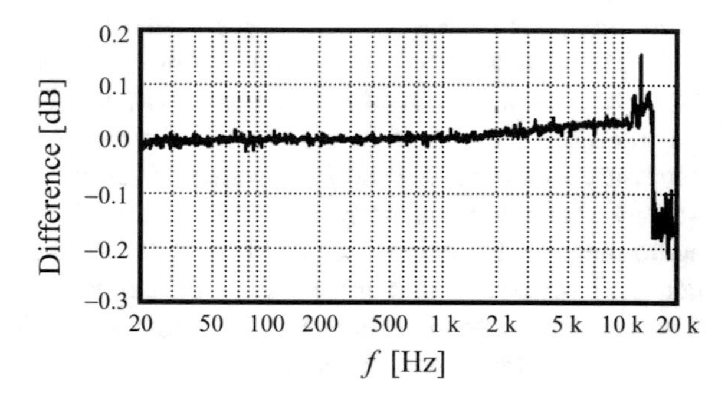

Fig. 9.3 Difference between the frequency response of the Zoom H4 obtained at T = 40 °C and at T = 24 °C

As it can be noted, the frequency response is acceptable since even in the worst case it is far below the tolerance limits. The sudden fall at about 16 kHz is due to the high sensitivity caused by the steep negative slope of the response at such frequency.

9.3.2 Noise

The self-generated or internal noise is composed of analog noise (due to miscellaneous sources such as thermal noise or shot noise) and digital noise, including discretization noise and dither. In order to measure the internal noise, the input has been short-circuited and recorded. Then RMS and Fast Fourier Transform (FFT) analysis has been performed on the .WAV file by software. A-weighting has

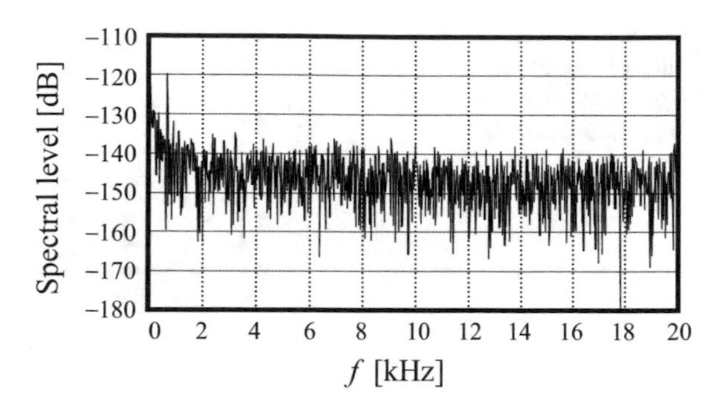

Fig. 9.4 Example of the worst-case spectral level of the internal noise of the Zoom H4 digital recorder, referenced to the maximum digital output. The signal was recorded with short-circuited input with a resolution of 16 bit and sample rate of 44.1 kHz

also been applied to the noise signal by software. Figure 9.4 shows the unweighted spectral density of the noise detected in the noisiest H4 unit, with a sampling rate of 44.1 kHz and a 16-bit resolution. While there are differences across units, sampling rates, and resolutions, these seem to be random and no significant effect is detected due to any single factor.

The worst case of the unweighted signal to noise ratio was 91.7 dB, while the A-weighted signal to noise ratio was 95.9 dBA, in both cases referred to an undistorted sine wave of maximum amplitude. This means that if a 94.0 dB calibration tone is recorded at maximum amplitude, it is possible to record signals down to about 20 dB with an uncertainty due to noise of about 0.1 dB. Note that most sound meters have an inherent noise equivalent to 20 dB.

9.3.3 Linearity

Linearity is tested in two different ways. First, by comparing its effect on the recorded signal for several input levels. This is the standard test specified in IEC 61672-2 (2003). Second, by computing the total harmonic distortion (THD) for a range of levels. Both tests have been performed using a tone modulated by a slowly varying exponential decay (duration $T = 20$ s; time constant $\tau = 8.69$ s; dynamic range, 80 dB).

In the first case (Fig. 9.5), the energy envelope has been computed with a first-order digital filter with $\tau = 0.05$ s and compared in a logarithmic scale to the theoretical, undistorted amplitude response. It has been expressed as the difference in dB between the real response and the ideal one.

An unexpected phenomenon can be observed close to −32 dB. There is an abrupt change in the error of about 0.2 dB. This does not affect the compliance with the IEC standard, since the maximum difference after a change of level between 1 dB and 10 dB should be less than ±0.6 dB. Once more, for levels below −70 dB the error increases quickly as a consequence of digital distortion. Besides, there is a spurious effect of inherent noise.

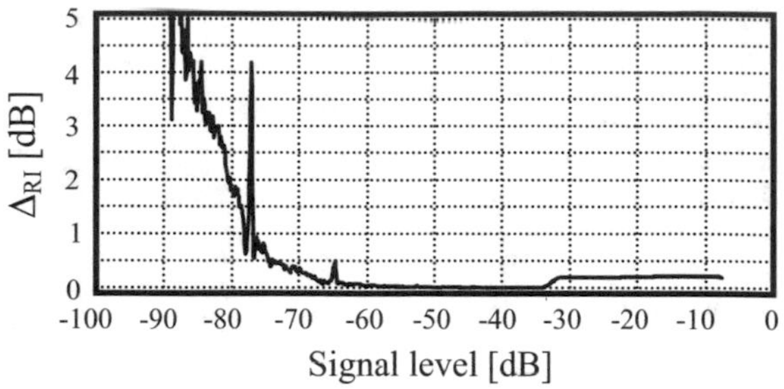

Fig. 9.5 Real response minus ideal response represented as $\Delta_{RI} = 20 \log (V_{RMS\text{-real}}/V_{RMS\text{-ideal}})$

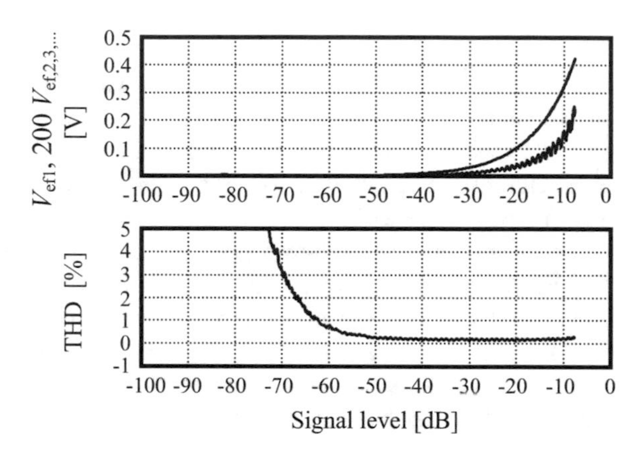

Fig. 9.6 Linearity analysis. *Top* First harmonic and higher harmonics (the latter scaled by a factor 200 for comparative purposes). *Bottom* Total harmonic distortion. In both cases the frequency is 1 kHz. The digital audio reference is the maximum digital signal

In the second case (Fig. 9.6), the fundamental and the remaining harmonics have been isolated by means of a highly selective multiband FFT filter in order to reduce the effect of wide band residual noise. Then both RMS values have been computed and compared. As it can be seen, harmonics 2, 3, … decrease more quickly than the fundamental, that is why the THD decreases slightly to about 0.02 %. As the signal level falls down, the THD starts to increase again due to digital distortion at very low levels. At −73 dB (referred to the maximum digital level), THD is around 5 %, which is equivalent to an error of 0.42 dB.

At the upper end of the spectrum (above 10 kHz) an aliasing effect has been observed, indicating the presence of some distortion after the anti-aliasing filter (perhaps due to the filter itself). This effect disappears altogether for signal levels below −20 dB respect to the digital maximum so it is recommended working below −20 dB. This somewhat reduces the dynamic range, but we have still about 70 dB available dynamic range, which is sufficient for most noise assessment applications.

9.3.4 Transient Response

In order to test the transient response, a square wave has been used. It has been found that the high frequency response is not exactly the same across units of model

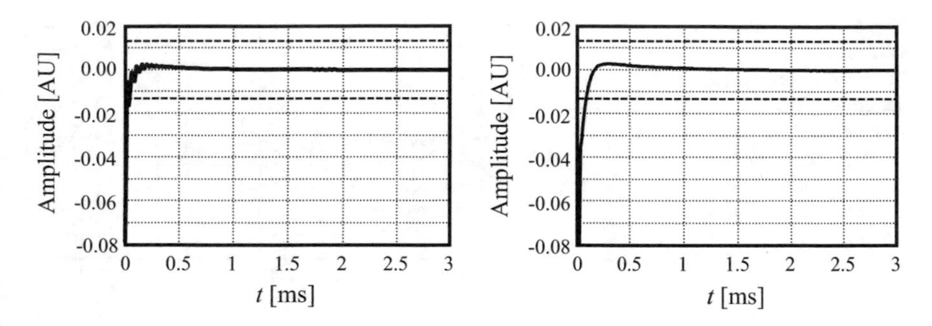

Fig. 9.7 Step response for 44.1 kHz (*left*) and 96 kHz (*right*). In *dashed line*, the 5 % limits

H4, but an upper bound for the 5 % settling time can be found. Responses are oscillating only for a sampling rate of 44.1 kHz, as shown in Fig. 9.7. Settling times were found to be at most 104 μs for 96 kHz and 136 μs for 44.1 kHz.

9.3.5 *Uncertainty*

In order to conform to the IEC 61672-1 (2002) Standard, the recorder should add an uncertainty of at most 0.1 dB. Test signals have been generated using a Stanford Research Systems DS 345 synthesized function generator. According to the specifications (Stanford Research Systems, undated), the sine wave accuracy is only ±0.4 dB. Since the use of the recorder requires the recording of a calibrated signal, this would be unimportant as long as the frequency response is sufficiently flat. However, no specification is provided regarding frequency response flatness, so it was checked using a Hewlett-Packard HP 974A digital multimeter (Hewlett-Packard Company 1995), whose true RMS total accuracy is given in Table 9.1 (including percentage of reading + counts). As shown in Fig. 9.8, the deviation of the DS 345 generator from flat response spans at most ±0.045 dB (96 % confidence), which combined with the HP 974A accuracy yields a 0.103 dB uncertainty. This uncertainty can be reduced to 0.095 dB taking into account the systematic deviation of the DS 345 at each frequency.

The distortion of the DS345 is specified as −55 dBc (decibels relative to carrier, i.e., to the fundamental). This represents a maximum error of $20 \log (1 + 10^{-55/20}) = 0.015$ dB when the peaks of harmonics are in phase with the main signal.

Table 9.1 True RMS accuracy of the multimeter HP 974A according to the frequency range (computed from specifications)

Frequency range (Hz)	Accuracy (dB)
20–50	±0.092
50–10,000	±0.049
10,000–20,000	±0.093

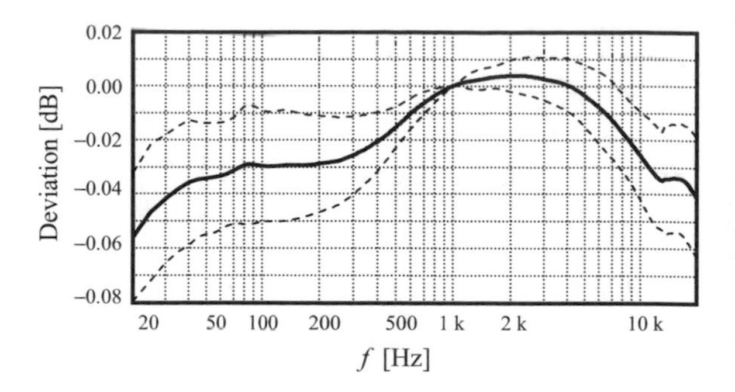

Fig. 9.8 Deviation from flatness of the frequency response of the signal generator SRS DS345. In *dashed lines* are indicated the 96 % (±2σ) confidence interval limits. An exact calibration is assumed at 1 kHz

9.3.6 Conclusion

Most parameters confirm that the tested units of model H4 are suitable for recording audio signals with measurement purposes. There is no significant difference between the various recording formats (16 bit or 24 bit). For instance, noise seems to have a lower bound due to analog noise and not to resolution. The model exhibits some distortion which causes, at the high frequency end (over 10 kHz), some noticeable aliasing. It is recommended to adjust the maximum recording level at about −20 dB in order to minimize distortion, which leaves about 60 dB of available dynamic range before reaching a level where the error due to noise grows to 0.1 dB.

Problems

Note: In order to work out some of these problems, you will need a true RMS voltmeter. Note that cheap multimeters may not have true RMS detectors. Most university electronic laboratories are equipped with a suitable instruments.

(1) (a) Simulate a sine wave of 20 kHz at a sample rate of 44.1 kHz. Find its RMS value and compare it with its theoretical value (b) Plot several cycles.

(2) (a) Generate a long exponential sine sweep from 20 Hz to 22 kHz with a sample rate of 44 100 Hz. Hint: See problem (1) from Chap. 8. (b) Find the moving maximum and the moving RMS over a sliding interval of N samples, for several values of N and plot the result as a function of the center of the interval, both versus time and versus frequency. (c) The frequency response should be flat. Investigate possible dips in the frequency response. (d) Try to find a theoretical criterion involving the center frequency, the rate of frequency increase and the size of the sliding interval.

(3) In order to calibrate the sound device of a computer connect the line output to the line input of the recorder to be tested. Use a specially assembled connector that provides access to the voltage through alligator clips. Make sure that the system mixer of the computer has its master and wave volume faders set at the maximum. If there is a software equalizer, it should be set flat at 0 dB. Panning should be at the center. If the sound driver has an automatic gain control (AGC), it should be disabled. Generate sine waves of different frequencies and half the full-scale digital amplitude. (a) While you record the input signal,

measure the true RMS voltage for each frequency. This voltage is representative of the computer sound device frequency response. (b) Calculate the recorded signal RMS value for each frequency. Find the frequency response of the recorder correcting for the frequency response of the sound device of the computer. (c) Repeat using a sine sweep instead of a series of discrete frequencies. Notes: (i) As a recorder you can use the input device of the computer. In full-duplex systems, Audacity will allow to record a signal while playing another. (ii) It is important that the frequency response be measured with the load (the recorder's input) connected.

(4) In order to measure the self-generated noise of the recorder, short-circuit the input of the recorder and record several seconds. Calculate the average FFT. As the number of averaged frames increases, the noise spectrum will tend to smooth out. Investigate the effect of several windows (Hann, Blackman, Blackman–Harris) and compare with the case of white Gaussian noise, considering the theoretical spectral density of a band-limited noise up to the Nyquist frequency.

(5) In order to test the nonlinear effects of the recorder, find the total harmonic distortion for 100 Hz, 1000 Hz, and 5000 Hz and three input amplitudes (adjusted from the software): Maximum amplitude (full scale), −10 dBFS and −20 dBFS.

(6) Test the transient response of the recorder. To this end you can use any external signal generator, which can be found in an electronics laboratory. Generate a square wave and look at the waveform with an oscilloscope in order to check that its settling time is much shorter than that of the recorder. Apply it to the recorder's input. Be sure not to exceed the maximum input signal of the recorder. If in doubt, start with an amplitude of 0.1 V and increase it gradually until the recorded signal is between 0.5 and 0.8 times the full scale amplitude. Write a Scilab script that finds automatically the 5 % settling time.

References

Hewlett-Packard Company (1995) HP 974A multimeter user's guide. http://cp.literature.agilent.com/litweb/pdf/00974-90002.pdf. Accessed 26 Feb 2016

IEC 61672-1 (2002) Electroacoustics—sound level meters—Part 1: specifications. International Electrotechnical Commission

IEC 61672-2 (2003) Electroacoustics—Sound level meters—Part 2: Pattern evaluation tests. International Electrotehcnical Commission

ISO 1996-2 (2007) Acoustics—description, measurement and assessment of environmental noise —Part 2: determination of environmental noise levels. Geneva, Switzerland

Miyara F, Accolti E, Pasch V, Cabanellas S, Yanitelli M, Miechi P, Marengo F, Mignini E (2010) Suitability of a consumer digital recorder for use in acoustical measurements. Internoise 2010. http://www.fceia.unr.edu.ar/acustica/biblio/consumer_recorders_for_noise_measurement_INTERNOISE_2010_993.pdf. Accessed 26 Feb 2016

Stanford Research Systems (no date) Synthesized function generators—DS345 specifications. http://www.thinksrs.com/downloads/PDFs/Manuals/DS345m.pdf. Accessed 27 Feb 2016

Zoom Corporation (no date) Handy recorder H4 operation manual. https://www.zoom.co.jp/sites/default/files/products/downloads/pdfs/E_H4v2_0.pdf. Accessed 27 Feb 2016

Further Readings

Anon (no date) FFT window and overlap. Accessed 2013/08/06, http://www.katjaas.nl/FFTwindow/FFTwindow.html

Audacity team (no date) File formats. Audacity Project website. Accessed 2013/07/02, http://audacity.sourceforge.net/xml/

Birch K (2003) Measurement good practice guide no. 36. estimating uncertainties in testing. An intermediate guide to estimating and reporting uncertainty of measurement in testing. British Measurement and Testing Association. Teddington, Middlesex, United Kingdom, 2003. http://publications.npl.co.uk/npl_web/pdf/mgpg36.pdf. Accessed 26 Feb 2016 http://www.dit.ie/media/physics/documents/GPG36.pdf

COVENIN 2552:1999 Vocabulario internacional de términos básicos y generales de metrología (in Spanish). http://www.sencamer.gob.ve/sencamer/normas/2552-99.pdf. Accessed 26 Feb 2016

Cox MG, Forbes AB, Harris PM (2004) Best practice guide no. 11. Numerical analysis for algorithm design in metrology. Technical report, National Physical Laboratory, Teddington, UK. http://resource.npl.co.uk/docs/science_technology/scientific_computing/ssfm/documents/ssfmbpg11.pdf. Accessed 26 Feb 2016

Cox MG, Harris PM (2006) Best practice guide no. 6. Software support for metrology. Uncertainty evaluation. NPL Report DEM-ES-011. Teddington, UK. http://resource.npl.co.uk/docs/science_technology/scientific_computing/ssfm/documents/ssfmbpg6.pdf. Accessed 26 Feb 2016

EA (2013) EA-4/02 M:2013 evaluation of the uncertainty of measurement in calibration. European co-operation for Accreditation. http://www.european-accreditation.org/publication/ea-4-02-m-rev01-september-2013. Accessed 26 Feb 2016

ISO/IEC Guide 99:2007 International vocabulary of metrology—basic and general concepts and associated terms (VIM)

ISO/TR 389-5:1998 Acoustics—reference zero for the calibration of audiometric equipment—Part 5: reference equivalent threshold sound pressure levels for pure tones in the frequency range 8 kHz to 16 kHz

ISO 11904-2:2004 Acoustics—determination of sound immission from sound sources placed close to the ear—Part 2: technique using a manikin

F. Miyara, *Software-Based Acoustical Measurements*, Modern Acoustics and Signal Processing, DOI 10.1007/978-3-319-55871-4

Janssen E, van Roremund A (2011) Look-ahead based sigma-delta modulation. Analog circuits and signal processing (Chapter 2). Springer. http://www.springer.com/cda/content/document/cda_downloaddocument/9789400713864-c2.pdf. Accessed 26 Feb 2016

Jarman D (1995) A brief introduction to sigma delta conversion. Intersil, May, 1995. http://www.intersil.com/content/dam/Intersil/documents/an95/an9504.pdf

Johnson R, Miller I, Freund J (2010) Miller and Freund's probability and statistics for engineers, 8th edn. Pearson

Klug A, Demmel H (no date) Digital editing with audacity 2.0.0. Pädagogische Hochschule Freiburg. http://www.mediensyndikat.de/grafik/material/audacity_reader_200_en.pdf. Accessed 29 Feb 2016

National Semiconductor (1986) An introduction to the sampling theorem. Application Note AN-236. Linear application handbook. Santa Clara, 1986. http://www.datasheetarchive.com/dlmain/Datasheets-21/DSA-418768.pdf. Accessed 26 Feb 2016. http://www.ti.com/lit/an/snaa079c/snaa079c.pdf. Accessed 26 Feb 2016

Park S (no date) Principles of sigma-delta modulation for analog to digital converters. Motorola APR8/D Rev. 1. http://xanthippi.ceid.upatras.gr/people/psarakis/courses/DSP_APL/demos/APR8-sigma-delta.pdf. Accessed 26 Feb 2016

Randall RB (1977) Application of B&K equipment to frequency analysis. Brüel & Kjær. Naerum, Denmark. http://www.ingelec.uns.edu.ar/pds2803/Materiales/LibrosPDF/Randall/TOC.htm. Accessed 26 Feb 2016

Schroder C (2011) The book of audacity—record, edit, mix, and master with the free audio editor. No Starch Press, San Francisco

Silva-Martinez J (no date) Sigma-delta modulators design issues. Texas A&M University Electrical and Computer Engineering. Accessed 2014/04/20 http://amesp02.tamu.edu/~jsilva/610/Lecture%20notes/Sigma-Delta-1.3.pdf

UKAS (2012) M3003—The expression of uncertainty and confidence in measurement, 2nd edn. United Kingdom Accreditation Service. https://www.ukas.com/download/publications/publications-relating-to-laboratory-accreditation/M3003_Ed3_final.pdf. Accessed 27 Feb 2016

Appendix A
Glossary and Definitions on Metrology[1]

Quantity: A property of a phenomenon or system susceptible of being represented numerically relative to a reference.

Note: Sometimes it is called *magnitude*.

Measurand: Quantity subject to measurement.

Note: The definition of a particular measurand requires to specify a series of conditions that allow a unique specification. Often the conditions are incompletely specified since it is impossible or impractical to take into account all the factors that could influence the measurand value. In general it is intended to consider at least those factors that presumably have a significant impact on the measurand value.

Realization of the measurand: A physical instance of the measurand compatible with its definition and susceptible of being subject to the measurement process.

Measurement: Process by means of which an approximation or estimate of the measurand value is obtained, as well as some expression of its uncertainty.

Measurement principle: Physical principle(s) on which the measurement is based.

Note: For example, electrostatic transduction of sound pressure to voltage in a microphone.

Measurement method: General description of the measurement steps.

Measurement procedure: Detailed description of the measurement process. It includes all aspects to take into account, from the type and features of the instruments to the calculations and formulas to apply, including the adequate selection of the samples and a survey of all variables that can influence the measurement result.

Measurement model: Mathematical relationship between the variables involved in a measurement. It may be an explicit or implicit formula or even an algorithm.

True value: Value of a measurand that would be obtained, in specified conditions, if an ideal instrument were used. In general the true value is not accurately known. It is only estimated or approximated.

---

[1]In general these definitions are conceptually in agreement with the ideas expressed in the ISO Guide to the Expression of Uncertainty of Measurements, GUM (ISO-GUM 1993). When there is no agreement or other alternatives are presented, it is advised. In many cases there are notes with examples and comments.

F. Miyara, *Software-Based Acoustical Measurements*, Modern Acoustics and Signal Processing, DOI 10.1007/978-3-319-55871-4

Note 1: Some exceptions are the quantities involved in the definition of the base units of the International System of Units (SI). For instance, the velocity of light in the vacuum is exactly 299,792,458 m/s.

Note 2: The ISO *Guide for the Expression of Uncertainty in Measurement* (ISO-GUM) avoids the use of the expression "true value". Instead it uses "value of the measurand", considering that the adjective "true" is redundant.

Note 3: The GUM also clarifies that in general there is not a single value of the measurand, as the ambiguity of the measurand definition due to the impossibility of specifying all the conditions that might influence its value, there could be more than one value that satisfies the (incomplete) definition. This gives rise to an *intrinsic uncertainty* that is not attributable to the instrument or the measurement procedure. An example is the measurement of the equivalent level of a white noise extended to a time interval of duration T. Equation (C.6) of Appendix C shows the statistical dispersion of the mean square pressure that in turn propagates to the equivalent level causing an intrinsic uncertainty. This uncertainty arises because the definition is ambiguous, since we are providing the duration T but not the exact location of the interval, such as would be if the measurand were defined as the equivalent level of the white noise between t_o and $t_o + T$ with a perfectly specified t_o.

Note 4: In some situations the true value is interpreted as a conventional reference value. An example is in the calibration context, where a reference standard is available and it is assumed that it is more accurate that the instrument to calibrate. This reference value has been measured with an uncertainty that is much lower than the expected uncertainty of the regular measurement procedure.

Precision: Degree of agreement of a series of measurements of a measurand with its mean value. It is generally expressed as the standard deviation of the values measured in repeatability conditions, i.e., the same procedure, location, operator, instrument, and ambient conditions.

Note: For example, a sound level meter has a precision of 0.1 dB if several measurements of the same sound in identical conditions have a standard deviation of 0.1 dB.

Accuracy: Degree of agreement of a measurement with the true value.

Note 1: For example, a sound level meter has an accuracy of 0.1 dB if any measurement differs from the true value in less than 0.1 dB. The accuracy depends on the calibration state of the instrument. However, the ISO (International Organization for Standardization) and other associated organizations such as the NIST (National Institute of Standards and Technology) ascribe to accuracy a qualitative nature not susceptible to numerical representation.

Note 2: A reason for this is that the uncertainty theory accepted by ISO, and other organizations implies some residual probability (though vanishingly small) that the measured value differs from the true value in an arbitrarily large quantity.

Resolution: For a given instrument, the minimum nonzero difference that is possible between two different readings.

Note 1: Resolution does not imply accuracy, but it does impose a limit to the attainable accuracy. For instance, a sound level meter with a resolution of 0.1 dB cannot guarantee accuracy better than 0.05 dB. Indeed, if the true value of a sound pressure level is 62.35 dB, the measured value will be, at best, 62.3 or 62.4 dB, which differs from the true value in 0.05 dB.

Note 2: In general the resolution of any instrument is much smaller than the uncertainty component due to the instrument, so the problem mentioned in Note 1 is not important.

Note 3: In the case of the sound level meter the resolution can be improved if instead of taking the reading directly it is obtained by software from a calibrated digital recording, since digitization is performed with many bits (in general 16 or even 24). This does not imply that the general uncertainty of the measurement will improve significantly, since the resolution factor has a very low impact on the measurement result.

Tolerance: Maximum acceptable dispersion between the true value and the specified value of a representative statistical parameter of a product.

Note 1: The dispersion is expressed statistically as the standard deviation multiplied by an adequate factor to guarantee a given confidence level, typically 95 %. For instance, if a batch of shafts has a specified diameter of (50 ± 0.06) mm, then the tolerance is achieved if the standard deviation of the diameter is 0.03 mm, since ± 2 standard deviations around the nominal value encompass 95 % of the units.

Note 2: If the tolerance were defined as the *maximum difference between the true value and the specified value of the parameter*, there would be no way to verify, in a particular case, if the product satisfies the tolerance, because of the measurement uncertainty.

Repetition: Replication within a short time interval of the measurement of the same measurand in identical conditions, such as procedure, operator, location, instruments, and environmental conditions.

Note 1: In the case of acoustical measurements the conditions include the relative position with respect to the sound source and surrounding objects and the orientation of the microphone.

Note 2: The repetition of a measurement and the subsequent averaging of the results allow to reduce the measurement uncertainty.

Note 3: Conceptually, in the case of acoustical measurements it is not possible in general to repeat the measurement because the measurand is time variable. An exception is laboratory conditions where a sound source can be controlled to produce essentially the same signal. Even in such a case, it is impossible to completely remove the residual acoustic noise.

Repeatability: Degree of agreement between the different measurements of the same measurand under the same conditions (procedure, location, operator, instruments, and ambient conditions) expressed in terms of the statistical dispersion of the results.

Repeatability conditions: Conditions in which the repeatability of a measurement process is assessed (the same procedure, location, operator, instruments, and ambient conditions).

Replication: Reiteration of a measurement of a realization of the measurand.

Note 1: The replication differs from the repetition in that the latter is performed on the same measurand. For instance the measurement of the length of several different rods produced by the same manufacturing process is a case of replication

Note 2: Replication allows to estimate the dispersion due to the manufacturing process, not to be confused with the measurement uncertainty.

Reproducibility: Degree of agreement between different measurements of the same measurand or different realizations of the measurand, varying some stated conditions (for instance the laboratory, the instruments, the operator), expressed in terms of the statistical dispersion of the results.

Note: The purpose of reproducibility is to have a knowledge of how robust a measurement procedure is faced with changes on some conditions that are expected to change when replicating a measurement. For instance, in interlaboratory comparisons, in the verification and confirmation of scientific or technical experimental results, in legal expert reports, in drafting specification sheets and in their independent verification. In all these cases changes are expected at least as regards the instrument (including brand, model and series number, even if they comply with the same technical standard), the operator, the location, the research material, and even the ambient conditions (temperature, atmospheric pressure and humidity). There could also be changes in the measurement protocol and sequence. The more the conditions are different, the larger is the expected dispersion of measurement values.

Reproducibility conditions: Conditions in which the reproducibility of a measurement process is assessed, specifying which conditions are allowed to change (for instance the operator, the instruments and the laboratory) and which not.

Measurement error: Difference between the measured value and the true value:

$$\varepsilon = x_{\text{measured}} - x_{\text{true}}.$$

Note 1: The error is inherently unknown, but in some cases it may be bounded, or specified statistically through the uncertainty concept. However, the generally accepted uncertainty theory does not allow an absolute error bound but a statistical specification. See the glossary entry for *Error bound*.

Note 2: The measurement error is the sum of the systematic error and the random error.

Note 3: Sometimes the measurement error is defined with respect to a conventional reference value and not the true value.

Observation error: Measurement error.

Systematic error: Statistical mean (mathematical expectancy) of the measurement errors when measuring the same measurand in repeatability condition:

$$\varepsilon_{\text{systematic}} = E(x_{\text{measured}} - x_{\text{true}}).$$

Note 1: The systematic error may be caused by an uncalibrated instrument or by measurement procedure defects. For instance, if a measurement of a low-frequency noise is performed very close to a reflecting surface, the measurement will systematically result in values higher than would result if measuring far away from any surface.

Note 2: The general concept of systematic error may include that of systematic effects such as gain errors or nonlinear transducer response.

Random error: Difference between a measurement value and the statistical mean (mathematical expectancy) of all possible measurements

$$\varepsilon_r = x_{\text{measured}} - E(x_k) = x_{\text{measured}} - (x_{\text{true}} + \varepsilon_{\text{systematic}}).$$

Note: The random error may be caused by limitations of the measurement instruments, by uncontrolled conditions within the measurement process (for instance, ambient conditions, measurement location, operator), or by the inherent variability of the measurand or its random nature. Of importance in acoustical metrology is the variability of the measurand over time.

Error bound: Maximum absolute measurement error under specified conditions.

Note: The ISO Guide to the Expression of Uncertainty of Measurement, (ISO-GUM 1993), as well as the NIST Guide (Taylor et al. 1994), tacitly presume that it is not possible to provide an error bound, since they only allow the use of statistical parameters such as the expanded uncertainty (see below the corresponding glossary entry), which stipulates an interval around the measured value within which the true value will lie with a stated probability (such as 95 %). The exception would be if somehow it were possible to accept that the error has a truncated distribution (such as the truncated normal distribution).

Probability density function: Function $f(x)$ such that the probability that a random variable X be between x_1 and x_2 is

$$P(x_1 < X < x_2) = \int_{x_1}^{x_2} f(x)\mathrm{d}x.$$

Mean (expected value): Mean value of all possible values of a random variable X:

$$\mu = \int_{-\infty}^{\infty} x f(x)\mathrm{d}x.$$

where $f(x)$ is the probability density function.

Note 1: The mean is also known as *mathematical expectancy* and, sometimes, *population mean*.

Standard deviation: Square root of the mean value of the squared difference between the random variable and the mean:

$$\sigma = \sqrt{\int_{-\infty}^{\infty} (x - \mu)^2 f(x) \mathrm{d}x}$$

where $f(x)$ is the probability density function.

Variance: It is the square of the standard deviation of a random variable.

Joint probability density function: Given two random variables X, Y, it is a function $f(x, y)$ such that

$$P(x_1 < x < x_2, y_1 < y < y_2) = \int_{x_1}^{x_2} \int_{y_1}^{y_2} f(x, y) \, \mathrm{d}x\mathrm{d}y.$$

Note: As an alternative differential version of the definition, it is a function such that

$$P(x < X < x + \mathrm{d}x, y < Y < y + \mathrm{d}y) = f(x, x)\mathrm{d}x\mathrm{d}y.$$

Covariance: Given two random variables X, Y, it is the mean value of the product of the differences between each variable and its mean:

$$\mathrm{cov}(X, Y) = \int_{-\infty}^{\infty} \int_{-\infty}^{\infty} (x - \mu_x)(y - \mu_y) f(x, y)\mathrm{d}x\mathrm{d}y$$

where $f(x, y)$ is the joint probability density function of X and Y.

Note 1: The covariance of a random variable with itself is equal to its variance. In such case the probability density function is

$$f(x, y) = \delta(y - x) f(x),$$

where δ is the Dirac impulse function.

Note 2: When two random variables X, Y are independent, i.e., such that the joint probability density function f_{XY} is the product of the probability density functions f_X and f_Y:

$$f_{XY}(x, y) = f_X(x) f_Y(y)$$

then

$$\mathrm{cov}(X, Y) = 0.$$

The larger is the covariance the more alike are the variables (i.e., simultaneous measurements of these variables are similar to each other).

Sample mean: Given n instances $x_1, \ldots, x_n$ of a random variable X, the sample mean is:

$$\bar{x} = \frac{1}{n}\sum_{k=1}^{n} x_k.$$

Note 1: This parameter approximates the mean or expected value. The *sample* mean is, therefore, an *estimate* of the *population* mean. Since $\bar{x}$ is, in turn, a random variable, it is possible to compute its mean $\mu_{\bar{x}}$ and its standard deviation $\sigma_{\bar{x}}$ resulting

$$\mu_{\bar{x}} = \mu, \quad \sigma_{\bar{x}} = \frac{\sigma}{\sqrt{n}}.$$

Note 2: Since the standard deviation decreases as n increases, it is possible to reduce the statistical dispersion in the determination of μ repeating the measurement several times.

Sample standard deviation: Given n samples $x_1, \ldots, x_n$ of a random variable X, it is:

$$s = \sqrt{\frac{1}{n-1}\sum_{k=1}^{n}(x_k - \bar{x})^2}.$$

Note: This parameter approximates the standard deviation of X. It is, therefore, an *estimate* of the population standard deviation.

Sample variance: The square of the sample standard deviation.

Note: The sample variance is an unbiased estimate of the variance (see Appendix B).

Sample covariance: Given n pairs of samples $(x_1, y_1), \ldots, (x_n, y_n)$ of the random variables X, Y, it is

$$m_{xy} = \frac{1}{n-1}\sum_{k=1}^{n}(x_k - \bar{x})(y_k - \bar{y})$$

Note 1: It is assumed that each pair (x_k, y_k) is the result of the same observation or replication in which both values have been measured *simultaneously*. For instance, the ambient temperature and the sound pressure level.

Note 2: This parameter approximates the covariance of X and Y. It is, therefore, an *estimate* of the covariance.

Note 3: The sample covariance of a variable with itself is equal to its sample variance.

Note 4: The sample covariance of two independent random variables is close to 0 but not necessarily equal to 0, since the independence is a population property.

Uncertainty: Qualitative concept or quantitative expression of the degree of ignorance, after a measurement, of the difference between the measured (estimated) value and the true value of the measurand (i.e., the measurement error). It is generally made up of a series of components attributable to different aspects of the measurement process (instruments, operator, nature of the observed phenomenon). It may be expressed as a statistical parameter.

Standard uncertainty: Standard deviation of a series of measurements of a measurand under specified conditions:

$$u = \sigma = \sqrt{\int_{-\infty}^{\infty} (x - \mu)^2 f(x) \mathrm{d}x}$$

Note: The standard uncertainty is usually composed of several components associated with different causes (see *Combined standard uncertainty*).

Type A uncertainty: Any uncertainty component obtained by statistical methods from the measured data, in general after several repetitions or replications of the measurement in specified conditions of repeatability. In this case it is possible to use generally accepted statistical methods:

$$u = s = \sqrt{\frac{1}{n-1} \sum_{k=1}^{n} (x_k - \bar{x})^2}$$

Type B uncertainty: An uncertainty component obtained by any method other than the statistical treatment of measurement results. Examples include the knowledge or independent estimation by scientific or technical consensus (such as the published uncertainty in physical constants or in measurement procedures), information provided in standards, instrument specifications, information provided in calibration certificates, and software simulations.

Combined standard uncertainty: The total uncertainty attributable to all known relevant sources of uncertainty. Given N random variables $X_1, \ldots, X_N$ on which measurand Y functionally depends according to

$$Y = f(X_1, \ldots, X_N),$$

let us suppose we wish to get y for a combination of measured data $x_1, \ldots, x_N$. The combined standard uncertainty can be computed as

$$u_C = \sqrt{\sum_{h=1}^{N}\sum_{k=1}^{N}\frac{\partial f}{\partial x_h}\frac{\partial f}{\partial x_k}\,u(x_h, x_k)}$$

where $\partial f/\partial x_h$ and $\partial f/\partial x_k$ are the partial derivatives of f calculated at the observed (estimated) values $x_1, \ldots, x_N$ and $u(x_h, x_k)$ are the estimated covariances:

$$u(x_h, x_k) = m_{x_h y_k} = \frac{1}{n-1}\sum_{i=1}^{n}(x_{hi} - \bar{x}_h)(x_{ki} - \bar{x}_k)$$

Note 1: The formula for u_c is derived applying a multivariate first-order Taylor approximation centered at the observed values of variables $x_1, \ldots, x_N$ (see Appendix H). It is also known as *error propagation formula.*

Note 2: For $h = k$ we have

$$u(x_h, x_h) = \sigma_{x_h}^2.$$

Note 3: If all variables are statistically independent, the combined standard uncertainty is

$$u_C = \sqrt{\sum_{h=1}^{N}\left(\frac{\partial f}{\partial x_h}\right)^2 \sigma_{x_h}^2}.$$

since the covariances between different variables vanish (the *covariance matrix* is diagonal).

Note 4: The function f may contain the physical model that allows to indirectly compute a desired quantity, for instance the volume of a rectangular parallelepiped from the lengths of its sides, a, b, c:

$$V = abc.$$

In this case it is a function of three variables. It could also describe the measurement process. For instance, if we want to determine the equivalent level of a stationary noise by making 10 measurements, we shall have

$$L_{eq} = 10\log_{10}\left(\frac{10^{L_{p1}/10} + \cdots + 10^{L_{p10}/10}}{10}\right).$$

In this case it is a function of 10 variables.

Expanded uncertainty: It is the combined standard uncertainty multiplied by a coverage factor k

$$U = k\,u_C$$

Note 1: The coverage factor is such that it allows to guarantee, with a specified probability P, that the measurand value is within $\pm U$ from the estimated value y, i.e.,

$$y - U \leq Y \leq y + U$$

If the probability density function is normal (see Appendix B) and $k = 2$, then $P = 0.96$, which means that there is a probability of only 4 % that the estimated measurand value differs in more than U from the true value.

Note 2: The expanded uncertainty, even with a rather high coverage factor, does not allow to define an interval where the true value will lie with probability $P = 1$. For instance, for a normal distribution, a coverage factor $k = 5$ gives a confidence of only 0.99999944. The main reason to adopt this approach in the expression of uncertainty is that the statistical formalism behind the uncertainty concept allows a relatively rigorous calculation of the effects of the different uncertainty sources, which happens to be, besides, rather simple in the case in which it can be safely assumed that the variables have normal distribution. But it drops any certainty in measurement. For instance, when one measures the length of a 1.2 m rod, there exists a nonzero probability that the result is 2 m, 10 km or even −1 m. This situation has given rise to some controversy. See *Error bound*.

Note 3: In some cases one can circumvent the problem stated in Note 2 by using a *truncated normal distribution*, i.e., one that is normal below some absolute value of the variable and 0 above it.

Traceability: A chain of known uncertainties that connect the result of a measurement (estimate of the measurand) with a national or international reference accepted by the metrological community.

Note 1: In the absence of traceability there is no way to quantify the uncertainty of a measurement. For instance, a precision instrument could be uncalibrated and in such case we would know neither its systematic error nor its possible statistical distribution.

Note 2: The traceability allows to link a measurement to a recognized measurement standard with maximum technical quality.

Measurement standard: Materialization or realization of a unit (or a multiple) corresponding to a quantity type (for instance 1 m if the quantity is a length; 1 Pa if the quantity is a sound pressure), or an instrument capable of measuring it, with the purpose of defining, keeping or reproducing it. A measurement standard has a specified associated uncertainty.

Primary standard (Primary measurement standard): Measurement standard (corresponding to a quantity type) of the maximum possible technical quality. Typically each country keeps a primary standard with traceability to the international prototype when it does exist.

Secondary standard (Secondary measurement standard): Measurement standard (corresponding to a quantity type) calibrated and traceable to a primary standard.

Reference standard (Laboratory standard): Measurement standard used for the calibration of other measurement standards in a laboratory and traceable through a traceability chain to a primary standard.

Working standard: Measurement standard (corresponding to a quantity type) with traceability to a primary standard used to calibrate measurement instruments.

Calibration: Process by means of which the uncertainty of an instrument is determined in several operating conditions with respect to a standard reference with traceability to a national or international reference accepted by the metrological community.

Note: In the case of a sound level meter, the operating conditions may include testing it at different frequencies and in several ranges.

Renard series: Series of preferred values introduced by Charles Renard in 1870. They are distributed approximately uniformly within a decade in logarithmic scale. They are designated by the letter *R* followed by the number of values in a decade, which can be 5, 10, 20 or 40. For instance, the *R*10 introduces the following 10 values in a decade: 1.00 1.25 1.60 2.00 2.50 3.15 4.00 5.00 6.30 8.00. The series are extended to other decades multiplying and dividing by powers of 10. The values are approximations of the series $10^{k/n}$, where $k = 0, \ldots, n - 1$ and n is the number of figures per decade.

Note 1: Since $10^{3/10} \cong 2$, the series *R*10, *R*20 and *R*40 have the property that the values also allow to divide exactly or almost exactly the octave. That is why they have been adopted as the standard series for the octave and one-third octave filters.

Note 2: The complete Renard series are

*R*5 : 1.00 1.60 2.50 4.00 6.30

*R*10 : 1.00 1.25 1.60 2.00 2.50 3.15 4.00 5.00 6.30 8.00

*R*20 : 1.00 1.12 1.25 1.40 1.60 1.80 2.00 2.24 2.50 2.80
3.15 3.55 4.00 4.50 5.00 5.60 6.30 7.10 8.00 9.00

*R*40 : 1.00 1.06 1.12 1.18 1.25 1.32 1.40 1.50 1.60 1.70
1.80 1.90 2.00 2.12 2.24 2.36 2.50 2.65 2.80 3.00
3.15 3.35 3.55 3.75 4.00 4.25 4.50 4.75 5.00 5.30
5.60 6.00 6.30 6.70 7.10 7.50 8.00 8.50 9.00 9.50

References

ISO-GUM (1993) Guide to the expression of uncertainty in measurement, prepared by ISO Technical Advisory Group 4 (TAG 4). Working Group 3 (WG 3), Oct 1993.

JCGM 100:2008 Evaluation of measurement data—guide to the expression of uncertainty in measurement. Working Group 1 of the Joint Committee for Guides in Metrology. http://www.bipm.org/utils/common/documents/jcgm/JCGM_100_2008_E.pdf. Accessed 26 Feb 2016

Taylor B, Kuyayt C (1994) NIST technical note 1297. Guidelines for evaluating and expressing the uncertainty of NIST measurement results. Physics Laboratory, National Institute of Standards and Technology, Gaithersburg, MD. http://www.nist.gov/pml/pubs/tn1297/. Accessed 17 Mar 2016

Appendix B
Fundamentals of Statistics

B.1 Descriptive Statistics

Statistical analysis is an important part of any measurement process, so it is worth reviewing some statistical concepts (Johnson et al. 2010).

First of all, one of the aims of this discipline is to derive global properties of a set of entities (called *population*) whose individual properties cannot be obtained accurately and exhaustively for all the entities. An example is the height of the people of a given geographic region. If one picks a person at random, it is not possible to predict his/her height. But it is possible to get a *global descriptor*, such as the *average height* of those people, and it is also possible to provide an idea of the height dispersion, i.e., how different can be the real height of an individual from the average height. The most interesting part of statistical methods is the possibility to *estimate* those global properties through the examination of a proper *sample*[2] of the population, i.e., a small subset of the whole population. This task is called *statistical inference*.

The properties associated with the entities included in the population are expressed in general as *random variables*, i.e., variables that take some value (in general unknown) for each population element. In the preceding example, the height of the individuals of the region is a random variable. In the case of acoustical metrology, the random variable of interest is the *result of a measurement*.

The general behavior of a random variable is governed by its *statistical distribution*. In order to clarify this concept, suppose we are interested in the hearing threshold at 4000 Hz of a group of 50 people. First, we split the total range of such variable (for instance, from 0 dB to 40 dB) into 5 dB intervals (usually called *bins*). Then we just count how many individuals have their threshold within each bin.

[2]Notice that in Statistics, a *sample* has quite a different meaning than in discrete signal processing. In Statistics it corresponds to a certain number n of instances of a random variable, while in signal processing it is a single value of the discrete signal (taken from the analog signal to be represented by the discrete signal).

F. Miyara, *Software-Based Acoustical Measurements*, Modern Acoustics and Signal Processing, DOI 10.1007/978-3-319-55871-4

Table B.1 An example of frequency distribution: hearing threshold for a group of 50 persons

Bin (dB)	0–5	5–10	10–15	15–20	20–25	25–30	30–35	35–40
Number	2	3	9	13	10	7	4	2
Percent (%)	4	6	18	26	20	14	8	4

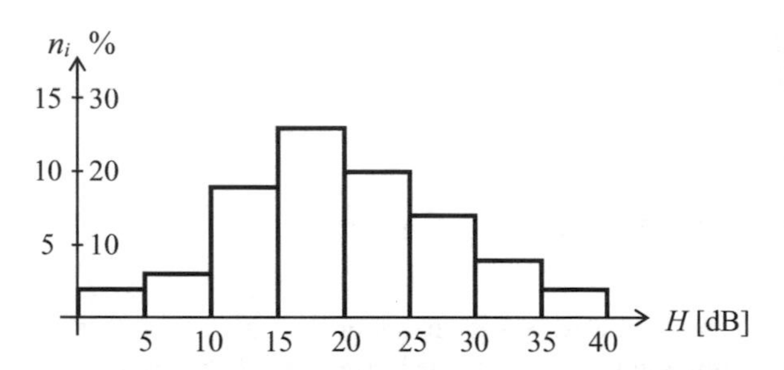

Fig. B.1 Histogram: A graphical representation of the frequency and percent distributions corresponding to the example of Table B.1

Table B.1 is an example, including the number of individuals in each bin as well as the percentage.

The number of elements in each bin is called *frequency*,[3] and the classification process yields the *frequency distribution*. If the same information is presented in terms of percentages we get the *percent distribution*. Both can be represented graphically as shown in Fig. B.1. This graph is called *histogram*.

The percent distribution is a more convenient representation of the statistical distribution of a random variable since on the one hand it allows to compare the behavior of different data sets and on the other hand it allows *statistical inference*. Indeed, if according to some criterion we can consider that the sample (in this case the threshold levels of the 50 people) is *representative* of the population, it will be possible to extrapolate the results to the whole population and, hence, to other samples taken from such population.

We can also introduce the *cumulative frequency distribution* as the cumulative sum of the frequencies corresponding to successive bins, as shown in Fig. B.2. If instead of the frequencies we add the percentages corresponding to each bin, we get the *cumulative percent distribution*.

We can associate two types of parameters with each random variable: the *measures of central tendency* and the *measures ofvariability or dispersion*. Measures of central tendency describe, according to different criteria, a "central" value around which the values of the random variable gather. Measures of variability, also according to several criteria, indicate how far can be the individual values from the central value.

[3]The *frequency of appearance* of an interval of values must not be confused with the physical concept of *frequency* of a tone or any other periodic phenomenon.

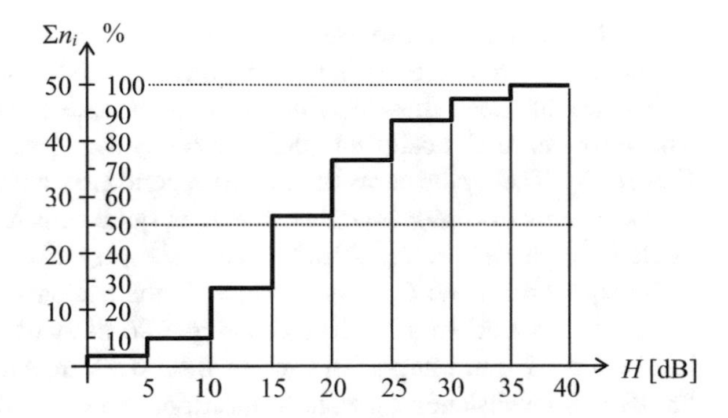

Fig. B.2 Graphical representation of the cumulative frequency distribution and the cumulative percent distribution corresponding to the example of Table B.1

The three more widely used measures of central tendency are the *mean* (also called *average*, or *mean value*), the *median* and the *mode*. The *mean* $\bar{x}$ of a random variable x is obtained as

$$\bar{x} = \frac{1}{N}\sum_{k=1}^{N} x_k, \tag{B.1}$$

where N is the total number of elements of the population and x_k the value of the random variable corresponding to the k-th element. In the case in which the total range of the random variable has been split into a relatively small number m of bins (as in the Example 1), the Eq. (B.1) can be rewritten as

$$\bar{x} = \frac{1}{N}\sum_{i=1}^{m} n_i x_i, \tag{B.2}$$

where x_i is the value of the random variable associated with the i-th bin and n_i the corresponding frequency. Applying this equation to the frequencies of Table B.1, and assuming that for each bin the variable takes a value equal to the average of its limits (for instance in the case of the interval 20–25, we will assume x_i = 22.5), we have $\bar{x} = 19.8$ dB.

The *median*, med(x) is such that 50 % of the population presents smaller values than med(x) and the remaining 50 % have larger values. It is easy to find the median from the cumulative percent distribution looking for the value that is closer to 50 %. In the preceding example, from Fig. B.2 we can see that it is a value between 15 and 20. Conventionally we take the central value of the interval, 17.5 dB.

Finally, the *mode* of a distribution, mode(x) is the value of the random variable where the frequency distribution reaches a maximum. In the previous example, from Fig. B.1 it can be seen that it is also the central value of the interval between 15 dB and 20 dB, i.e., 17.5 dB. Note that a distribution could have more than one mode. This happens when the variable values gather around two or more values. An example may be the hearing threshold of the workers of a manufacturing plant: office employees, who work in quiet rooms, will have different thresholds but generally lower than those who work in production activities using noisy tools and machinery. In this case we have a *bimodal* distribution.

The use of one of these measures of central tendency depends on several factors. The mean is analytically simpler, allowing the systematic use of results from Probability Theory because of its correspondence with the *mathematical expectancy*. The main disadvantage is that it is quite sensitive to extreme values, particularly in the case of variables that can take very dissimilar values. A typical case is the equivalent level L_{eq}. For instance, if the background noise is about 55 dBA, the pass of a truck during 10 s causing a 90 dBA noise will rise the equivalent level over a 1 min integration time to 82.3 dBA, much closer to 90 dBA than to 55 dBA, notwithstanding its short duration. When the distribution is symmetrical, both the mean and the median are equal. In strongly asymmetric distributions the median is usually more representative. The mode is a good descriptor for communicating results, but it is not well suited for their processing. It has the drawback that it completely ignores what is happening with all values other that the most frequent one.[4]

The most widely used measure of variability is the *standard deviation*, σ_x, defined as the square root of the mean square deviation with respect to the mean:

$$\sigma_x = \sqrt{\frac{\sum_{k=1}^{N} (x_k - \bar{x})^2}{N}}, \tag{B.3}$$

where x_k are the population values of the random variable, $\bar{x}$ the mean value of x and N the size of the population. If the variable has been split into bins, it can also be computed as

$$\sigma_x = \sqrt{\frac{\sum_{i=1}^{m} n_i (x_i - \bar{x})^2}{N}}. \tag{B.4}$$

where x_i is the value corresponding to the i-th bin and n_i the frequency of the values within that bin, and $N = \sum n_i$. As an example, we can compute the standard deviation corresponding to the example of Table B.1. The result is $\sigma_x = 8.2$ dB. In general, a substantial part of all the values of a distribution are found to be between $\bar{x} - 2\sigma_x$ and $\bar{x} + 2\sigma_x$. The larger the standard deviation, the more diffuse, and hence the less concentrated, the distribution of the random variable.

It is also interesting in Statistics the concept of *variance*, defined as the square of the standard deviation, since it has additive properties with respect to independent

[4]The mode is also useful when analyzing the behavior of nominal (non-numerical) variables, for instance when a survey is conducted to find out which kind of noise is the most annoying from a descriptive list (for instance, motorbikes without exhaust silencer, discotheques or pneumatic drills). There is no way to "average" the three noises or to order them on a low to high scale.

causes of variability of a random variable. A drawback is that it is not dimensionally compatible with the random variable, since its units are squared.

Both the mean and the standard deviation of a random variable associated with the population can be estimated from a properly selected *sample*. In general, for a sample to be representative of the population it must be selected at random[5] and its size must be large enough to guarantee with a high probability that the error in the estimation will not exceed a specified value (for instance that the error will be smaller than 2 dB with probability 95 %). The detailed discussion of the selection criteria and the preliminary tests needed exceed the introductory purpose of this Appendix.

The *population mean* may be directly estimated by the *sample mean*, which can be computed with the same expression as the former, i.e.,

$$\bar{x} = \frac{1}{n}\sum_{k=1}^{n} x_k, \tag{B.5}$$

where n is the *sample size*. The *population standard deviation* may be estimated by the *sample standard deviation*, which is computed with a slightly different formula:

$$s = \sqrt{\frac{\sum_{k=1}^{n} \left(x_k - \bar{x}\right)^2}{n-1}}. \tag{B.6}$$

The difference in the denominator tends to be small for large-sized samples.[6] For instance, for samples larger than 10 elements, the difference is smaller than 5 %, and for samples with more than 50 elements, it is smaller than 1 %.

Although the random variable that we have been considering in the examples (the hearing threshold) is inherently *continuous*, since their values may be in principle any real number, for its statistical treatment we have transformed it into a *discrete variable*, since from each interval (bin) we have taken its central value as representative. There are two reasons for that. The first one is that in general an accurate determination of the threshold is not possible, since even for the same individual the threshold might change over time and the audiometric test offers

[5]For instance, if it is desired to find the mean hearing threshold (at a given frequency) of the workers of a manufacturing plant, the sample should contemplate similar demographic variables (age, working area, years of service) as those in the whole population (all the employees working in the plant). It would be incorrect to select all the workers belonging to the same area or of the same etary segment, unless the study is meant to analyze those particular groups.

[6]The reason to divide by $n - 1$ instead of n is that in this way we get an *unbiased estimate* of the variance. An estimate is unbiased if the mean of all estimations from different samples is equal to the population parameter. Note that the sample standard deviation is not an unbiased estimate of the populational standard deviation.

several difficulties.[7] The second is that reducing the number of possible values of the random variable simplifies the statistical data handling.

B.2 Probability

For the purpose of modeling and theoretical analysis, it is more convenient to work with continuous distributions, especially when dealing with large populations of indeterminate size. In this case we will consider a concept akin to the percent distribution. Since there are infinite possible values, it is unlikely that one of them corresponds to a specific element of the population. For example: How many people have a hearing threshold at 1 kHz exactly equal to 4.572349 dB? Probably, no one. It is more reasonable to ask how many people have a hearing threshold between 4 dB and 5 dB. More generally, we can ask what fraction of the population will have a hearing threshold between x and $x + \Delta x$. In other words we are asking what is the *probability* that an individual taken at random has a hearing threshold between x and $x + \Delta x$. It is evident that the smaller Δx the smaller such probability. But if we divide the probability by Δx we get a quantity that tends (for each x) to a well defined value that we shall call *probability density*, i.e.,

$$f(x_0) = \lim_{\Delta x \to 0} \frac{P(x_0 < X < x_0 + \Delta x)}{\Delta x}, \tag{B.7}$$

where $P(A)$ denotes the probability that condition A holds.[8] If we consider all possible values of x_0 as the domain we will have the *probability density function*, shown for a particular case in Fig. B.3.

We define the *cumulative probability function*, or *cumulative distribution function* $F(x_0)$, as the probability that x is smaller than x_0. It can be computed as

$$F(x_0) = \int_{-\infty}^{x_0} f(x)\mathrm{d}x. \tag{B.8}$$

Figure B.4 shows the cumulative probability function corresponding to the probability density function of Fig. B.3. The probability density function is always normalized so that $F(\infty) = 1$.

In the case of continuous random variables, the mean can be computed with the formula

[7] For instance, the headset position, the attention or the fatigue of the subject under test may affect significantly the measurement results.

[8] The probability is 0 when the condition is never true, for instance, that when throwing a die we get 7, and is 1 when the condition is true, for instance that the result of throwing a die is between 1 and 6.

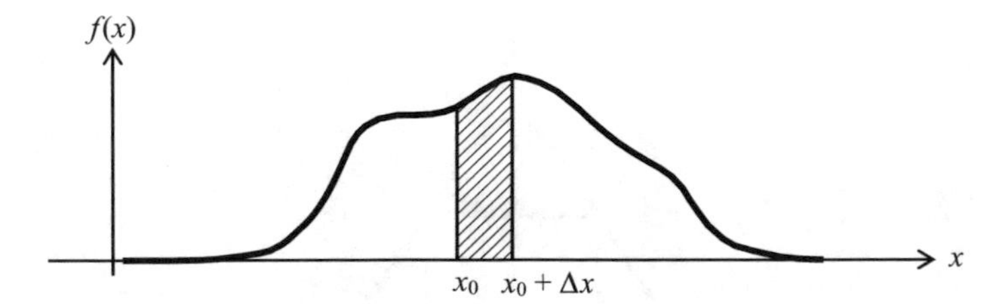

Fig. B.3 An example of probability density function. The *shaded area* represents the probability that the random variable is between x_0 and $x_0 + \Delta x$

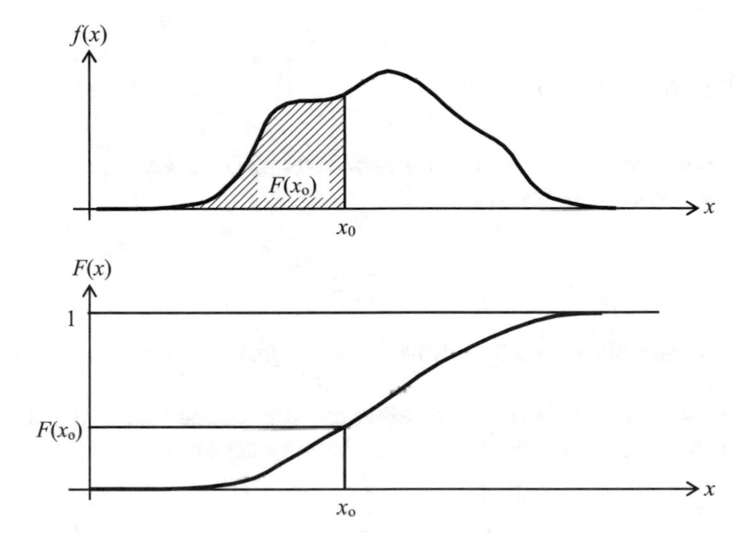

Fig. B.4 Probability density function $p(x)$ and its cumulative probability function. The *shaded area* represents the probability that the random variable is smaller than x_0

$$\mu = \int_{-\infty}^{\infty} x f(x)\mathrm{d}x \tag{B.9}$$

Likewise, the standard deviation is computed as

$$\sigma = \sqrt{\int_{-\infty}^{\infty} (x - \mu)^2 f(x)\mathrm{d}x} \tag{B.10}$$

These formulas are formally equivalent to those corresponding to a discrete random variable, where the sums have been replaced by integrals.

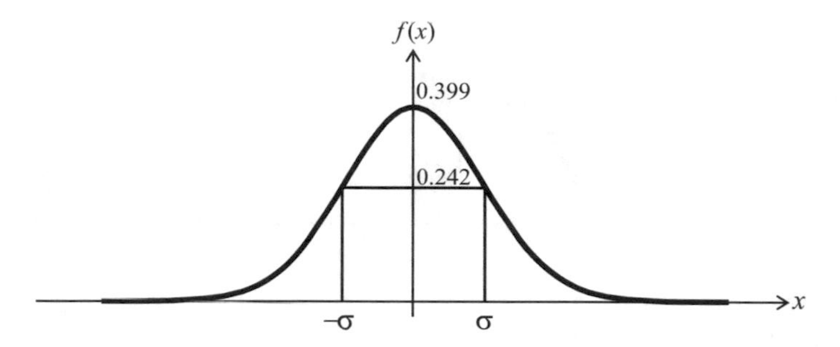

Fig. B.5 Probability density function of the normal distribution (Gaussian distribution) with $\mu = 0$ and $\sigma = 1$

B.3 Normal Distribution

A very important case is the *normal distribution* or *Gaussian distribution*, whose probability density function is

$$f(x) = \frac{1}{\sqrt{2\pi}} e^{-\frac{x^2}{2}}. \tag{B.11}$$

This function, shown in Fig. B.5, has a mean equal to 0 and a standard deviation equal to 1.

It is possible to have a normal distribution with arbitrary mean μ and standard deviation σ by means of the following probability density function:

$$f(x) = \frac{1}{\sqrt{2\pi}\,\sigma} e^{-\frac{(x-\mu)^2}{2\sigma^2}}. \tag{B.12}$$

One of the most interesting properties of the normal distribution is that the means of samples of size n of whatever distribution tend to be normally distributed when $n \rightarrow \infty$. This property, known as the *central limit theorem*, has important applications in error theory.

B.4 Uniform Distribution

Another distribution of interest in metrology is the *uniform distribution*. In this distribution the random variable may take values in a bounded interval $a < X < b$ and it does so in such a way that all values are equally probable, meaning that the probability density function is constant in that interval and 0 in all remaining values:

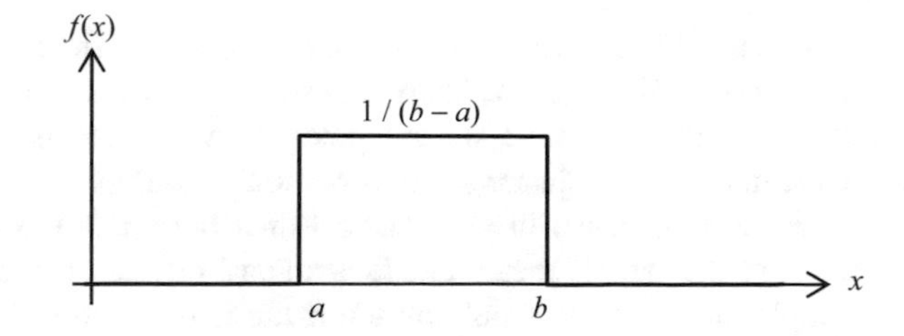

Fig. B.6 Probability density function of a uniform distribution

$$f(x) = \begin{cases} \frac{1}{b-a} & a \leq x \leq b \\ 0 & x < a \text{ or } b < x \end{cases} \tag{B.13}$$

The value $1/(b - a)$ has been adopted so that the total probability is 1. Figure B.6 shows this distribution.

The uniform distribution appears naturally in measurements made with any finite-resolution instrument, in particular, instruments with a digital display, but also analog or quasi-analog instruments with a scale with divisions. When observing an indication between two divisions, in the absence of an interpolating device such as the Vernier system it will only be possible to record the value corresponding to the closest division. Since there is, in principle, no reason to consider that the measurand might be at any distance in particular from a given division, the distribution of all possible values is uniform.

The mean of a uniform distribution is half the sum of the interval limits,

$$\mu = \frac{a+b}{2}. \tag{B.14}$$

The standard deviation is

$$\sigma = \sqrt{\int_a^b \frac{1}{b-a}\left(x - \frac{b+a}{2}\right)^2 \mathrm{d}x} = \frac{b-a}{\sqrt{12}}. \tag{B.15}$$

B.5 Distribution of the Sample Mean

If we take a sample of size n from a population, the mean of the n instances of the random variable is, in turn, a new random variable called *sample mean*:

$$\bar{x} = \frac{1}{n}\sum_{k=1}^{n} x_k, \tag{B.16}$$

It is a random variable because it changes with the particular sample we take. In other words, if we repeat many times the operation of extracting a random sample of size n and each time we compute its sample mean, those sample mean values constitute different instances of a random variable.

We are interested in its statistical distribution. We need to know its mean (the mean of the sample mean) and its standard deviation (the standard deviation of the sample mean). It turns out that the mean is equal to the original population mean:

$$\mu_{\bar{x}} = \mu. \tag{B.17}$$

The standard deviation is equal to the population standard deviation divided by $\sqrt{n}$:

$$\sigma_{\bar{x}} = \frac{\sigma}{\sqrt{n}}. \tag{B.18}$$

In other words, the larger the sample size, the smaller the expected "error" when estimating the population mean with the sample mean. This is the basis of statistical inference.

These properties hold, in general, for any random variable, either normal or other distributions.

Now, in the preceding discussion we assumed that we knew μ and σ, but in general they are unknown and we can only estimate them. In fact, μ is estimated by the sample mean $\bar{x}$ computed through Eq. (B.5) and σ is estimated by the sample standard deviation using Eq. (B.6)

$$s = \sqrt{\frac{\sum_{k=1}^{n} (x_k - \bar{x})^2}{n - 1}}. \tag{B.19}$$

When $n \gg 1$, the estimate is quite good, not so for small sample sizes. The lack of knowledge of the real dispersion from which we start implies that the dispersion that would be derived from (B.18) rises.

It can be shown that the auxiliary normalized variable

$$t = \frac{\bar{x} - \mu}{s/\sqrt{n}} \tag{B.20}$$

has a statistical distribution known as Student t distribution with $\nu = n - 1$ degrees of freedom. Figure B.7 shows the t distribution for three different degrees of freedom and their comparison with the normal distribution. For $\nu \geq 30$ it is generally accepted that it can be replaced by a normal distribution with mean 0 and standard deviation 1.

The probability density function for the Student t distribution is given by (Smirnov et al. 1978)

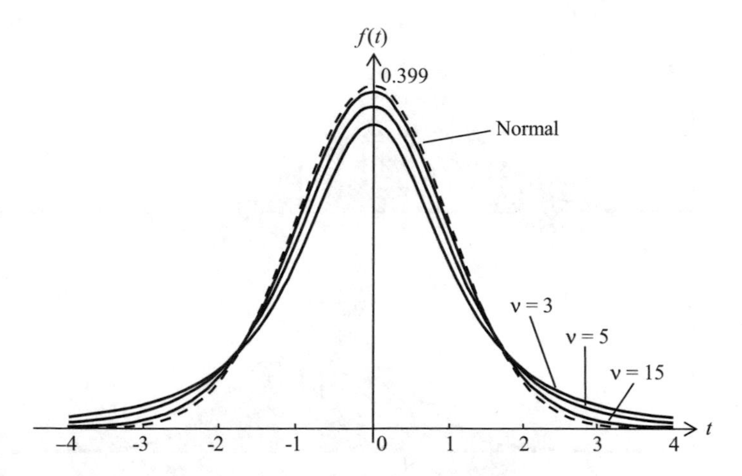

Fig. B.7 Probability density function of the Student t distribution for three different numbers of degrees of freedom, compared with the normal distribution (in *dashed line*)

$$f(t,\nu) = \frac{\Gamma\left(\frac{\nu+1}{2}\right)}{\sqrt{\pi\nu}\,\Gamma\left(\frac{\nu}{2}\right)}\left(1+\frac{t^2}{\nu}\right)^{-\frac{\nu+1}{2}}, \tag{B.21}$$

where $\Gamma(x)$ is the gamma function, defined as

$$\Gamma(x) = \int_0^\infty z^{x-1}\mathrm{e}^{-z}\mathrm{d}z. \tag{B.22}$$

The t distribution has mean 0 and standard deviation, for $\nu > 2$,

$$\sigma_t = \sqrt{\frac{\nu}{\nu - 2}}. \tag{B.23}$$

This standard deviation ranges from 1.73 for $\nu = 3$ down to 1 for $\nu = \infty$.

B.6 Inference Relative to the Population Mean

The sample mean $\mu_{\bar{x}}$ may be used to estimate the population mean; and its distribution, particularly its standard deviation, to estimate the estimation error. Now, if the distribution of the sample mean is, for instance, a Student t or a normal distribution, it would imply that any value is actually possible, even values very different from the true population mean. Does this mean that the estimate is

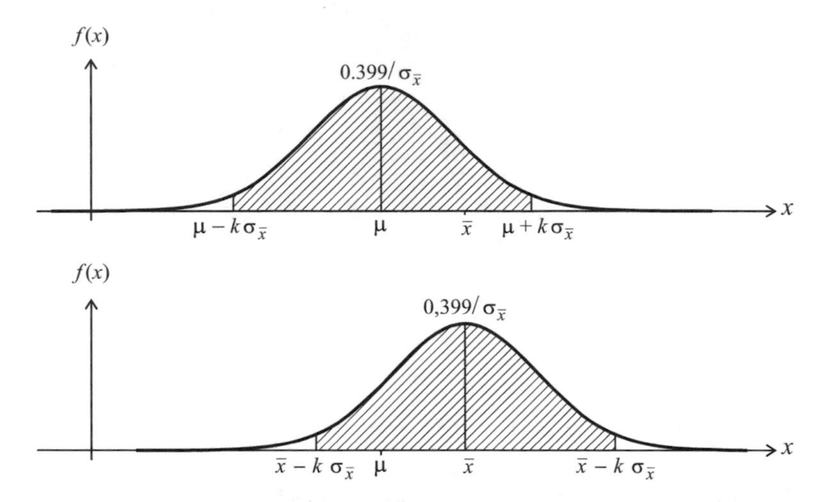

Fig. B.8 Two interpretations of the proximity between the population mean μ and the sample mean $\bar{x}$

completely useless? Not at all, if we can find a *confidence interval*, i.e., an interval in which the population mean is with a specified probability. If this probability is chosen to be reasonably high (for instance, 95 % or 99 %), this would imply that the fact that our estimate fell outside that interval would be a very rare event. Of course, the higher the probability, the wider the confidence interval.

Let us consider, in the first place, the case in which σ is previously known (or it can be accurately estimated). Suppose we want to get an interval around our estimate $\bar{x}$ that contains the population mean with probability $1 - \alpha$, or, in other words, so that the probability that it is *not* in that interval (residual probability) is α. The problem is equivalent to finding an interval of the same amplitude around μ that contains the estimated mean $\bar{x}$ with probability $1 - \alpha$.[9] In particular, this means that we can use exactly the same distribution replacing the mean as shown in Fig. B.8.

Calling $z_{\alpha/2}$ the value of the variable of the standard normal distribution ($\mu = 0$, $\sigma = 1$) such that $P(x > z_{\alpha/2}) = \alpha/2$, because of the symmetry the confidence interval such that $P(|x| < z_{\alpha/2}) = 1 - \alpha$ can be expressed as

$$\bar{x} - z_{\alpha/2}\sigma_{\bar{x}} < \mu < \bar{x} + z_{\alpha/2}\sigma_{\bar{x}},$$

[9] Indeed, let us see that μ and $\bar{x}$ are interchangeable:

$$\begin{aligned} P(\bar{x} - L < \mu < \bar{x} + L) &= P(-L < \mu - \bar{x} < L) \\ &= P(-\mu - L < -\bar{x} < -\mu + L) \\ &= P(\mu - L < \bar{x} < \mu + L) \end{aligned}$$

Table B.2 Values of $z_{\alpha/2}$ and $t_{\nu,\ \alpha/2}$ for several values of residual probability α and number ν of degrees of freedom for the determination of confidence intervals for the population mean

α	0.1	0.05	0.025	0.010	0.005	0.001
$z_{\alpha/2}$	1.6449	1.9600	2.2414	2.5758	2.8070	3.2905
$t_{2,\ \alpha/2}$	2.9200	4.3027	6.2053	9.9248	14.089	31.599
$t_{3,\ \alpha/2}$	2.3534	3.1824	4.1765	5.8409	7.4533	12.924
$t_{5,\ \alpha/2}$	2.0150	2.5706	3.1634	4.0321	4.7733	6.8688
$t_{10,\ \alpha/2}$	1.8125	2.2281	2.6338	3.1693	3.5814	4.5869
$t_{15,\ \alpha/2}$	1.7531	2.1314	2.4899	2.9467	3.2860	4.0728
$t_{20,\ \alpha/2}$	1.7247	2.0860	2.4231	2.8453	3.1534	3.8495
$t_{25,\ \alpha/2}$	1.7081	2.0595	2.3846	2.7874	3.0782	3.7251
$t_{30,\ \alpha/2}$	1.6973	2.0423	2.3596	2.7500	3.0298	3.6460
$t_{40,\ \alpha/2}$	1.6839	2.0211	2.3289	2.7045	2.9712	3.5510

or, considering (B.18), as

$$\bar{x} - z_{\alpha/2}\frac{\sigma}{\sqrt{n}} < \mu < \bar{x} + z_{\alpha/2}\frac{\sigma}{\sqrt{n}}. \tag{B.24}$$

where σ is the population standard deviation.

Consider now the case in which σ is not known but it is estimated by the sample standard deviation of a size n sample. The distribution is, in this case, the Student t distribution with $n - 1$ degrees of freedom. Calling $t_{n-1,\ \alpha/2}$ the value of t such that $P(x > t_{n-1,\ \alpha/2}) = \alpha/2$, the confidence interval results

$$\bar{x} - t_{n-1,\alpha/2}\frac{s}{\sqrt{n}} < \mu < \bar{x} + t_{n-1,\alpha/2}\frac{s}{\sqrt{n}}. \tag{B.25}$$

Table B.2 gives the values of $z_{\alpha/2}$ and $t_{\nu,\ \alpha/2}$ for different values of the residual probability α and number ν of degrees of freedom.

Example 1 Suppose we have measured the following values of the sound pressure level:

73.6	80.2	76.7	75.5	76.7	75.3	75.4	77.8	77.7	77.7

The mean value is 76.66 dB and the sample standard deviation, 1.83 dB. The confidence interval for a probability of 95 % can be determined using the t distribution with 9 degrees of freedom and $\alpha = 0.05$. Interpolating in Table B.2 between 5 and 10 degrees of freedom we have

$$t_{9,0.05/2} \cong 2.2966,$$

from where the confidence interval that contains the population mean with a probability of 95 % will be

$$76.66 - 2.2966\frac{1.83}{\sqrt{10}} < \mu < 76.66 + 2.2966\frac{1.83}{\sqrt{10}}$$

i.e.,

$$75.33 \text{ dB} < \mu < 77.99 \text{ dB}.$$

B.7 Distribution of the Sample Variance

The sample variance, i.e., the square of the sample standard deviation, given by

$$s^2 = \frac{\sum_{k=1}^{n} (x_k - \bar{x})^2}{n - 1}, \tag{B.26}$$

is an *unbiased estimate* of the population variance. Unbiased means that the mean over a large number of samples tends to the population variance. We are interested in how it is statistically distributed. Since s^2 is an unbiased estimate, its mean is, precisely, σ^2. If the population is normally distributed, then the auxiliary variable χ^2, defined as

$$\chi^2 = \frac{(n-1)\, s^2}{\sigma^2}, \tag{B.27}$$

obeys a *chi-squaredistribution*, with $\nu = n - 1$ degrees of freedom, whose probability density function is

$$f(x, \nu) = \begin{cases} \frac{1}{2^{\frac{\nu}{2}}\Gamma\left(\frac{\nu}{2}\right)} x^{\frac{\nu}{2}-1} \mathrm{e}^{-\frac{x}{2}} & x \geq 0 \\ 0 & x < 0 \end{cases} \tag{B.28}$$

The null value for negative values of the argument is because the sample variance is a positive variable.

The mean of the chi-square distribution is ν and its variance is 2ν. Figure B.9 shows the probability density function of this distribution for selected values of ν. As can be observed, the distribution is asymmetric and it has a mode (maximum) at $\nu - 2$.

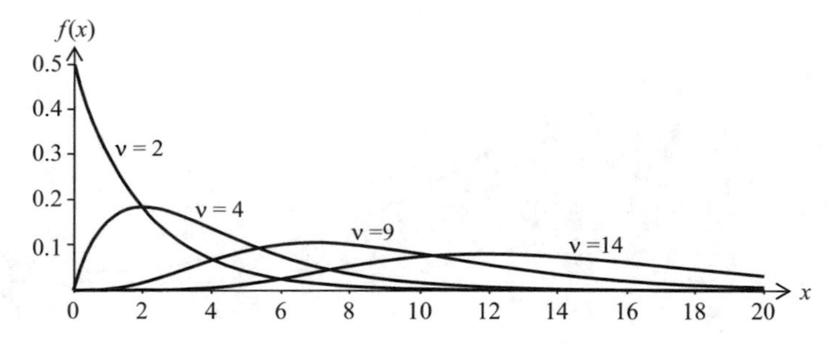

Fig. B.9 Probability density function of the chi-square distribution for selected values of ν

B.8 Inference Relative to the Population Variance

The sample variance s^2 can be used to estimate the population variance σ^2, and its distribution allows to study the estimation error. As in the case of the mean, we want to find a confidence interval around the estimate s^2 that contains the population variance σ^2 with probability $1 - \alpha$.

Since the chi-square distribution is asymmetric, we should get two values, $\chi^2_{1-\alpha/2}$ and $\chi^2_{\alpha/2}$, which define the limits towards both sides of the mean, as shown in Fig. B.10. Indeed, $\chi^2_{1-\alpha/2}$ leaves to its right a fraction $1 - \alpha/2$ of instances of the variable χ^2, while $\chi^2_{\alpha/2}$ leaves to its right a fraction $\alpha/2$. The difference is the value $1 - \alpha$ that we were looking for.

We can express the confidence interval for the normalized variable as

$$\chi^2_{1-\alpha/2} < \frac{(n-1)s^2}{\sigma^2} < \chi^2_{\alpha/2}.$$

Solving for σ^2, we get

$$\frac{(n-1)s^2}{\chi^2_{\alpha/2}} < \sigma^2 < \frac{(n-1)s^2}{\chi^2_{1-\alpha/2}}. \tag{B.29}$$

Table B.3 provides the values of $\chi^2_{1-\alpha/2}$ and $\chi^2_{\alpha/2}$ for selected values of the parameters α and ν.

Example 2 With the data of Example 1, let us find the confidence interval for a 95 % probability that it includes the population standard deviation. In this case $n = 10$, so $\nu = 9$. Besides, $\alpha = 1 - 0.95 = 0.05$. The interpolated values between the ones corresponding to $\nu = 5$ and $\nu = 10$ of the table are

$$\chi^2_{1-\alpha/2} = 3.3813$$
$$\chi^2_{\alpha/2} = 16.86$$

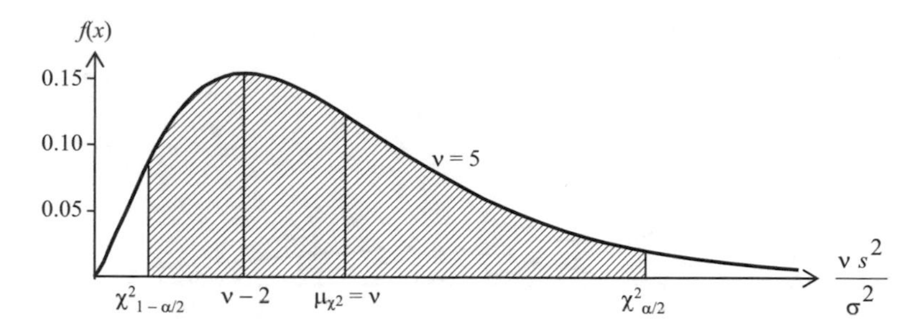

Fig. B.10 Determination of a confidence interval for the normalized variance from the chi-square distribution. The normalized variance is between $\chi^2_{1-\alpha/2}$ and $\chi^2_{\alpha/2}$ with probability $1 - \alpha$. The example is for $\nu = 5$ and $\alpha = 0.1$. The mean is $s^2 = \sigma^2$

Table B.3 Values of $\chi^2_{1-\alpha/2}$ and $\chi^2_{\alpha/2}$ for several values of the residual probability α and the number of degrees of freedom ν for the determination of the confidence intervals for the population variance

ν	α	0.1	0.05	0.025	0.010	0.005	0.001
3	$\chi^2_{\alpha/2}$	7.8147	9.3484	10.8610	12.8380	14.3200	17.7300
	$\chi^2_{1-\alpha/2}$	0.3519	0.2158	0.1338	0.0717	0.0449	0.0153
5	$\chi^2_{\alpha/2}$	11.0700	12.8330	14.5440	16.7500	18.3860	22.1050
	$\chi^2_{1-\alpha/2}$	1.1455	0.8312	0.6109	0.4117	0.3075	0.1581
10	$\chi^2_{\alpha/2}$	18.3070	20.4830	22.5580	25.1880	27.1120	31.4200
	$\chi^2_{1-\alpha/2}$	3.9403	3.2470	2.7072	2.1559	1.8274	1.2650
15	$\chi^2_{\alpha/2}$	24.9960	27.4880	29.8430	32.8010	34.9500	39.7190
	$\chi^2_{1-\alpha/2}$	7.26090	6.2621	5.4569	4.6009	4.0697	3.1075
20	$\chi^2_{\alpha/2}$	31.4100	34.1700	36.7610	39.9970	42.3360	47.4980
	$\chi^2_{1-\alpha/2}$	10.8510	9.5908	8.5563	7.4338	6.7228	5.3981

from where

$$\frac{9 \times 1.83^2}{16.86}\ \mathrm{dB}^2 < \sigma^2 < \frac{9 \times 1.83^2}{3.3813}\ \mathrm{dB}^2.$$

It turns out, therefore, that with a 95 % probability

$$1.377\ \mathrm{dB} < \sigma < 2.986\ \mathrm{dB}.$$

References

Johnson R, Miller I, Freund J (2010) Miller and Freund's probability and statistics for engineers, 8th edn. Pearson

Smirnov NV, Dunin-Barkowski I (1978) Cálculo de probabilidades y estadística matemática. Ed. Paraninfo. Madrid, España.

Appendix C
Statistical Dispersion of the RMS Value of a Stationary Noise as a Function of the Integrating Time

Let us consider a spectrally limited white noise $p(t)$ of RMS value P_{ef}. The probability density function of its instant values is

$$f(p) = \frac{1}{\sqrt{2\pi}\,\sigma} e^{-\frac{p^2}{2\sigma^2}}. \tag{C.1}$$

Because of ergodicity, the temporal average is equal to the mean, for a given instant, extended to all possible realizations or instances of the noise

$$P_{ef}^2 = \lim_{T\to\infty} \frac{1}{T} \int_{-T/2}^{T/2} p^2(t)\mathrm{d}t = \int_{-\infty}^{\infty} p^2 f(p)\mathrm{d}p$$

$$P_{ef}^2 = \frac{1}{\sqrt{2\pi}\,\sigma} \int_{-\infty}^{\infty} p^2 e^{-\frac{p^2}{2\sigma^2}} \mathrm{d}p \tag{C.2}$$

This can be integrated by parts, rewriting p^2 as the product of p by itself. One is left alone and the other multiplies the exponential, yielding an easily integrable function. Finally, we get $P_{ef}^2 = \sigma^2$, from which

$$P_{ef} = \sigma. \tag{C.3}$$

Let us now calculate the variance of the square pressure. To this end, consider that by the very definition of standard deviation, the mean of p^2 is, precisely, σ^2. We have, then

F. Miyara, *Software-Based Acoustical Measurements*, Modern Acoustics and Signal Processing, DOI 10.1007/978-3-319-55871-4

$$
\begin{aligned}
\sigma_{p^2}^2 &= \int_{-\infty}^{\infty} \left(p^2 - \sigma^2\right)^2 f(p)\mathrm{d}p \\
&= \frac{1}{\sqrt{2\pi}\,\sigma} \int_{-\infty}^{\infty} \left(p^2 - \sigma^2\right)^2 \mathrm{e}^{-\frac{p^2}{2\sigma^2}}\mathrm{d}p \\
&= \frac{1}{\sqrt{2\pi}\,\sigma} \int_{-\infty}^{\infty} \left(p^4 - 2p^2\sigma^2 + \sigma^4\right)\mathrm{e}^{-\frac{p^2}{2\sigma^2}}\mathrm{d}p
\end{aligned}
$$

This integral can be split into three integrals. The first two are integrated by parts and the third is the normal probability density function. It results

$$\sigma_{p^2}^2 = 2\sigma^4,$$

from where,

$$\sigma_{p^2} = \sqrt{2}\,\sigma^2 = \sqrt{2}P_{\mathrm{ef}}^2. \tag{C.4}$$

This is the statistical dispersion of the square pressure (notice that the square pressure is not normally distributed).

We are now interested in the statistical dispersion of the effective or RMS value extended to a time T, as a function of T. Suppose that the white noise is band-limited to f_{max}. Then it is possible to sample it at a sample rate $F_{\mathrm{s}} = 2\,f_{\mathrm{max}}$. Calling $K = [F_{\mathrm{s}}T]$, where [ ] is the integer part, we can calculate the square RMS pressure as

$$P_{\mathrm{ef},T}^2 = \frac{1}{K}\sum_{k=1}^{K} p^2(k).$$

This is interesting since it allows to express the square RMS pressure as a discrete sum of random variables $p^2(k)$. Moreover, it is indeed the mean of a sample of size K. The standard deviation of the sample mean is equal to the standard deviation of the individual variables divided by the square root of the number of variables:

$$\sigma_{P_{\mathrm{ef},T}^2} = \sqrt{\frac{2}{K}}P_{\mathrm{ef},\infty}^2, \tag{C.5}$$

where $P_{\mathrm{ef},\infty}^2$ is the long-term RMS value. In other words,

$$\sigma_{P^2_{\mathrm{ef},T}} \cong \sqrt{\frac{1}{f_{\max} T}} P^2_{\mathrm{ef},\infty}. \tag{C.6}$$

We can see that the statistical dispersion of the square RMS value is reduced as the square root of the integrating time. This phenomenon is known as the *stabilization* of the effective or RMS value of pressure and is useful to find the minimum measuring time required to guarantee a given error bound with a given confidence level.

Appendix D
Statistical Dispersion of the RMS Value of a Nonstationary Noise as a Function of the Integrating Time

Consider the example of a traffic noise $p(t)$. This noise is the result of the superposition of a series of events consisting in the pass-by of Q individual vehicles per unit time (traffic flow). If the maximum pressure measured at a distance d of the vehicle trajectory is P_{max} and its velocity is v, it can be shown (Miyara 1999) that the square RMS pressure extended to a time interval T centered at the instant of maximum proximity is given by

$$P_{ef,T}^2 = \frac{P_{max}^2}{T}\frac{d}{v}2\arctan\frac{vT}{2d}. \tag{D.1}$$

when $T \gg 2d/v$, the argument of the arctangent is very large and it tends to $\pi/2$, from where we have

$$P_{ef,T}^2 \cong P_{max}^2 \frac{\pi d}{vT}. \tag{D.2}$$

As we can see, for large T the square RMS pressure is inversely proportional to T. This is because virtually all the acoustic energy contributed by the vehicle at the receiver has been already radiated but it is distributed over an increasing time interval.

The square RMS value differs from one vehicle to the next one because of statistic dispersion. We can assume that the square RMS pressure corresponding to the k-th vehicle is

$$P_{ef,k,T}^2 = K_k P_{ef,o,T}^2 \tag{D.3}$$

where $P_{ef,o,T}^2$ is the mean value and K_k, a dimensionless random variable that is independent of the integration time[10] whose mean is equal to 1. We can express this in terms of the equivalent level dividing by the square reference pressure and applying 10 log:

[10] K_k depends essentially on the acoustic power, the velocity, and the distance, which are random variables.

F. Miyara, *Software-Based Acoustical Measurements*, Modern Acoustics and Signal Processing, DOI 10.1007/978-3-319-55871-4

$$L_{\mathrm{eq},k,T} = L_{\mathrm{eq,o},T} + 10\log K_k = L_{\mathrm{eq,o},T} + \Delta L_{\mathrm{eq},k},$$

where

$$\Delta L_{\mathrm{eq},k} = 10\log K_k.$$

Then we can present (D.3) as

$$P^2_{\mathrm{ef},k,T} = P^2_{\mathrm{ef,o},T}10^{\Delta L_{\mathrm{eq},k}/10} \tag{D.4}$$

The exponential can be approximated linearly when the exponent is small:

$$P^2_{\mathrm{ef},k,T} = P^2_{\mathrm{ef,o},T}\left(1 + \frac{\ln 10}{10}\Delta L_{\mathrm{eq},k}\right) \tag{D.5}$$

This equation allows to get the square RMS pressure caused by the $Q \cdot T$ vehicles that passed during time T by energy superposition of incoherent sources:

$$P^2_{\mathrm{ef},T} = \sum_{k=1}^{Q\,T} P^2_{\mathrm{ef},k,T} = QTP^2_{\mathrm{ef,o},T} + P^2_{\mathrm{ef,o},T}\frac{\ln 10}{10}\sum_{k=1}^{QT}\Delta L_{\mathrm{eq},k}$$

We replace Eq. (D.2) in this equation, getting

$$\begin{aligned} P^2_{\mathrm{ef},T} &= QTP^2_{\mathrm{max,o}}\frac{\pi d_\mathrm{o}}{v_\mathrm{o}T} + P^2_{\mathrm{max,o}}\frac{\pi d_\mathrm{o}}{v_\mathrm{o}T}\frac{\ln 10}{10}\sum_{k=1}^{QT}\Delta L_{\mathrm{eq},k} \\ &= QP^2_{\mathrm{max,o}}\frac{\pi d_\mathrm{o}}{v_\mathrm{o}} + QP^2_{\mathrm{max,o}}\frac{\pi d_\mathrm{o}}{v_\mathrm{o}QT}\frac{\ln 10}{10}\sum_{k=1}^{QT}\Delta L_{\mathrm{eq},k} \\ &= P^2_{\mathrm{ef},\infty}\left(1 + \frac{1}{QT}\frac{\ln 10}{10}\sum_{k=1}^{QT}\Delta L_{\mathrm{eq},k}\right) \\ &= P^2_{\mathrm{ef},\infty}\left(1 + \frac{1}{QT}\frac{\ln 10}{10}\Delta L_{\mathrm{eq}}\right) \end{aligned}$$

where $P_{\mathrm{ef},\,\infty}$ is the long-term RMS pressure. The global increment ΔL_{eq} has a standard deviation $\sqrt{QT}$ times larger than the individual increments $\Delta L_{\mathrm{eq},\,k}$ of each vehicle with respect to an average vehicle. We can rewrite the preceding equation in terms of the equivalent level:

$$L_{\mathrm{eq},T} = L_{\mathrm{eq},\infty} + \frac{\Delta L_{\mathrm{eq}}}{QT} = L_{\mathrm{eq},\infty} + \Delta L.$$

We can set the following bound with a confidence level of 99.7 %:

$$\left|\Delta L_{\text{eq}}\right| < 3\ \sigma_{\Sigma L_{\text{eq},k}} = 3\sqrt{QT}\,\sigma_{L_{\text{eq},k}}$$

hence,

$$|\Delta L| = \left|\frac{\Delta L_{\text{eq}}}{QT}\right| < \frac{3}{\sqrt{QT}}\,\sigma_{L_{\text{eq},k}} \tag{D.6}$$

We can conclude that the error when integrating during a finite time T with respect to the value that would be obtained integrating during an infinite time (long-term value) decreases with the square root of the integration time T and with the square root of the number of vehicles per unit time, Q.

Finally, we can calculate the *stabilization time* to have an error smaller than $\Delta L_{\max}$ considering the inequality

$$\frac{3}{\sqrt{QT}}\,\sigma_{L_{\text{eq},k}} < \Delta L_{\max}.$$

We get

$$T_{\text{stabil}} = \frac{9}{Q}\left(\frac{\sigma_{L_{\text{eq},k}}}{\Delta L_{\max}}\right)^2. \tag{D.7}$$

Reference

Miyara F (1999) Modelización del ruido del tránsito automotor (in Spanish: Traffic noise modelization). INGEACUS 99. Valdivia, Chile, 1999. http://www.fceia.unr.edu.ar/acustica/biblio/MRT.pdf. Accessed 26 Feb 2016

Appendix E
Statistical Dispersion of Percentiles

We shall consider a random variable X with normal distribution of mean μ and standard deviation σ, i.e., whose probability density function is

$$f(X) = \frac{1}{\sqrt{2\pi}\,\sigma} \mathrm{e}^{-\frac{(X-\mu)^2}{2\sigma^2}}. \tag{E.1}$$

We can find the percentiles from the cumulative probability function

$$F(X) = \int_{-\infty}^{X} \frac{1}{\sqrt{2\pi}\,\sigma} \mathrm{e}^{-\frac{(x-\mu)^2}{2\sigma^2}} \mathrm{d}x. \tag{E.2}$$

The p percentile is the value X_p such that

$$F(X_p) = \frac{p}{100}, \tag{E.3}$$

i.e.,

$$X_p = F^{-1}\left(\frac{p}{100}\right). \tag{E.4}$$

The inverse function F^{-1} can be obtained from the *inverse error function*, which is available in many mathematical software packages. The error *function* is defined as

$$\operatorname{erf}(x) = \frac{2}{\sqrt{\pi}} \int_{0}^{x} \mathrm{e}^{-x^2} \mathrm{d}x. \tag{E.5}$$

F. Miyara, *Software-Based Acoustical Measurements*, Modern Acoustics and Signal Processing, DOI 10.1007/978-3-319-55871-4

After some algebraic steps we can express $F(X)$ in terms of erf(x), from which

$$X_p = \sqrt{2}\,\sigma\,\mathrm{erf}^{-1}\left(2\frac{p}{100} - 1\right) + \mu. \tag{E.6}$$

X_p is the population $p$percentile of variable X, i.e., the p percentile extended to the whole population. What happens if instead of analyzing the whole population we analyze a sample of size n, i.e., $\{X_1, \ldots, X_n\}$? A different p percentile will be obtained for each sample. The mean of all sample p percentiles is an estimate of the population p percentile, and its variability, expressed in terms of the standard deviation, can be obtained from the following equation (Kendall and Stuart 1977):

$$\sigma_{X_p} = \sqrt{\frac{\frac{p}{100}\left(1-\frac{p}{100}\right)}{n}}\frac{1}{f(X_p)}. \tag{E.7}$$

This formula indicates that for a given p, the standard deviation of the sample p percentile decreases with the square root of the sample size n. It is also observed that in the case of the extreme percentiles (close to 0 and 100), although the numerator tends to 0, the denominator tends to 0 more quickly through $f(X_p)$, so the statistical dispersion is larger.

In particular, the minimum (0‰) and the maximum (100‰) are subject to an infinite variability. That is why, whatever the sample size, the minimum and maximum are not representative.

Reference

Kendall M, Stuart A (1977) The advanced theory of statistics, vol 1. Charles Griffin & Co., London

Appendix F
Envelope of a Filtered Noise

F.1 Theoretical Time Dependence of a Filtered Noise

Let $n(t)$ be a Gaussian white noise and suppose we filter it with an ideal bandpass filter with cut-off frequencies f_1 and f_2. The response $x(t)$ can be obtained as the convolution with the impulse response $h(t)$ of the filter:

$$x(t) = n(t) * h(t) \tag{F.1}$$

The impulse response is

$$h(t) = \frac{1}{2\pi}\int_{-\infty}^{\infty} H(\omega)e^{j\omega t}\mathrm{d}t = \frac{1}{2\pi}\left(\int_{2\pi f_1}^{2\pi f_2} e^{j\omega t}\mathrm{d}t + \int_{-2\pi f_2}^{-2\pi f_1} e^{j\omega t}\mathrm{d}t\right).$$

Integrating we get

$$h(t) = 2(f_2 - f_1)\cos\left(2\pi\frac{f_2+f_1}{2}t\right)\frac{\sin\left(2\pi\frac{f_2-f_1}{2}t\right)}{2\pi\frac{f_2-f_1}{2}t}. \tag{F.2}$$

If $f_2 - f_1 \ll (f_1 + f_2)/2$, the impulse response is a sinusoidal wave of frequency $(f_1 + f_2)/2$ modulated by a slowly varying sinc function. Since the convolution in (F.1) is but a weighted sum of time-displaced impulse responses, the filtered noise resembles a sine wave of frequency close to $(f_1 + f_2)/2$ modulated by a slowly and randomly varying envelope, as can be seen in Fig. F.1.

This sine-wave behavior makes it possible to obtain the peak envelope multiplying the energy envelope[11] by $\sqrt{2}$:

[11]The energy envelope is in turn obtained by filtering $x^2(t)$ with a first-order lowpass filter with a suitable time constant τ.

F. Miyara, *Software-Based Acoustical Measurements*, Modern Acoustics and Signal Processing, DOI 10.1007/978-3-319-55871-4

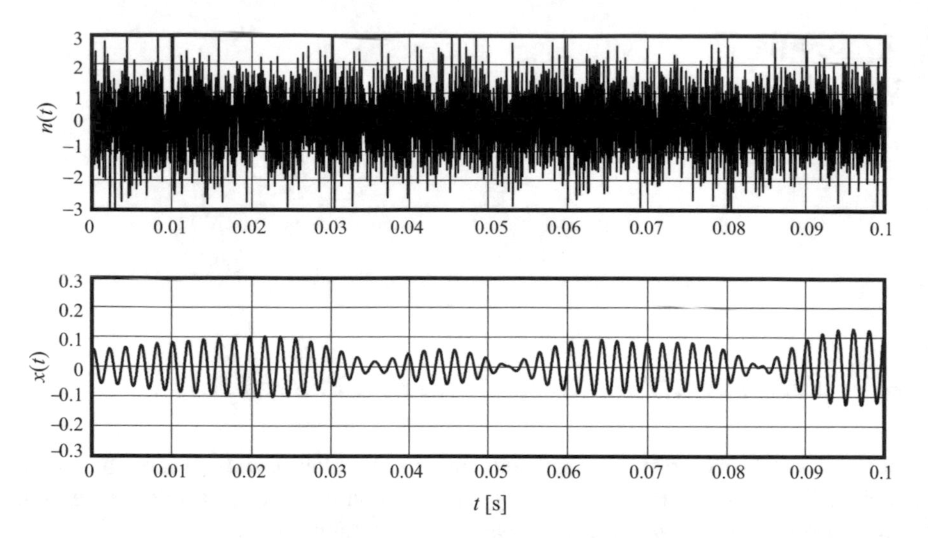

Fig. F.1 *Upper panel* White noise signal with an effective value equal to 1. *Lower panel* The signal after being filtered with an ideal bandpass filter with f_1 = 500 Hz and f_2 = 550 Hz

$$Y(t) = \sqrt{2 \int_{-\infty}^{t} x^2(\theta) e^{\frac{\theta - t}{\tau}} d\theta} \tag{F.3}$$

F.2 Statistical Description of the Envelope of a Filtered Noise

Under the previous assumption that $f_2 - f_1 \ll (f_1 + f_2)/2$, the peak envelope of $x(t)$ has a distribution that approximates the Rayleigh distribution (Rice 1944), whose probability density function is

$$f(Y) = \frac{Y}{\sigma^2} e^{-\frac{Y^2}{2\sigma^2}}, \tag{F.4}$$

where σ is the standard deviation of the filtered noise. The standard deviation is equivalent to the effective or RMS value, which can be computed by integration of the spectral density (which for a white noise filtered with an ideal filter is constant in the pass band):

$$\sigma = \sqrt{\int_{f_1}^{f_2} \overline{n^2}(f) df} = \sqrt{(f_2 - f_1)\overline{n^2}} \tag{F.5}$$

If, prior to going through the filter, the white noise is band-limited between 0 and f_{max} (this is to prevent aliasing) and its effective value is N_{ef}, then its spectral density is N_{ef}/f_{max}, hence,

$$\sigma = N_{ef}\sqrt{\frac{f_2 - f_1}{f_{max}}}. \tag{F.6}$$

The Rayleigh distribution has a mean

$$\mu_Y = \sqrt{\frac{\pi}{2}}\sigma = 1.2533\ \sigma, \tag{F.7}$$

and a standard deviation

$$\sigma_Y = \sqrt{\frac{4-\pi}{2}}\sigma = 0.6551\ \sigma. \tag{F.8}$$

Reference

Rice SO (1944) Mathematical analysis of random noise. Bell Syst Tech J 23:282–332

Appendix G
Transient Response of a Third-Order Bandpass Filter

The transfer function of a third-order bandpass filter is given by

$$G(s) = K\prod_{k=1}^{3}\frac{a_{1k}s}{s^2 + a_{1k}s + a_{2k}}, \tag{G.1}$$

so when it is excited by a sine wave input of angular frequency ω_o and amplitude 1 the Laplace transform of the response will be

$$R(s) = K\frac{\omega_o}{s^2 + \omega_o^2}\prod_{k=1}^{3}\frac{a_{1k}s}{s^2 + a_{1k}s + a_{2k}}. \tag{G.2}$$

This expression may be expanded in partial fractions as follows:

$$R(s) = \frac{A_{1o} + A_{2o}s}{s^2 + \omega_o^2} + \sum_{k=1}^{3}\frac{A_{1k} + A_{2k}s}{s^2 + a_{1k}s + a_{2k}} \tag{G.3}$$

In order to compute the coefficients A_{nk} we add all the fractions of (G.3) to form a common denominator and then we equate the numerator to that of Eq. (G.2). We get a system of 8 equations with 8 unknowns which we can solve for the coefficients A_{nk}.

It is then easy to find the inverse transform from (G.3):

$$r(t) = A_o \sin(\omega_o t + \varphi) + \sum_{k=1}^{3} A_k e^{-t/\tau_k}\ \sin(\omega_k t + \psi_k) \tag{G.4}$$

where

$$\omega_k = \sqrt{a_{2k} - \left(\frac{a_{1k}}{2}\right)^2} \tag{G.5}$$

F. Miyara, *Software-Based Acoustical Measurements*, Modern Acoustics and Signal Processing, DOI 10.1007/978-3-319-55871-4

and

$$\tau_k = \frac{2}{a_{1k}} \tag{G.6}$$

The amplitudes A_k and phase shifts φ and ψ_k depend on the constants A_{nk}. A_o and φ correspond to the amplitude and phase frequency response of the filter, and can be found substituting $s = j\omega$ in (G.1). Using the notation of Eq. (1.60) and rearranging,

$$\omega_k = \frac{\omega_{ok}}{2Q}\sqrt{4Q_k - 1}, \tag{G.7}$$

and

$$\tau_k = \frac{2Q_k}{\omega_{ok}}. \tag{G.8}$$

Appendix H
Combined Uncertainty

Let Y be a random variable that depends functionally of other N independent random variables:

$$Y = g(X_1, \ldots, X_N). \tag{H.1}$$

Given a value or instance (x_{1o}, ..., x_{No}) of the N-tuple (X_1, ..., X_N), we can express Y in the neighborhood of (x_{1o}, ..., x_{No}) by means of an N-dimensional first-order Taylor approximation:

$$Y \cong g(x_{1o}, \ldots, x_{No}) + \sum_{i=1}^{N} \frac{\partial g}{\partial x_i}(x_{io})(X_i - x_{io}). \tag{H.2}$$

We shall analyze two cases. In the first case we shall work with population parameters and in the second with sample parameters.

H.1 Population combined uncertainty

We can define an N-dimensional joint probability density function $f(x_1, \ldots, x_N)$ such that

$$P(x_{io} < x_i < x_{io} + \mathrm{d}x_i, \;\; i = 1, \ldots N) = f(x_{1o}, \ldots, x_{No})\mathrm{d}x_1 \ldots \mathrm{d}x_N. \tag{H.3}$$

The mean of the random variable Y defined in Eq. (H.1) is

$$\mu_Y = \int_{-\infty}^{\infty} \cdots \int_{-\infty}^{\infty} g(x_1, \ldots, x_N)\, f(x_1, \ldots, x_N)\mathrm{d}x_1 \ldots \mathrm{d}x_N. \tag{H.4}$$

F. Miyara, *Software-Based Acoustical Measurements*, Modern Acoustics and Signal Processing, DOI 10.1007/978-3-319-55871-4

Replacing the approximate value from Eq. (H.2),

$$\mu_Y \cong \int_{-\infty}^{\infty} \cdots \int_{-\infty}^{\infty} \left(g(x_{1o}, \ldots, x_{No}) + \sum_{i=1}^{N} \frac{\partial g}{\partial x_i}(x_{io})(x_i - x_{io}) \right) f(x_1, \ldots, x_N) \mathrm{d}x_1 \ldots \mathrm{d}x_N.$$

It might seem that the approximation (H.2) does not hold since x_i takes values very distant from x_{io}. However, often f decreases very quickly as x_i grows.

Since $g(x_{1o}, \ldots, x_{No})$ is constant, it is equal to its mean. Commuting the sum and the integration,

$$\begin{aligned}
\mu_Y &\cong g(x_{1o}, \ldots, x_{No}) + \sum_{i=1}^{N} \frac{\partial g}{\partial x_i}(x_{io}) \int_{-\infty}^{\infty} \cdots \int_{-\infty}^{\infty} (x_i - x_{io}) f(x_1, \ldots, x_N) \mathrm{d}x_1 \ldots \mathrm{d}x_N \\
&= g(x_{1o}, \ldots, x_{No}) + \sum_{i=1}^{N} \frac{\partial g}{\partial x_i}(x_{io})(\mu_i - x_{io})
\end{aligned}$$

If we select the coordinates of the point around which the Taylor expansion is centered equal to the mean of each variable, i.e., $x_{io} = \mu_i$, it turns out that the mean is equal to the function g calculated at the means of the independent variables:

$$\mu_Y \cong g(\mu_1, \ldots, \mu_N). \tag{H.5}$$

Similarly, its variance (square standard deviation) is

$$\begin{aligned}
\sigma_Y^2 &= \int_{-\infty}^{\infty} \cdots \int_{-\infty}^{\infty} (y(x_1, \ldots, x_N) - \mu_Y)^2 f(x_1, \ldots, x_N) \mathrm{d}x_1 \ldots \mathrm{d}x_N \\
&\cong \int_{-\infty}^{\infty} \cdots \int_{-\infty}^{\infty} \left(\sum_{i=1}^{N} \frac{\partial g}{\partial x_i}(\mu_i)(x_i - \mu_i) \right)^2 f(x_1, \ldots, x_N) \mathrm{d}x_1 \ldots \mathrm{d}x_N \\
&= \int_{-\infty}^{\infty} \cdots \int_{-\infty}^{\infty} \left(\sum_{i=1}^{N} \sum_{j=1}^{N} \frac{\partial g}{\partial x_i}(\mu_i) \frac{\partial g}{\partial x_j}(\mu_j)(x_i - \mu_i)(x_j - \mu_j) \right) f(x_1, \ldots, x_N) \mathrm{d}x_1 \ldots \mathrm{d}x_N
\end{aligned}$$

Reversing the sum and integration order,

$$\sigma_Y^2 \cong \sum_{i=1}^{N} \sum_{j=1}^{N} \frac{\partial g}{\partial x_i}(\mu_i) \frac{\partial g}{\partial x_j}(\mu_j) \int_{-\infty}^{\infty} \cdots \int_{-\infty}^{\infty} (x_i - \mu_i)(x_j - \mu_j) f(x_1, \ldots, x_N) \mathrm{d}x_1 \ldots \mathrm{d}x_N$$

The integrals are, by definition, the population covariances of X_i and X_j, so

$$\sigma_Y^2 \cong \sum_{i=1}^{N}\sum_{j=1}^{N}\frac{\partial g}{\partial x_i}(\mu_i)\frac{\partial g}{\partial x_j}(\mu_j)\mathrm{cov}(X_i,X_j) \tag{H.6}$$

The population analysis is useful for the calculation of type B combined uncertainty, when the covariance matrix is previously known by some method that does not depend on the samples, for instance, from a previous technical or scientific study.

H.2 Sample Combined Uncertainty

Suppose now that we have made n measurements of each of the N variables, i.e., that we have n instances of the N-tuple, (x_{1k}, …, x_{Nk}), where k = 1, …, n. For the k-th instance we can express y_k as a first-order Taylor expansion,

$$y_k \cong g(x_{1o},\ldots,x_{No}) + \sum_{i=1}^{N}\frac{\partial g}{\partial x_i}(x_{io})(x_{ik}-x_{io}). \tag{H.7}$$

The sample mean of $\{y_k\}$ is, therefore,

$$\bar{y} = \frac{1}{n}\sum_{k=1}^{n} y_k \cong g(x_{1o},\ldots,x_{No}) + \frac{1}{n}\sum_{k=1}^{n}\sum_{i=1}^{N}\frac{\partial g}{\partial x_i}(x_{io})(x_{ik}-x_{io}). \tag{H.8}$$

We can reverse the summation order and rearrange,

$$\bar{y} \cong g(x_{1o},\ldots,x_{No}) + \sum_{i=1}^{N}\frac{\partial g}{\partial x_i}(x_{io})\frac{1}{n}\sum_{k=1}^{n}(x_{ik}-x_{io}),$$

$$\bar{y} \cong g(x_{1o},\ldots,x_{No}) + \sum_{i=1}^{N}\frac{\partial g}{\partial x_i}(x_{io})(\bar{x}_i - x_{io}). \tag{H.9}$$

If we select $x_{io} = \bar{x}_i$ (i.e., we consider the Taylor expansion around the population means of the measured arguments), it results,

$$\bar{y} \cong g(\bar{x}_1,\ldots,\bar{x}_N). \tag{H.10}$$

In other words, the function evaluated at the estimated measurands is an estimate of the expected value of y.

Now let us determine the sample variance, s_y^2, of $\{y_k\}$:

$$s_y^2 = \frac{1}{n-1}\sum_{k=1}^{n}(y_k - \bar{y})^2 = \frac{1}{n-1}\sum_{k=1}^{n}\left(\sum_{i=1}^{N}\frac{\partial g}{\partial x_i}(\bar{x}_i)(x_{ik} - \bar{x}_i)\right)^2$$
$$= \frac{1}{n-1}\sum_{k=1}^{n}\sum_{i=1}^{N}\sum_{j=1}^{N}\frac{\partial g}{\partial x_i}(\bar{x}_i)\frac{\partial g}{\partial x_j}(\bar{x}_j)(x_{ik} - \bar{x}_i)(x_{jk} - \bar{x}_j)$$

Introducing the sum respect to k inside the other two we can write

$$s_y^2 = \sum_{i=1}^{N}\sum_{j=1}^{N}\frac{\partial g}{\partial x_i}(\bar{x}_i)\frac{\partial g}{\partial x_j}(\bar{x}_j)\frac{1}{n-1}\sum_{k=1}^{n}(x_{ik} - \bar{x}_i)(x_{jk} - \bar{x}_j)$$

The sum

$$\frac{1}{n-1}\sum_{k=1}^{n}(x_{ik} - \bar{x}_i)(x_{jk} - \bar{x}_j) \tag{H.11}$$

is the sample covariance c_{ij} of variables x_i and x_j, i.e., an estimate of the population covariance $\mathrm{cov}(X_i, X_j)$, so we can write

$$s_y^2 = \sum_{i=1}^{N}\sum_{j=1}^{N}\frac{\partial g}{\partial x_i}(\bar{x}_i)\frac{\partial g}{\partial x_j}(\bar{x}_j)c_{ij}, \tag{H.12}$$

This formula is identical to (H.6) for the population case, except that the covariances are replaced by their estimates. It is useful for the treatment of type A uncertainty components, if the sample covariances can be computed from the measured data.

If the variables are independent, only the covariances of each variable and itself, i.e., the variances, are nonzero, so

$$s_y^2 = \sum_{i=1}^{N}\left(\frac{\partial g}{\partial x_i}(\bar{x}_i)\right)^2 s_i^2. \tag{H.13}$$

Note: Even if the variables are independent, when computing the sample covariances of pairs of different variables the result will probably be nonzero. The reason is the same for which if we take a sample of white noise during a short time, its sample mean will most likely be nonzero.

Appendix I
Example of Uncertainty Calculation in the Case of a Nonlinear Systematic Error

Consider a data acquisition system that records sound pressure values p obtained from the analog/digital conversion of the microphone output voltage v applying the equation

$$p = v/K_o, \tag{I.1}$$

where K_o is the nominal sensitivity of the microphone. We will assume that the sensitivity is 0.040 V/Pa and its standard uncertainty is 0.002 V/Pa, and that the sound pressure level measurements have a standard uncertainty of 0.4 dB (this implies that the uncertainty in the sound pressure is proportional to the pressure).

We will assume that the relationship between p and v is nonlinear and can be represented by the equation

$$v = Kp + K_1 p^2 + K_2 p^3. \tag{I.2}$$

The measurement model is shown in the block diagram of Fig. I.1.

Suppose that we have three nominal reference values L_{pk} of 74.0 dB, 94.0 dB and 114.0 dB available, with a standard uncertainty of 0.2 dB in all cases which includes any systematic error. Table I.1 shows these values converted into sound pressure, P_k, along with the observed values P_{ko} and the same expressed as sound pressure level L_{pko}.

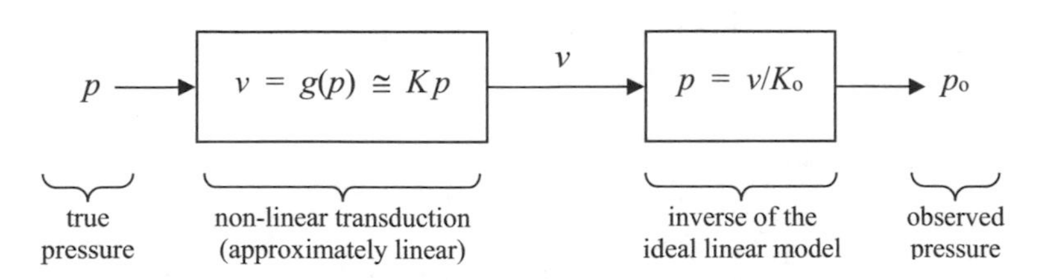

Fig. I.1 Measurement model of a nonlinear transduction process that can be linearly approximated through a constant K

F. Miyara, *Software-Based Acoustical Measurements*, Modern Acoustics and Signal Processing, DOI 10.1007/978-3-319-55871-4

Table I.1 Nominal reference sound pressure level and its corresponding sound pressure; measured sound pressure and the corresponding sound pressure level

k	L_{pk} (dB)	P_k (Pa)	P_{ko} (Pa)	L_{pko} (dB)
1	74.0	0.10024	0.10029	74.004
2	94.0	1.00240	1.00660	94.036
3	114.0	10.02400	9.72040	113.730

Instead of attempting to find the parameters of the model (I.2) that would allow to find p from the intermediate variable v, we propose a similar cubic model to find p directly from the observed value p_o:

$$p \cong ap_o + bp_o^2 + cp_o^3. \tag{I.3}$$

We shall now apply the system of Eq. (2.73):

$$\begin{bmatrix} P_{1o} & P_{1o}^2 & P_{1o}^3 \\ P_{2o} & P_{2o}^2 & P_{2o}^3 \\ P_{3o} & P_{3o}^2 & P_{3o}^3 \end{bmatrix} \begin{bmatrix} a \\ b \\ c \end{bmatrix} = \begin{bmatrix} P_1 \\ P_2 \\ P_3 \end{bmatrix}. \tag{I.4}$$

Calling A the system matrix, we have

$$A \begin{bmatrix} a \\ b \\ c \end{bmatrix} = B.$$

Replacing the numerical values,

$$A = \begin{bmatrix} 0.10029 & 0.010057 & 0.0010086 \\ 1.0066 & 1.0132 & 1.0199 \\ 9.7204 & 94.486 & 918.45 \end{bmatrix}$$

$$B = \begin{bmatrix} 0.10024 \\ 1.0024 \\ 10.024 \end{bmatrix}.$$

Solving for a, b and c

$$\begin{bmatrix} a \\ b \\ c \end{bmatrix} = A^{-1}B = \begin{bmatrix} 1 \\ -0.0050166\ \text{Pa}^{-1} \\ 0.00084634\ \text{Pa}^{-2} \end{bmatrix}. \tag{I.5}$$

We have already the model to compensate the nonlinear systematic effects. Now we shall calculate the uncertainty. To that end we need first to compute the 9 partial derivatives of a, b and c with respect to P_{1o}, P_{2o}, P_{3o}. Note that a, b and c are implicit functions of those variables that are defined by the system of Eq. (I.4). Instead of attempting to find them analytically and then compute their derivatives, which would be tedious, we can find the derivatives by implicit differentiation. The system (I.4) can be written as

$$\begin{cases} f_1(a,b,c,P_{1o}) = 0 \\ f_2(a,b,c,P_{2o}) = 0 \\ f_3(a,b,c,P_{3o}) = 0 \end{cases} \tag{I.6}$$

where, for k = 1, 2, 3,

$$f_k = P_{ko}a + P_{ko}^2 b + P_{ko}^3 c - P_k. \tag{I.7}$$

Since

$$\begin{cases} a = a(P_{1o}, P_{2o}, P_{3o}) \\ b = b(P_{1o}, P_{2o}, P_{3o}) \\ c = c(P_{1o}, P_{2o}, P_{3o}) \end{cases} \tag{I.8}$$

we can differentiate each f_k as a composite function with respect to the three variables P_{1o}, P_{2o}, P_{3o}, applying the chain rule. For instance, to find the derivatives with respect to P_{1o}, we get the following system of three equations and three unknowns that allow to solve for the derivatives $\partial a/\partial P_{1o}$, $\partial b/\partial P_{1o}$, $\partial c/\partial P_{1o}$:

$$\begin{cases} \frac{\partial f_1}{\partial a}\frac{\partial a}{\partial P_{1o}} + \frac{\partial f_1}{\partial b}\frac{\partial b}{\partial P_{1o}} + \frac{\partial f_1}{\partial c}\frac{\partial c}{\partial P_{1o}} + \frac{\partial f_1}{\partial P_{1o}} = 0 \\ \frac{\partial f_2}{\partial a}\frac{\partial a}{\partial P_{1o}} + \frac{\partial f_2}{\partial b}\frac{\partial b}{\partial P_{1o}} + \frac{\partial f_2}{\partial c}\frac{\partial c}{\partial P_{1o}} = 0 \\ \frac{\partial f_3}{\partial a}\frac{\partial a}{\partial P_{1o}} + \frac{\partial f_3}{\partial b}\frac{\partial b}{\partial P_{1o}} + \frac{\partial f_3}{\partial c}\frac{\partial c}{\partial P_{1o}} = 0 \end{cases} \tag{I.9}$$

The last term of the first equation is present because f_1 (but not f_2 or f_3) depends on P_{1o}. The system may be rewritten in matrix form as

$$\begin{bmatrix} \frac{\partial f_1}{\partial a} & \frac{\partial f_1}{\partial b} & \frac{\partial f_1}{\partial c} \\ \frac{\partial f_2}{\partial c} & \frac{\partial f_2}{\partial b} & \frac{\partial f_2}{\partial c} \\ \frac{\partial f_3}{\partial a} & \frac{\partial f_3}{\partial b} & \frac{\partial f_3}{\partial c} \end{bmatrix} \begin{bmatrix} \frac{\partial a}{\partial P_{1o}} \\ \frac{\partial b}{\partial P_{1o}} \\ \frac{\partial c}{\partial P_{1o}} \end{bmatrix} = \begin{bmatrix} -\frac{\partial f_1}{\partial P_{1o}} \\ 0 \\ 0 \end{bmatrix} \tag{I.10}$$

The derivatives that appear in the system matrix are directly computed from the data:

$$\frac{\partial f_1}{\partial a} = P_{1o} \tag{I.11}$$

$$\frac{\partial f_1}{\partial b} = P_{1o}^2 \tag{I.12}$$

$$\frac{\partial f_1}{\partial c} = P_{1o}^3 \tag{I.13}$$

As we can see, it turns to be the same system matrix of the system (I.4) for the calculation of a, b and c. The derivative in the independent term is

$$\frac{\partial f_1}{\partial P_{1o}} = a + 2P_{1o}b + 3P_{1o}^2 c, \tag{I.14}$$

Besides P_{1o}, we need here a, b, and c, which had been computed numerically in (I.5). It is now possible to find the derivatives needed for the uncertainty calculations through the numerical solution of system (I.10) and similar systems for the derivatives with respect to P_{2o} and P_{3o}. The results, computed by software, are included in Table I.2.

We can now compute the corrected value and its uncertainty in any measurement. Suppose, for instance, that we have measured a sound pressure $P_o = 8$ Pa with a standard uncertainty for the sound pressure level of 0.4 dB. The corrected value is

$$P_c = 8 \times 1 - 0.0050166 \times 8^2 + 0.00084634 \times 8^3 = 8.1123 \text{ Pa}.$$

In order to compute the uncertainty we shall use Eq. (2.78) combined with (2.79) and (2.80). Eq. (2.79) has the form

$$\frac{\partial P_c}{\partial P_o} = a + 2bP_o + 3cP_o^2 = 1.0822.$$

Equation (2.80),

$$\begin{aligned}
\frac{\partial P_c}{\partial P_{1o}} &= \frac{\partial a}{\partial P_{1o}} P_o + \frac{\partial b}{\partial P_{1o}} P_o^2 + \frac{\partial c}{\partial P_{1o}} P_o^3 = 109.97 \\
\frac{\partial P_c}{\partial P_{2o}} &= \frac{\partial a}{\partial P_{2o}} P_o + \frac{\partial b}{\partial P_{2o}} P_o^2 + \frac{\partial c}{\partial P_{2o}} P_o^3 = -13.574 \\
\frac{\partial P_c}{\partial P_{3o}} &= \frac{\partial a}{\partial P_{3o}} P_o + \frac{\partial b}{\partial P_{3o}} P_o^2 + \frac{\partial c}{\partial P_{3o}} P_o^3 = -0.6278
\end{aligned}$$

The uncertainties in (2.78) can be determined from the data. As we have the uncertainty in decibels, we must convert it into pascals. Since

$$L_p = 20 \log \frac{P}{P_{ref}} = \frac{20}{\ln 10} \ln \frac{P}{P_{ref}}, \tag{I.15}$$

Table I.2 Numerical values needed to calculate the uncertainty

Derivative	a	b	c
$\frac{\partial}{\partial P_{1o}}$	−11.179	12.256	−1.1425
$\frac{\partial}{\partial P_{2o}}$	0.1217	−1.2261	0.12485
$\frac{\partial}{\partial P_{3o}}$	−0.00014341	0.0015725	−0.0014206

we have

$$u_{L_P} = \frac{20}{\ln 10}\frac{1}{P}u_P. \tag{I.16}$$

Solving for u_p,

$$u_p = \frac{\ln 10}{20} P\, u_{L_P} \tag{I.17}$$

The uncertainty in the measurement of the reference sound pressure level has two components: the uncertainty of the reference itself (0.2 dB) and the measurement uncertainty (0.4 dB):

$$u^2_{P_{io}} = \left(\frac{\ln 10}{20} P_1 u_{L_{p1}}\right)^2 + \left(\frac{\ln 10}{20} P_{1o} u_{L_{p1o}}\right)^2. \tag{I.18}$$

Therefore,

$$u_{P_{1o}} = 0.005163 \text{ Pa}$$
$$u_{P_{2o}} = 0.051783 \text{ Pa}$$
$$u_{P_{3o}} = 0.50364 \text{ Pa}$$

The uncertainty in the observed value P_o of the measurand is only the measurement uncertainty (0.4 dB),

$$u_{P_o} = \frac{\ln 10}{20} P_o u_{L_{po}}, \tag{I.19}$$

hence, for $P_o = 8$ Pa,

$$u_{P_o} = 0.36841 \text{ Pa}. \tag{I.20}$$

Finally, we apply the Eq. (2.78), which in this case is:

$$u_{P_c}^2 = \left(\frac{\partial P_c}{\partial P_o}\right)^2 u_{P_o}^2 + \left(\frac{\partial P_c}{\partial P_{1o}}\right)^2 u_{P_{1o}}^2 + \left(\frac{\partial P_c}{\partial P_{2o}}\right)^2 u_{P_{2o}}^2 + \left(\frac{\partial P_c}{\partial P_{3o}}\right)^2 u_{P_{3o}}^2. \quad (I.21)$$

The result is

$$u_{P_c} = 1.037 \text{ Pa}.$$

It is interesting to note that the uncertainty in the corrected value is larger (almost a threefold increase) than the uncertainty of the original uncorrected measurement (I.20). This is because the search for the parameters of a model such as (I.3) adds new uncertainty components. If, as in this case, the nonlinearity is not too significant, it is not convenient to correct the result.

Appendix J
The Sampling Theorem

Suppose we want to sample an arbitrary signal $v(t)$ with the only restriction of some bandwidth limit. An example is any audio signal, whose maximum significant frequency is near 20 kHz.

A central issue in these cases is the selection of an adequate *sample rate*, i.e., the number of samples per unit time. We shall see that if we wish to retrieve the signal later or to digitally process it in some way that resembles an analog device (such as amplification, filtering, compression) the sample rate should be greater than twice the maximum frequency present in the signal.

To that end, consider the Fourier transform $V(\omega)$ of $v(t)$:

$$V(\omega) = \int_{-\infty}^{+\infty} v(t)\mathrm{e}^{-j\omega t}\mathrm{d}t. \tag{J.1}$$

and its inverse

$$v(t) = \frac{1}{2\pi}\int_{-\infty}^{+\infty} V(\omega)\mathrm{e}^{j\omega t}\mathrm{d}t. \tag{J.2}$$

Suppose now that we multiply the signal by a periodic sampling function $s(t)$ given by (see Fig. J.1)

$$s(t) = \begin{cases} 1/\tau & \text{if } -\tau/2 < t < \tau/2 \\ 0 & \text{if } \tau/2 < t < T - \tau/2 \\ s(t-T) & \text{for every other } t \end{cases} \tag{J.3}$$

The result is the sampled signal

$$v_{\mathrm{s}}(t) = v(t)\cdot s(t). \tag{J.4}$$

F. Miyara, *Software-Based Acoustical Measurements*, Modern Acoustics and Signal Processing, DOI 10.1007/978-3-319-55871-4

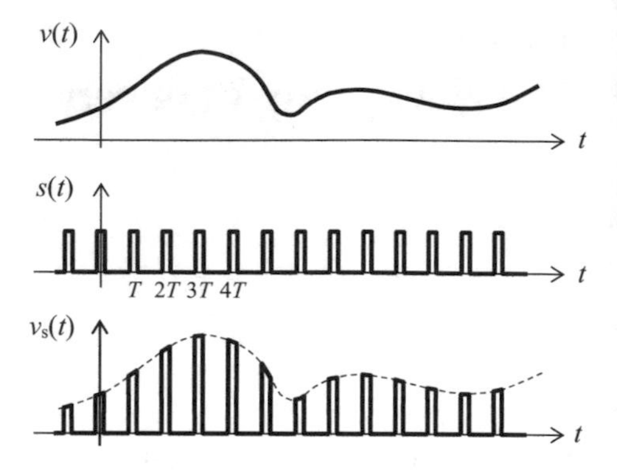

Fig. J.1 Sampling of a time-varying signal with a pulse train of period T and width τ

Although any signal can be described from the point of view either of its time evolution or its spectrum, often the spectral description provides a better idea of the effect of a given process on the signal. This is the case for sampling, so we shall calculate the spectrum of $v_s(t)$. Let us obtain first the spectrum of the sampling signal itself. As it is a periodic function, it has a Fourier series expansion. We will prefer the complex form

$$s(t) = \sum_{n=-\infty}^{\infty} c_n e^{jn\omega_s t}. \tag{J.5}$$

where $\omega_s = 2\pi f_s = 2\pi/T$, and

$$c_n = \frac{1}{T} \int_{-T/2}^{+T/2} m(t) e^{-jn\omega_m t} dt = \frac{1}{T} \int_{-T/2}^{+T/2} \frac{1}{\tau} e^{-jn\omega_m t} dt.$$

i.e.,

$$c_n = \frac{1}{T} \frac{\sin n\omega_s \tau/2}{n\omega_s \tau/2}. \tag{J.6}$$

These coefficients have a behavior with n according to the unnormalized sinc function, $\text{sinc}(x) = \text{sen } x/x$. For $\tau \to 0$, all the coefficients tend to be equal to $1/T$. Multiplying the expression (J.5) for $s(t)$ by $v(t)$ we have

$$v_s(t) = \sum_{n=-\infty}^{\infty} c_n v(t) e^{jn\omega_s t}. \tag{J.7}$$

Its Fourier transform is calculated applying the definition, and introducing the exponential and the integral inside the sum

$$V_s(\omega) = \int_{-\infty}^{+\infty} \left(\sum_{n=-\infty}^{\infty} c_n v(t) e^{jn\omega_s t} \right) e^{-jn\omega t} dt$$
$$= \sum_{n=-\infty}^{\infty} c_n \int_{-\infty}^{\infty} v(t) e^{-j(\omega - n\omega_s)t} dt \tag{J.8}$$
$$= \sum_{n=-\infty}^{\infty} c_n V(\omega - n\omega_s)$$

This result is very interesting, since it indicates that the sampled signal $v_s(t)$ is formed by infinite replicas of the bilateral spectrum of the original signal which are periodically distributed along the frequency axis with a "period" equal to the sample rate. The replicas are weighted by the coefficients c_n of the sampling signal spectrum (see Fig. J.2b). In the ideal case in which $\tau \to 0$, i.e., the sampling pulses are Dirac deltas, all replicas have the same amplitude (Fig. J.2c).

In the example of Fig. J.2 the spectrum of the signal is limited to a band $\pm\omega_{max}$, and the angular sample rate ω_s has been chosen so that

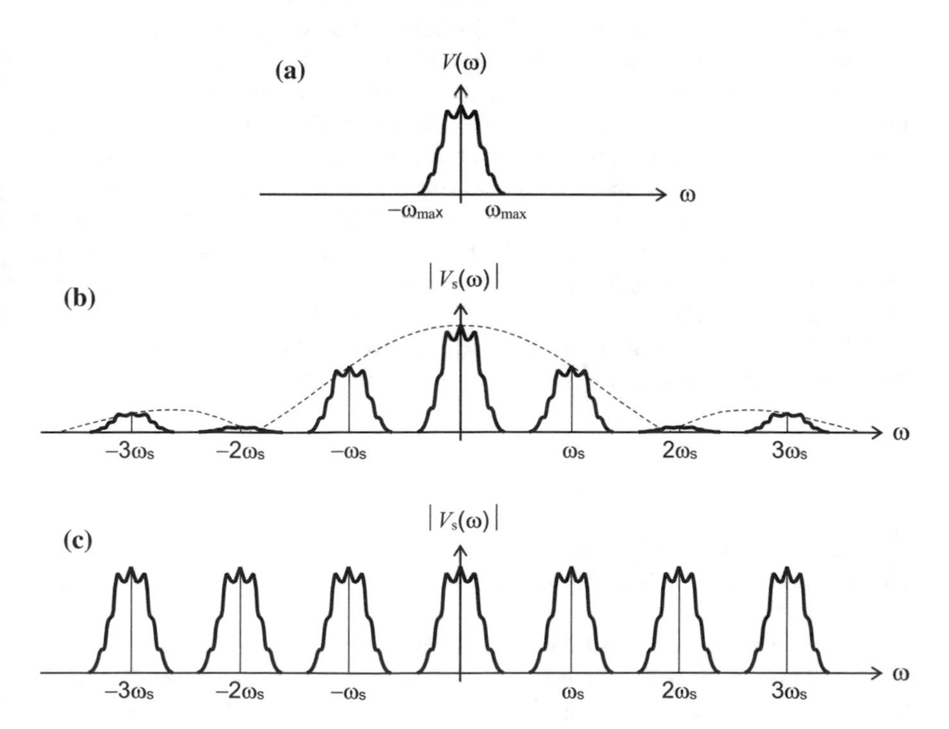

Fig. J.2 **a** Spectrum of a band-limited signal, which is only nonzero within the range $[-\omega_{max}, \omega_{max}]$. **b** Spectrum of the sampled signal with a sample rate $\omega_s > 2\omega_{max}$. **c** The same as (**b**) when $\tau \to 0$

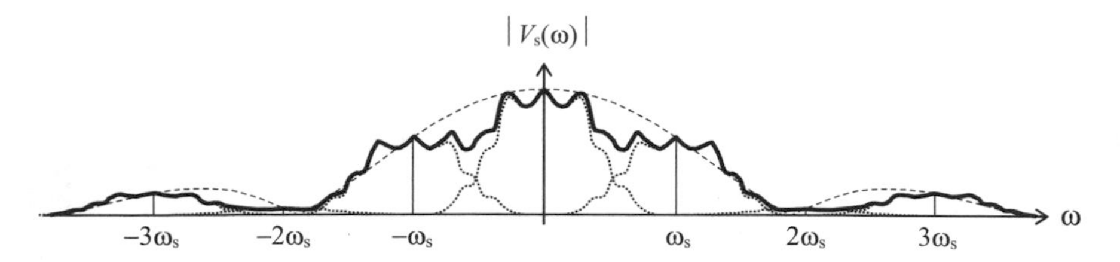

Fig. J.3 Spectrum of a signal that has been sampled at sample rate $\omega_s < 2\omega_{max}$

$$\omega_s > 2\omega_{max}. \tag{J.9}$$

If this condition, called *Nyquist condition*, is fulfilled, the replicas of the original spectrum are separated and it is then easy to recover the original signal by means of a sufficiently selective lowpass filter, i.e., capable of removing the replicas centered at $\pm\omega_s$, $\pm 2\omega_s$, $\pm 3\omega_s$, etc. This result is known as the *sampling theorem*.

If, instead, the Nyquist condition were not fulfilled, as in the example of Fig. J.3, the spectral replicas would be overlapped and it would not be possible to recover the base band $[-\omega_{max}, \omega_{max}]$ since it would contain residuals from the higher bands. If we tried to recover the signal with a lowpass filter, we would have spectral components "reflected" back at $\omega_s/2$. For instance, a frequency $\omega_s/2 + \Delta\omega$ would be reflected as the frequency $\omega_s/2 - \Delta\omega$. These frequencies are called *alias frequencies* and the phenomenon is known as *aliasing*. There is no method to remove the alias frequencies without removing at the same time frequencies that have been correctly sampled. If in the preceding example we remove the alias frequency $\omega_s/2 - \Delta\omega$, we would also remove any non-aliased component at that frequency.

The frequency $f_s/2$ is called *Nyquist frequency*. The sampling theorem can be alternatively stated as follows: It will be possible to perfectly recover a signal that has been sampled at a sample rate f_s only if its spectral content does not exceed the Nyquist frequency $f_s/2$.

If the sample rate complies with the requirement of the sampling theorem, there is a *reconstruction formula* for a sampled signal, given by

$$x(t) = \sum_{k=-\infty}^{\infty} x(kT) \frac{\operatorname{sen} \pi f_m(t - kT)}{\pi f_m(t - kT)}. \tag{J.10}$$

In the Fig. J.4 is shown the reconstruction based on this formula, also called *sinc interpolation*because it uses the sinc function to interpolate between the discrete sample instants.

The sinc interpolation is not computationally efficient, since although the series of Eq. (J.10) is convergent, it converges very slowly; so it is necessary for a large number of terms to attain an acceptable error. In some cases a truncated sinc is used, i.e., instead of calculating each sinc over the full signal duration, only the values corresponding to a few cycles before and after its central value are used. This

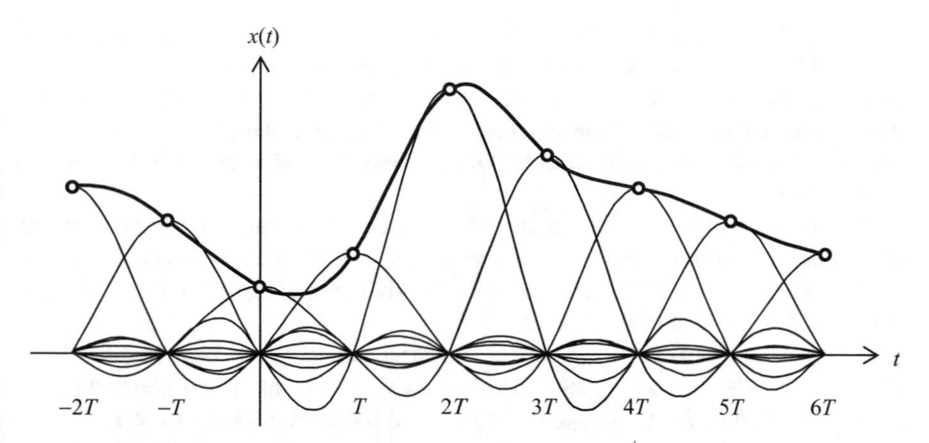

Fig. J.4 Sinc interpolation for the perfect reconstruction of a sampled signal that complies with the requisite of the sampling theorem

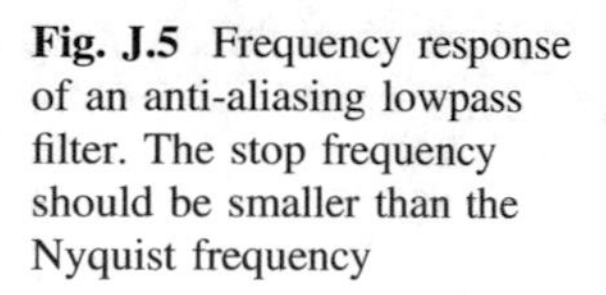

Fig. J.5 Frequency response of an anti-aliasing lowpass filter. The stop frequency should be smaller than the Nyquist frequency

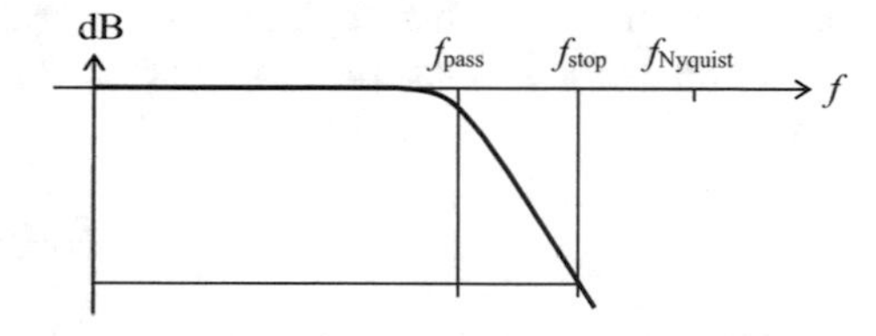

interpolation is useful for display purposes. It is used, for example, in digital oscilloscopes. In contrast, it has no serious applications to digital audio, where it is better to use much more efficient reconstruction or smoothing filters.

Getting back to the Nyquist condition, observe that it is not sufficient that the meaningful frequencies be limited by the Nyquist frequency, since aliasing does not distinguish between meaningful and meaningless spectral components. For instance audio signals do not contain relevant frequencies above 20 kHz, but they contain noise well above that limit, so it is not correct to use a sample rate of 40 kHz. Indeed, the original ultrasonic noise is inaudible, but when reflected back into the audio region because of aliasing, it becomes audible. As an example, a 38 kHz tone originating in the subcarrier of an FM signal exceeds by 18 kHz the Nyquist frequency, so an aliased component will appear at 20 kHz – 18 kHz = 2 kHz, a perfectly audible frequency.

The solution is to use a lowpass filter before sampling. This filter is called *anti-aliasing filter* and it is meant to reduce the out-of-band components to vanishingly small levels (see Fig. J.5). Since any real filter has a gradual transition between the pass frequency and the stop frequency, in practice the sample rate must be larger than twice the maximum relevant frequency. For instance, the standard sample rate for CD quality is $f_s = 44.1$ kHz instead of 40 kHz. The Nyquist frequency is, in this case, 22.05 kHz, allowing for a 2.05 kHz margin for the filter transition. This is quite a rapid transition, requiring high order filters that produce undesirable transients. That is why when there are no memory restrictions it is preferable to use $f_s = 48$ kHz or even $f_s = 96$ kHz, allowing the use of simpler filters, which have better transient response. Another technique is to use oversampling, such as in sigma-delta converters, where the oversampling factor may be as high as 128.

The attenuation of the anti-aliasing filter above the Nyquist frequency must be high enough to ensure that the noise created by residual aliasing is acceptable. In the typical case in which the samples will be subsequently digitized, it is sufficient that such noise is less than 0.5 LSB.

It is interesting to mention that the sampling theorem may be generalized for bandpass signals. It can be shown (Proakis et al. 1996) that if the spectrum of a signal is limited to the range $f_{\min} < f < f_{\max}$ and therefore its bandwidth is

$$B = f_{\max} - f_{\min}, \tag{J.11}$$

then it can be correctly sampled (i.e., without in-band aliasing) with a sample rate f_s that is much smaller than $2f_{\max}$ as would require the plain application of the Nyquist condition (J.9). Indeed, one can find a range of sample rates contained in the interval

$$2B \le f_s < 4B, \tag{J.12}$$

such that the replicas predicted by Eq. (J.8) do not overlap with the original signal and, consequently, they can be removed by filtering. This is particularly useful when the signal bandwidth B is very small compared with $f_{\max}$, since in this case the sample rate can be chosen far below the Nyquist condition. Sampling a signal at a smaller rate than required according to the standard version of the sampling theorem is called *undersampling*.

There are two cases that must be considered. The first one is when $f_{\min}$ is an integer multiple of B, i.e.,

$$f_{\min} = kB, \tag{J.13}$$

where k is a positive integer. In this case it is shown (Proakis et al. 1996) that the passband signal can be sampled at a sample rate

$$f_s = 2B. \tag{J.14}$$

without aliasing. Figure J.6 illustrates this situation. If, for instance, all spectral components of a signal lie between $f_{\min} = 15$ kHz and $f_{\max} = 20$ kHz, so that $B = 5$ kHz, then we can sample at a rate of 10 kHz, even if it is below twice the maximum frequency, i.e., 40 kHz (or even below the maximum signal frequency!).

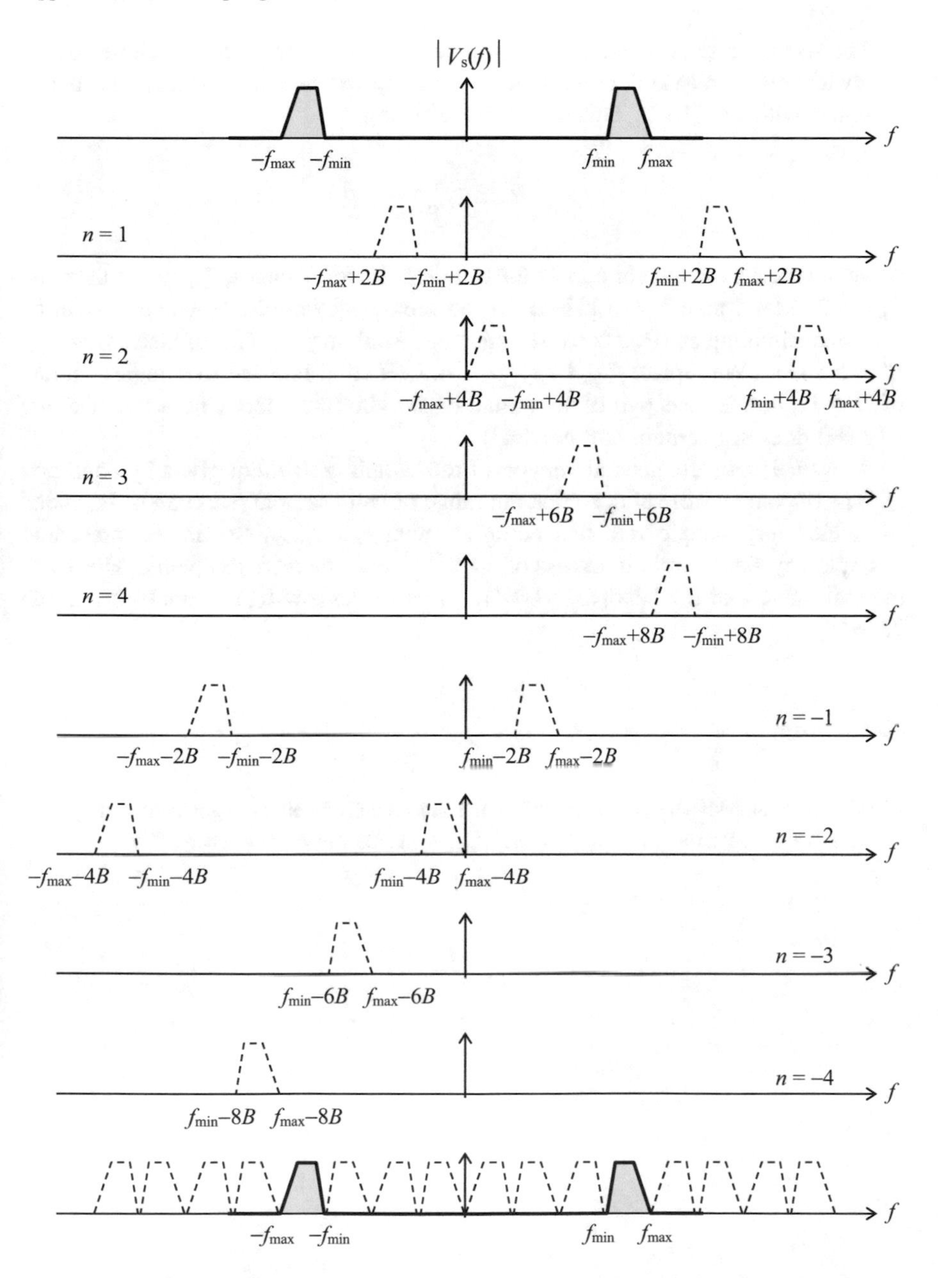

Fig. J.6 Sampling theorem for bandpass signals when $f_{\min} = kB$ with k integer. *Top* bilateral spectrum of a bandpass signal of bandwidth B. *Center* eight positive and negative replicas of the signal spectrum when sampled at a sample rate $f_s = 2B$, according to Eq. (J.8). *Bottom* spectrum of the sampled signal, including the replicas

The second case is when $f_{\min}$ and, hence, $f_{\max}$, are not integer multiples of the bandwidth. In this case it is possible to virtually extend the bandwidth so that it complies with Eq. (J.13). This can be done taking

$$B' = \frac{f_{\max}}{\left[\frac{f_{\max}}{B}\right]}. \tag{J.15}$$

where [ ] means the integer part of the quotient. If, for instance, $f_{\min} = 17$ kHz and $f_{\max} = 22$ kHz , then $B = 5$ kHz as in the previous example. However, it can be seen that sampling at 10 kHz causes aliasing. Applying Eq. (J.15), instead, we get $B' = 5.5$ kHz. We replace $f_{\min}$ by $f'_{\min} = 16.5$ kHz and proceed to sample at a rate of 11 kHz. In this case, part of the virtual bandwidth (the interval between 16.5 and 17 kHz) does not contain any power.

As a final note, the general version of the sampling theorem given by condition (J.9) is, of course, still valid, so that the range of valid sample rates can be extended to include any sample rate that complies with $f_s > 2f_{\max}$. So, in the preceding example any sample rate in excess of 44 kHz would be also acceptable. However, the real interest of the bandpass version arises for narrow band signals, for which $2B \ll f_{\max}$.

Reference

Proakis JG, Manolakis DG (1996) Digital signal processing. Principles, algorithms and applications, 3rd edn. Prentice-Hall International, Upper Saddle River, New Jersey, USA

Appendix K
Structure of a FLAC File

K.1 File and Data Stream

Even if we will be dealing mainly with FLAC files, it is worth mentioning that their structure is equally applicable to a *data stream*. The difference between a file and a data stream is that the file is static, i.e., data associated physically with storing media, while a data stream is associated with a communication channel within a data network (such as Internet or an intranet).

In fact, the FLAC format is conceived to allow its transmission in real time (i.e., at an information rate equal to or greater than that required for playback) over a network. The format is tolerant against loss of some information, allowing decoding from an intermediate point.

K.2 Data structure

The audio data structure of a FLAC file (Coalson 2008; Roveri 2011) is shown in Fig. K.1. It is comprised of *frames* and *subframes*, which code, respectively, the *blocks* and *subblocks* that contain PCM audio information. A block comprises a number N of multichannel samples (in a typical case, stereo samples) and is split into subblocks, each of which contains the N samples of a channel. In the reference codec,[12] the size N of each block is determined from the sample rate. For instance, for $F_s = 44\ 100$ Hz, $N \leq 4608 = 2^9 \times 3^2$.

Each frame has a *header* with metadata and a series of subframes. Each subframe has its own header and the encoded information corresponding to a subblock. Unlike WAV files, where the channels are interleaved, in this case the subframes appear successively. This simplifies the decoding process.

The file starts with the identifier "fLaC" (i.e., 0x6C4C6143 in hexadecimal ASCII coding) followed by one or more metadata blocks, followed by one or more frames.

---

[12]The FLAC project (https://xiph.org/flac/) provides a format specification and a reference codec. However, since it is a free and open format, any software developer can create their own codec.

F. Miyara, *Software-Based Acoustical Measurements*, Modern Acoustics and Signal Processing, DOI 10.1007/978-3-319-55871-4

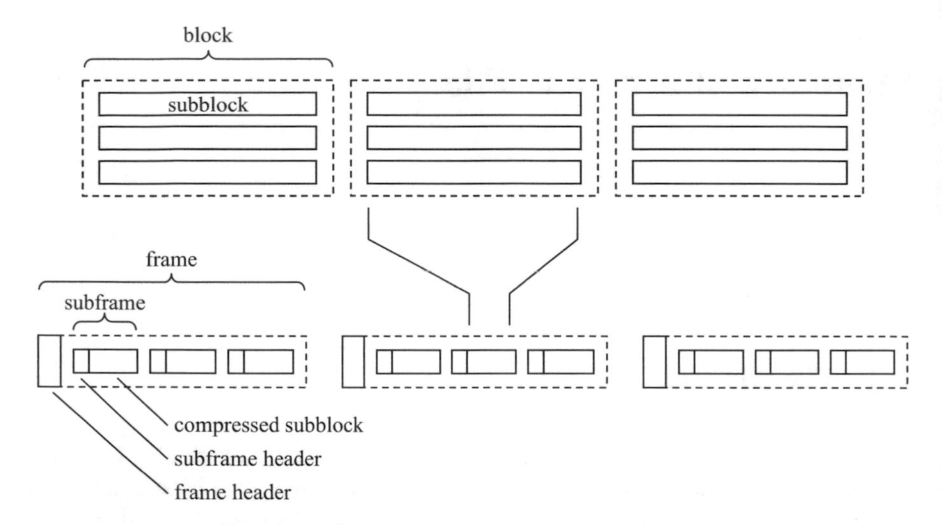

Fig. K.1 Example of structure of a FLAC audio file. *Top* the PCM signal corresponding to three audio channels before compression. *Bottom* the compressed signal with the addition of the frame and subframe headers Table K.1

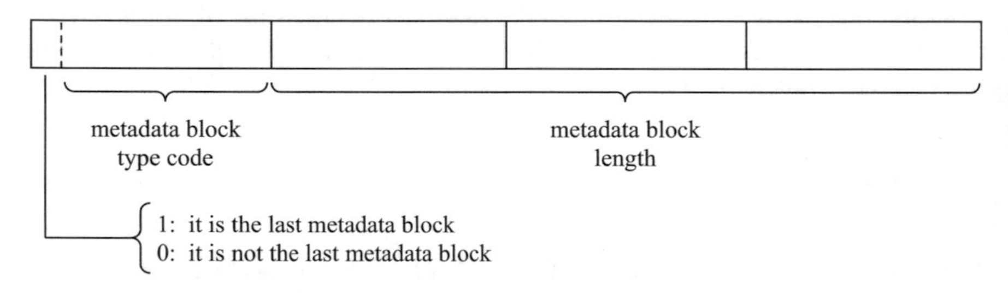

Fig. K.2 Header of each metadata block

All metadata blocks have a 4 byte header (Fig. K.2). The first bit of the first byte is 1 if it is the last metadata block, and 0 if it is not. The remaining 7 bits contain a code between 0 and 127 indicating the metadata type. In Table K.1 are shown the normalized block codes. The 3 bytes that follow contain the length of the rest of the block in *big-endian* format (i.e., the most significant byte comes first as in the conventional hexadecimal numbering system).

The first metadata block, called *streaminfo* is mandatory and contains data stream information (see Table K.2). It is identified with number 0.

As can be appreciated, the individual fields do not have necessarily an integer number of bytes, although the whole block has 34 bytes. The first couple of fields are a priori specifications defined by the encoder. The next two fields indicate the minimum and maximum sizes of the encoded frames in the file or data stream. The values are highly dependent on the redundancy of the different signal blocks. The most redundant blocks will have smaller frame length than the most random ones. The three fields that follow describe the characteristics of the signal and its sampling:

Table K.1 Numerical identifiers of metadata block types

No.	Metadata block
0	STREAMINFO
1	PADDING
2	APPLICATION
3	SEEK TABLE
4	VORBIS COMMENT
5	CUE SHEET
6	PICTURE
7 and 126	Reserved for future metadata
127	Invalid

Table K.2 Data structure of the *streaminfo* metadata block

Bits	Description
16	Minimum block size (in samples)[a]
16	Maximum block size (in samples)[b]
24	Minimum frame size (in bytes)[c]
24	Maximum frame size (in bytes)[c]
20	Sample rate (in Hz)[d]
3	Number of channels −1
5	Number of bits per sample −1[e]
36	Total number of samples[f]
128	MD5 digital signature of the unencoded data

[a]The minimum acceptable number is 16
[b]The maximum acceptable number is 65 535
[c]0 indicate it is unknown
[d]This number cannot be greater than 655 350 Hz
[e]The reference codec allows up to 24 bit per sample
[f]A multichannel sample is considered as a single sample

sample rate, number of channels and bits per sample. The eighth field records the number of samples of the whole file or data stream.

The last 16 bytes (128 bit) of the frame header constitute an MD5*digital signature*[13] computed from the original data before encoding. It is the result of the application of a *hash algorithm*[14] and its objective is to allow the decoder to check for data integrity on the decoded audio. To this end, the algorithm is applied again after decoding and the result is compared with the digital signature in the header. If they are identical, the data are validated.

[13]See, for instance, http://en.wikipedia.org/wiki/MD5, http://tools.ietf.org/html/rfc1321.
[14]A *hash* algorithm allows to obtain a fixed-length number (*hash code*) from an arbitrary number of bits. The simplest hash algorithm is the one that computes a parity bit, i.e., a bit that is 1 if the number of 1's is even and 0 in the opposite case. They are used as error detecting codes and in cryptography.

Although there have been detected cases where two different data sets give rise to the same digital signature (situation known as *collision*) they are extremely rare, so it is very unlikely that a corrupted result is validated. The validation by MD5 is convenient but it is not a mandatory function of a decoding algorithm, particularly in the handling of local files with a highly dependable hardware.

The second metadata block, called *padding* is optional and is meant to reserve file space for future use without having to change the size. It consists of an integer number of null bytes. A possible use may be to add later a metadata block not originally planned during encoding, or for any other particular information without a standard format. It could be useful in metrological applications to incorporate contextual information such as GPS data, date, description, observations, operator, etc. It could be argued that reserving an unused space is against the philosophy of data compression, but the required storage space is in general negligible in comparison with the audio data, even after data reduction.

The third block, *application*, is also optional and it contains information about the software with which the file has been created, for instance the reference codec from the FLAC project. The structure and length are defined by the software manufacturer. The first four bytes contain a software registration number that may be applied for at the FLAC project web site.

The fourth block, *seek table*, is optional. It deals with the problem of the efficient location of specific samples within the file, in a context in which every sample has a different length. The direct seek is very slow since it requires to count samples one by one. Instead, a series of *seek points* that allow to locate specific frames are included. The seek table has 18 bytes per seek point. The first 8 bytes indicate the correlative sample number of the first sample in the target frame. The following 8 bytes indicate the *offset* in bytes between the first header byte of the first frame and the first header byte of the target frame. The last 2 bytes indicate the number of samples in the target frame. Since a frame has relatively few samples, the final seek within the frame is fast. Figure K.3 explains the seek point addressing mechanism. If, for example, we are looking for the m-th sample, it will suffice to find the closest seek point. That seek point corresponds to the beginning of some frame, of which, from the second argument, we know in which byte it starts. From then, a sequential seek begins, but starting at a much closer position.

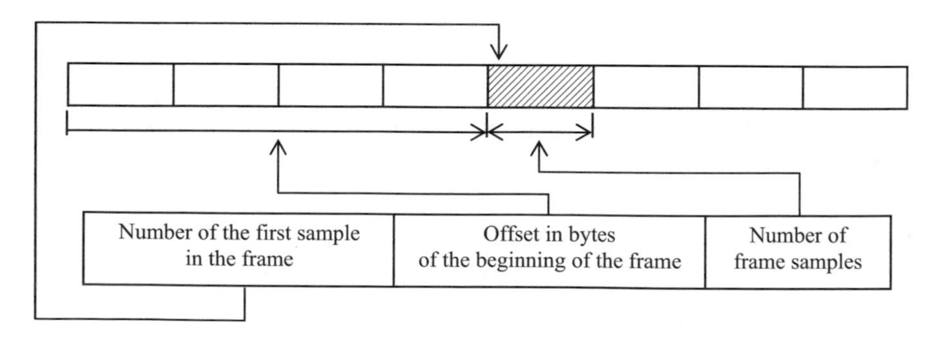

Fig. K.3 *Top* a sequence of frames. *Bottom* information associated to a seek point

Table K.3 Data structure of the metadata block *Vorbis comment*

Bytes	Type	Content
4	Little endian	Length N of a character string provided by the codec manufacturer
N	ASCII/UTF8	N character string provided by the codec manufacturer
4	Little endian	Number L of user comments
4	Little endian	Length M_1 of user comment #1
M_1	ASCII/UTF8	User comment #1 ($M1$ characters)
…	…	…
4	Little endian	Length M_L of user comment #L
M_L	ASCII/UTF8	User comment #L (M_L characters)

Unlike ASCII, UTF-8 is a character representation code that allows the use of values over 127, and even more than one byte, offering support for several languages

It is also possible to reserve space to insert a future seek point. This is accomplished adding, at the end, one or more placeholder seek points, characterized by a value 0xFFFFFFFFFFFFFFFF as first sample in the frame.

The fifth block, also optional, is a *Vorbis comment*. It includes a series of fields with a non-fully normalized weak structure. Vorbis comments are used in the lossy compressed format Vorbis-OGG. The structure of this metadata block is shown in Table K.3.

The string provided by the codec manufacturer contains in general information on the coded name, version and date. For instance, for the reference codec it corresponds to the $N = 32$ characters of the string

"reference libFLAC 1.2.120070917"

User comments must have a minimum structure of the type:

$$[\text{Field name}] = [\text{Comment}]$$

where [Field name] is an ASCII case-insensitive string that represents an English description of the field content and [Comment] is a UTF-8 string in any compatible language that represents the comment itself. In Table K.4 there are some examples of frequently used field names.

These fields are, evidently, designed for the music recording industry. While the Vorbis[15] specification includes a description and examples, it is not normative and there are no established or mandatory features for the field contents. In fact, it is possible to include ad hoc comments which have a greater interest for the documentation of a recording with metrological purposes. For instance, we could have a comment such as

$$\text{TITLE} = \text{Morning soundscape of Lopez Square}$$

[15]http://xiph.org/vorbis/doc/Vorbis_I_spec.html.

Table K.4 Usual field names in Vorbis comments

TITLE	ORGANIZATION
VERSION	DESCRIPTION
ALBUM	GENRE
TRACKNUMBER	DATE
ARTIST	LOCATION
PERFORMER	CONTACT
COPYRIGHT	ISRC
LICENSE	

Table K.5 Data structure of the cue sheet

Length	Type	Content
128 byte	ASCII	Media catalog number (MCN)
8 byte	Big-endian hex	Number of lead-in samples in the CD
1 bit	Bit	1 for CD, 0 in any other case
7 bit	Bit	Reserved. All bits are 0
258 byte	Bit	Reserved. All bits are 0
1 byte	Hex	Number *N* of tracks
Variable		Cue sheet track 1
…	…	…
Variable		Cue sheet track *N*

that would be preceded by the hex number 0x28000000, which in little endian format represents decimal 40, i.e., the byte count used by the comment (included the field name and any blank spaces).

The next metadata block, also optional, is the *cue sheet*. In principle it is thought for the case in which the FLAC file is used to compress the content of an audio compact disk (CD-DA), though it allows more general use. It contains information on *track* location, and within each track, the *index points*. Tracks are large divisions that usually correspond to individual musical pieces while the index points allow locating sections within a long piece. The structure of a cue sheet is shown in Tables K.5, K.6 and K.7.

Notes: (1) The Media Catalog Number for a CD is a 13 digit number. The remaining 115 bytes must be 0. (2) The lead-in samples correspond to track 0, which contains at least 2 s silence. (3) The International Standard Recording Code is a 12 digit number that is assigned to each track by the International Federation of the Phonographic Industry through national agencies.

The last metadata block corresponds to images, such as icons, cover, author's and performer's photographs, recording session photographs, etc. It is rather specific and it is of little use for our purposes, so we will not get into more detail.

After the end of the metadata block section, frames and subframes containing headers and encoded audio begin to appear.

Table K.6 Data structure of a cue sheet track

Length	Type	Content
8 byte	Big-endian hex	Offset in samples of the first index point relative to the beginning of the audio stream
1 byte	hex	Track number n
8 byte	ASCII	International Standard Recording Code
1 bit	bit	Track type. 0 for audio, 1 for any other
1 bit	bit	0: without pre-emphasis; 1: with pre-emphasis
6 bit	bit	Reserved. All bits are 0
13 byte	bit	Reserved. All bits are 0
1 byte	hex	Number M of index points in the track
12 byte		Index point 1
…	…	…
12 byte		Index point M

Table K.7 Data structure of a track index point

Length	Type	Content
8 byte	Big-endian hex	Offset in samples of the index point, relative to the track offset (for an audio CD it is divisible by 588)
1 byte	hex	Index point number L
3 byte	bit	Reserved. All bits are 0

The frame header begins with a synchronization code, “11111111111110”. This allows to start decoding information of a FLAC audio stream anytime, even if the metadata blocks are lost.[16] The sync code is followed by two reserved bits. The first one is 0 and the second one is 0 for a fixed block size (the strategy adopted by the reference codec) and 1 for a variable block size.

Then follows a 4 bit code indicating the size B of the block in samples. The idea is to use few bits for the most frequent block sizes, but also allow more general sizes, in which case the size can be found at the end of the frame header (Table K.8). For a sample rate of 44.1 kHz the reference codec uses 0101, i.e., $B = 576 \times 2^{5-2} = 4608$.

The next 4 bits indicate the sample rate. While it is constant throughout the file and it had already been declared in the streaminfo metadata block at the beginning of the file (see Table K.2), the information is repeated in each frame to enable the decoding of a data stream even if the metadata are lost. However, since this information appears in every frame, few bits are used for the most popular sample rates. Table K.9 shows each code’s meaning.

[16]This bit combination could also appear as a succession of Golomb–Rice codes, for instance, for a divisor $D = 4$, the end of an odd number, then four times the number 3 followed by a number greater than 4 would present the same pattern.

Table K.8 Block size codes in the frame header

Code	Block size B in samples
0000	Reserved
0001	192
$0010 \leq n \leq 0101$	$576 \times 2^{n-2}$
0110	Read 8 bits at the header end and add 1 ($1 \leq B \leq 256$)
0111	Read 16 bits at the header end and add 1 ($1 \leq B \leq 65{,}536$)
$1000 \leq n \leq 1111$	$256 \times 2^{n-8}$

Table K.9 Sample rate codes in the frame header

Code	Sample rate F_s (kHz)
0000	It is taken from *streaminfo*
0001	88.2
0010	176.4
0011	192
0100	8
0101	16
0110	22.05
0111	24
1000	32
1001	44.1
1010	48
1011	96
1100	F_s is an 8 bit value in kHz at the end of the header
1101	F_s is a 16 bit value in en Hz at the end of the header
1110	F_s is a 16 bit value in tens of Hz at the end of the header
1111	Invalid

Table K.10 Channel allocation codes in the frame header

Code	Content	
0000–0111	Number of channels −1	
1000	Channel 0: left	Channel 1: difference
1001	Channel 0: right	Channel 1: difference
1010	Channel 0: half sum	Channel 1: difference
1011–1111	Reserved	

The next 4 bits represent the channel allocation as shown in Table K.10. From 0000 to 0111 it indicates the number of channels C minus 1. The next three code values indicate, for the stereo case, how has been made the channel decorrelation, a necessary information in order to retrieve the original left–right channels.

Table K.11 Sample resolution code in the frame header

Code	Sample resolution n_b in bit
000	It is taken from the streaminfo block
001	8
010	12
011	Reserved
100	16
101	20
110	24
111	Reserved

The next 3 bits refer to the sample resolution in bit. The same as in the case of the sample rate, fewer bits are used for the most usual resolutions (Table K.11). Then there is a bit whose mandatory value is 0, being the 1 reserved for an unspecified future use.

Then follows the number of frame N if blocks have fixed block size L or the number of the first frame sample n_N if blocks are of variable length L_k. In the first case, the number of the first sample could be calculated simply as

$$n_N = (N - 1)L + 1. \tag{K.1}$$

In the second case it is necessary to sum the sizes of the $N - 1$ preceding blocks, which is a bit more complicated:

$$n_N = \sum_{k=1}^{N-1} L_k + 1 \tag{K.2}$$

Since the number of frames is much smaller than the number of samples, its representation requires few bytes.

The number of frame or the first frame sample (depending on the case) is codified in the file with a method similar to the one used by the UTF-8 code for the Unicode character coding. It is essentially a variable length code and it consists in allocating bits in the following way. If the number is not greater than 127, it is represented by a single byte whose most significant bit is 0. If it is greater than 127, two or more bytes are used, where all but the most significant byte start with the prefix 10 followed by 6 of the original bits that have not been yet allocated, starting by the least significant ones. The bits are allocated from right to left. The most significant byte starts with as many 1's as the number of total bytes (a unary code) followed by one or more 0's, followed by the rest of the bits not yet allocated. For instance, the decimal 11521, which in natural binary is

Table K.12 Codes associated with the predictive model of a subframe

Code	Subframe predictive model
000000	Constant
000001	Verbatim (identical)
00001x	Reserved
0001xx	Reserved
001000	Polynomial of order 0
001001	Polynomial of order 1
001010	Polynomial of order 2
001011	Polynomial of order 3
001100	Polynomial of order 4
001101–001111	Reserved
01xxxx	Reserved
1xxxxx	LPC of order xxxxx + 1

101101 00000001,

will be represented as

11100010 **10**110100 **10**000001,

where the prefixes have been highlighted in bold face and the original bits have been underlined. Decoding is also very easy, since the first prefix indicates the number of bytes and then it suffices to remove all prefixes.

If the block size code (see Table K.8) was 0110 or 0111, then the next 8 bits or 16 bits, respectively, represent the block size in samples. Similarly, if the sample rate code (see Table K.9) was 1100, 1101 or 1110, then the next 8 bits or 16 bits represent, respectively, the sample rate in kHz ($F_s \leq 128$ kHz), in Hz ($F_s \leq 65\,535$ Hz) or in tens of Hz ($F_s \leq 655\,350$ Hz).

Finally, the header of each frame ends with a byte that contains the *cyclic redundancy code*CRC-8, corresponding to all preceding bits in the header, including the initial synchronization code (Williams 1993).

After the frame header (see Fig. K.1) come the subframes, each of which, in turn, begins with a subframe header. This header begins with a bit 0, to prevent that the subframe can be confused with a new frame (that would begin with the synchronization code, whose first bit is 1). Then there is a 6 bit code that states what type of predictive model has been used in the present subframe.[17] The codes are summarized in Table K.12.

Then follows a bit that is 1 if in the original subblock there are wasted bits per sample and 0 if not. Wasted bits are the least significant bits that are systematically

[17]Note that the predictive model that optimizes compression might vary (and *do* vary) from block to block, that is why it must be stated on a subframe basis.

0 throughout the subblock. In the first case follows the number k of wasted bits per sample as a unary representation of $k - 1$. For instance, if there are 4 wasted bits (i.e., all samples in the subblock are multiple of $2^4 = 16$) we will have the bit string 10001, where the first 1 indicates that there are wasted bis, the three 0's indicate that there are $3 + 1 = 4$ wasted bits and the last 1 acts as ending symbol for the unary code.

Wasted bits are used by the FLAC encoder to reduce the bit depth since the signal has actually less bits. This is used in combination with some lossy preprocessors, such as *lossyWAV*, that work replacing, on a block basis, the least significant bits by 0's when that modification happens to be inaudible. Note that the subframe header has a variable length, which can be a number of bits not divisible by 8.

From here on comes the encoded audio itself. The content and how it shall be interpreted by the decoder depend on the predictive model used to encode the current subframe (the prediction strategy may change from one subblock to the next). The general structure is a list of model parameters followed, if appropriate, by a series of residuals. Note that the model parameters are the same throughout the whole subframe.

K.2.1 Constant Model

In this case, since the block size B is known, it is sufficient to specify the PCM value of the constant value with no compression, using a single word of as many bits as the resolution.[18] Since this predictor is selected only if the encoder has detected that the signal is constant throughout the subblock, the residuals are all 0 so it is pointless to include them. It is essentially a *run-length* code (one where the value and the number of times it repeats are given).

K.2.2 Verbatim Model

In the verbatim model, we have $n_b B$ bits corresponding to B samples in the subblock and n_b bits per sample. Note that in this case all the residuals are 0. As a model it is highly inefficient since it requires as many parameters as samples in the subblock. It is only justified when the signal is of maximum entropy (no redundancy), such as white noise, so any other model will produce residuals that are impossible to compress.

[18]Any attempt to compress a single value would be probably unsuccessful.

K.2.3 Polynomial Model

In principle, this model would need the order P of the polynomial model (Table 3.8), which is already included in the header (Table 3.20), since once defined the order the coefficients are fixed (Table 3.8). However, in order to predict a new value we need the values of the P previous samples. Therefore the first P samples (in uncompressed PCM format) are also included as parameters of the model. The prediction starts from sample $P + 1$.

It is interesting to note that the zero-order model is not equivalent to the constant model, as it could at first seem. Indeed, in the latter no residuals are calculated since we assume that the signal is constant. In the zero-order polynomial, which predicts that samples will be identically 0, the residuals are identical to the signal. An application example would be the case of an imperfect silence, i.e., a very low-level background noise. In this case probably the signal will be a high-entropy, low-redundancy one. Any model would yield residuals comparable to the signal, so the simplest model is preferable.

After the model parameters (the P initial samples) come the Golomb–Rice-encoded residuals, as shall be described later.

K.2.4 LPC Model

As in the polynomial model, the LPC model begins with P initial samples. The ideal LPC coefficients (i.e., those that minimize the total quadratic error throughout the subblock to which the subframe refers) have a potentially infinite precision, so it will be necessary to quantize them.[19]

The next 4 bits contain the value $n_{\mathrm{LPC}} - 1$, where n_{LPC} is the number of bits that are assigned to the coefficients. So, $n_{\mathrm{LPC}} = 8$, means that the coefficients will be specified with an 8 bit resolution. The combination 1111 is not allowed, so the maximum resolution, which corresponds to 1110, is 15 bit.

The next 5 bits indicate, in 2's complement format,[20] the number of bits m that the representation of the LPC coefficients must be shifted to the right (if m positive)

[19]Because of coefficient quantization, the model does not actually reach the minimum quadratic error, but it can get much closer than is attainable with a fixed-coefficient polynomial approximation.

[20]The two's complement is a numbering format allowing a representation of negative numbers. For n bits it represents numbers between -2^{n-1} and $2^{n-1} - 1$. Positive numbers are represented in conventional binary system, being its most significant bit 0. The code for a negative number is found complementing each bit, adding 1 and ignoring the carry if it exists. For instance, with 5 bit we can represent numbers between −16 and 15. Hence, the 6 is represented as 00110 an the −6 as 11001 + 1 = 11010.

or to the left (if m negative). That shift is equivalent to dividing by 2^m. If, for instance an LPC coefficient is represented with an 8 bit binary $a_h = 01110101$ (decimal 117) and $m = 4$, then the real value of the coefficient will be 0111.0101, i e., decimal 7.3125.

Then follow the P coefficients, each one represented with n_{LPC} bits, with a total bit count equal to $n_{\text{LPC}}P$.

After the parameters of the LPC model, appear the Golomb–Rice-encoded residuals described in the next section.

K.2.5 Residuals

In the polynomial and LPC models the FLAC format requires a list of the residuals (prediction error) of the successive samples. The Golomb–Rice code requires, as its only parameter, the base 2 logarithm of the divisor, or *Rice parameter*. FLAC allows representing it either with 4 bit or 5 bit. The next 2 bits indicate the resolution of the Rice parameter: 00 corresponds to 4 bits and 01 to 5 bits, while codes 10 y 11 are reserved.

The Rice parameter may be the same for all the subblock or, on the contrary, the block may be split into $Q = 2^q$ equal partitions and specify for each one a different Rice parameter. The next 4 bits are intended for this purpose, indicating the *partition order*, i.e., the exponent q.

The format of each partition includes, depending on the case, 4 bits or 5 bits that determine the Rice parameter and then the number n of residuals given by

$$n = \begin{cases} B - P & \text{if there is one partition} \\ B/2^q - P & \text{if there are } 2^q \text{ partitions and it is the first} \\ B/2^q & \text{if there are } 2^q \text{ partitions and it is not the first} \end{cases} \tag{K.3}$$

where B is the number of samples per block and P the model order. The first P residuals are unnecessary because they are replaced by the initial values.

If the Rice parameter is represented with 4 bits, 0000 corresponds to a divisor equal to 1 (it would be applied to a series of very small residuals). The value 1111 is an *escape code*, which means that the residuals are not encoded. The next 5 bits indicate, in that case, the bit resolution of the sample. Similarly if the Rice parameter is represented with 5 bits, in which case the escape code is 11111.

Although the Eq. (K.3) indicates the number of residuals, it is not a priori possible to predict how many bits are required, since the code is of variable length. The only way is to decode the successive samples until completion of the n residuals as indicated in (K.3).

K.3 Error Detection

Each frame ends with a cyclic redundancy code CRC-16 corresponding to all frame data from the synchronism code to the bit immediately preceding the CRC.

References

Coalson J (2014) FLAC—Format. Xiph.Org Foundation. https://xiph.org/flac/format.html. Accessed 26 Feb 2016

Williams RN (1993) A painless guide to CRC error detection algorithms. http://www.ross.net/crc/download/crc_v3.txt. Accessed 27 Feb 2016

Appendix L
Document Type Definition (DTD) for Audacity Project Files

L.1 Introduction

Audacity project files, .aup, are a particular kind of text documents that adhere to a generic structure based on XML (*extensible markup language*), a markup metalanguage[21] used to build specific markup languages for a variety of applications (Bray 2008). A markup language is, in general, a language that can include pieces of text delimited by tags that declare to what category belongs each piece of text. Categories may deal with structure, semantics, format, etc. One of the applications of markup languages is the possibility of automated parsing and extraction of information classified by categories.

Tags have the general form <xxx> and </xxx> when they delimit a non-empty text and <xxx/> when there is no associated text (the information is provided within the tag itself). Here xxx represents the name of the tag and possibly a list of attributes. There can be nested tags. The internal ones correspond to subcategories. Each tag may have attributes and each attribute adopts a value (numeric, text or alphanumeric) that modifies the tag.

L.2 Document Type Definition (DTD)

A *document type definition* (DTD) is a declaration of the syntax of a particular type of XML documents (Miloslav, n.d.); in this case, those corresponding to Audacity project files.

It contains a series of lines enclosed between delimiters, which can be of three types. Lines enclosed between are comments; those enclosed between <! ELEMENT and > define the *elements*, i.e., the main components of the syntax; and those enclosed between <!ATTLIST and > are attributes that modify and customize the elements.

[21]A metalanguage is a language whose object is to refer to other languages.

F. Miyara, *Software-Based Acoustical Measurements*, Modern Acoustics and Signal Processing, DOI 10.1007/978-3-319-55871-4

One element can have nested elements (*child elements*). In that case, those that are immediately subordinated are enclosed between normal parenthesis and within the parenthesis, they are separated by commas. The following list shows the resources for combining elements:

xxx* means that there may be 0 or more elements of type xxx
xxx+ means that there may be 1 or more elements of type xxx
xxx? indicates that xxx is an optional element (may appear at most once)
, separates elements within an ordered series
| separates alternative and exclusionary elements

So, xxx (yyy, (zzz | www)*) means that an element of type xxx contains a nested element of type yyy followed by any combination in any order of nested elements of types zzz and www, even their absence.

The following list indicates some key words as regards the content of elements and attributes:

#PCDATA indicates arbitrary text (including reserved characters such as < and >)
CDATA indicates a character text, where, for instance, "<" is represented by "<"
#REQUIRED indicates that it is a required item
#IMPLIED indicates that it is an optional attribute
EMPTY is an element that can only contain attributes, but not text.

L.3 DTD for Audacity Project Files

In what follows, we describe the contents of the document type definition (DTD) for the Audacity project files (whose extension is .aup), available at https://github.com/audacity/audacity/blob/master/src/xml/audacityproject.dtd.[22] The different entries are presented in bold face and large font, while a description and some comments (not present in the original document) are included in indented normal case font.

<!–DTD for Audacity Project files, as created by Audacity version 1.3.0–>
<!–For more information: http://audacity.sourceforge.net/xml/–>

<!ELEMENT project (tags, (wavetrack | labeltrack | timetrack)*)>

The project element contains a tags element (see below) and can contain 0 or more elements to be selected from the following ones, in any order: wavetrack, labeltrack, or timetrack.

[22]The declared version as of March 2017 is out of date. See at the end of this Appendix.

Its attributes are:

<!ATTLIST project projname CDATA #REQUIRED>

Project name (the file name followed by _data).

<!ATTLIST project version CDATA #REQUIRED>

DTD version.

<!ATTLIST project audacityversion CDATA #REQUIRED>

Audacity version with which the file has been edited.

<!ATTLIST project sel0 CDATA #REQUIRED>

Start of the last saved selection, in seconds.

<!ATTLIST project sel1 CDATA #REQUIRED>

End of the last saved selection, in seconds.

<!ATTLIST project vpos CDATA #REQUIRED>

Vertical position. (?)[23]

<!ATTLIST project h CDATA #REQUIRED>

Horizontal position. (?)

<!ATTLIST project zoom CDATA #REQUIRED>

Horizontal zoom with respect to a window duration of approximately 15 s

<!ATTLIST project rate CDATA #REQUIRED>

Sample rate in Hz.

<!ELEMENT tags EMPTY>

The tags element contains typical metadata of musical recordings.[24]
Its attributes are:

<!ATTLIST tags title CDATA #REQUIRED>

Title

<!ATTLIST tags artist CDATA #REQUIRED>

Artist

---

[23]No further information could be found or reverse engineered by this author for items marked with (?).

[24]In the project files generated by version 2.05 or later, the specific tags are child elements instead of attributes. See Note at the end of this Appendix.

<!ATTLIST tags album CDATA #REQUIRED>

Album

<!ATTLIST tags track CDATA #REQUIRED>

Track

<!ATTLIST tags year CDATA #REQUIRED>

Year

<!ATTLIST tags genre CDATA #REQUIRED>

Genre

<!ATTLIST tags comments CDATA #REQUIRED>

Comments

<!ATTLIST tags id3v2 (0|1) #REQUIRED>

ID3v2 is a *de facto* standard for the identification of musical recordings and their metadata, which include information such as the author, performer, title, etc. The standard can be downloaded from http://id3.org/id3v2.3.0

<!ELEMENT labeltrack (label*)>

The labeltrack element can contain 0 or more labels.
Its attributes are as follows:

<!ATTLIST labeltrack name CDATA #REQUIRED>

Label track name.

<!ATTLIST labeltrack numlabels CDATA #REQUIRED>

Number of labels.

<!ELEMENT label EMPTY>

Each label element refers to some text, in general a short one, associated to a time interval (which may have zero duration). It does not have any content; its information is expressed as attributes, which are as follows:

<!ATTLIST label t CDATA #REQUIRED>

Start instant in seconds.

<!ATTLIST label t1 CDATA #REQUIRED>

End instant in seconds.

<!ATTLIST label title CDATA #REQUIRED>

Title (the label text).

<!ELEMENT timetrack (envelope)>

The timetrack element contains time envelope information. The time envelope consists of control points and associated values (see below).
Its attributes are as follows:

<!ATTLIST timetrack name CDATA #REQUIRED>

Time track name.

<!ATTLIST timetrack channel CDATA #REQUIRED>

Channel to which it is applied.

<!ATTLIST timetrack offset CDATA #REQUIRED>

Offset of the time track from the initial instant.

<!ELEMENTwavetrack (waveclip*)>

Each wavetrack contains 0 or more waveclips. A wavclip is an audio fragment.
A wavetrack element has the following attributes:

<!ATTLIST wavetrack name CDATA #REQUIRED>

Name that is displayed in the track control panel.

<!ATTLIST wavetrack channel CDATA #REQUIRED>

Assigned channel.

<!ATTLIST wavetrack linked CDATA #REQUIRED>

(?)

<!ATTLIST wavetrack offset CDATA #REQUIRED>

Offset in seconds from the beginning of the wavetrack (seemingly this is not used at all)

<!ATTLIST wavetrack rate CDATA #REQUIRED>

Sample rate of the wavetrack (it could be different from track to track).

<!ATTLIST wavetrack gain CDATA #REQUIRED>

Gain to be applied to the wavetrack, including all associated wavclips.

<!ATTLIST wavetrack pan CDATA #REQUIRED>

Panning to be applied to the wavetrack, including all associated wavclips.

<!ATTLIST wavetrack mute CDATA "0">

Mute state (0 is the default value, corresponding to disabled mute).

<!ATTLIST wavetrack solo CDATA "0">

Solo state (0 is the default value, corresponding to disabled solo).

<!ELEMENT waveclip (sequence, envelope)>

Each wavclip element contains a sequence formed by a declared number of samples, and it is in turn split into waveblocks, i.e., segments of 2^{20} bytes (approximately 1 Mb), as well as an amplitude envelope that consists of control points and associated amplitude values.
A waveclip has a single attribute:

<!ATTLIST waveclip offset CDATA #REQUIRED>

Offset in seconds from the beginning of the track.

<!ELEMENT sequence (waveblock*)>

A sequence is formed by 0 or more waveblocks (segments of 2^{20} bytes).
It has the following attributes:

<!ATTLIST sequence maxsamples CDATA #REQUIRED>

Maximum number of samples of the waveblocks it contains. It is equal to 2^{20}/bps, where bps is the amount of bytes per sample.

<!ATTLIST sequence sampleformat CDATA #REQUIRED>

Number of samples plus eleven 4 byte words corresponding to the header.

<!ATTLIST sequence numsamples CDATA #REQUIRED>

Number of samples in the sequence.

<!ELEMENTwaveblock (simpleblockfile | silentblockfile | legacyblockfile | pcmaliasblockfile)>

A waveblock is formed by one and only one of four block types.
It has a single attribute:

<!ATTLIST waveblock start CDATA #REQUIRED>

Sample number in which the block begins. If there are more than one block they are multiple of 2^{20}/bps. For instance, if bps = 2 (equivalent to a 16 bit resolution), then they are multiple of $2^{20}/2 = 524\ 288$.

<!ELEMENT simpleblockfile EMPTY>

The most common block is of simpleblockfile type. It has no content besides its attributes.
It has the following attributes:

<!ATTLIST simpleblockfile filename CDATA #REQUIRED>

File name. It is something like "e000075b.au".

<!ATTLIST simpleblockfile len CDATA #REQUIRED>

Length of the waveblock in samples.

<!ATTLIST simpleblockfile min CDATA #REQUIRED>

Minimum decimal value of all waveblock samples, normalized between −1 and 1.

<!ATTLIST simpleblockfile max CDATA #REQUIRED>

Maximum decimal value of all waveblock samples, normalized between −1 and 1.

<!ATTLIST simpleblockfile rms CDATA #REQUIRED>

RMS value of the waveblock samples, normalized between −1 and 1.

<!ELEMENT silentblockfile EMPTY>

The element silentblockfile represents silence. It has no content except its only attribute:

<!ATTLIST silentblockfile len CDATA #REQUIRED>

Length of silence in samples.

<!ELEMENT legacyblockfile EMPTY>

The element legacyblockfile represents a block in a previous legacy format (corresponding to Audacity versions from 0.98 to 1.0 and from 1.1.0 to 1.1.2. It is possible to load, edit and save the modified version of a block with that format but not to create a new one (More information in https://github.com/audacity/audacity/blob/master/src/blockfile/LegacyBlockFile.cpp and in https://sourcecodebrowser.com/audacity/1.2.4b/_legacy_block_file_8h.html). It has the following attributes:

<!ATTLIST legacyblockfile name CDATA #REQUIRED>

File name.

<!ATTLIST legacyblockfile len CDATA #REQUIRED>

File length in samples. (?)

<!ATTLIST legacyblockfile summarylen CDATA #REQUIRED>

Length of the summary included in the format.

<!ATTLIST legacyblockfile norms CDATA "0">

Boolean variable. It is 1 if there is no RMS value.

<!ELEMENT pcmaliasblockfile EMPTY>

The element pcmaliasblockfile represents audio stored in an "alias" file outside the project data directory (such files are present when a file has been imported without saving a copy).
It has the following attributes:

<!ATTLIST pcmaliasblockfile summaryfile CDATA #REQUIRED>

Name of a file with extension .auf that contains a summary of the alias file. The summary consists of the maximum, minimum and RMS values for each 256 audio samples, and the maximum, minimum and RMS values for each 56 536 audio samples (a file of an approximate length 1/85 if the original one). Note: If the original file is very large, it is virtually split into fragments of up to 2^{20} bytes. The summary has, consequently up to $2^{20}/85$ samples, preceded by a short header (see Sect. 3.16).

<!ATTLIST pcmaliasblockfile aliasfile CDATA #REQUIRED>

Complete path and name of the alias file.

<!ATTLIST pcmaliasblockfile aliasstart CDATA #REQUIRED>

Sample number within the alias file where the audio fragment to which the .auf summary file refers. For the first file it is 0, then they are multiples of $2^{20}/4$.

<!ATTLIST pcmaliasblockfile aliaslen CDATA #REQUIRED>

Number of samples contained in the audio fragment that corresponds to the summary file. If there is more than one summary file, all except the last one have $2^{20}/4$ samples. The last one may have less samples.

<!ATTLIST pcmaliasblockfile aliaschannel CDATA #REQUIRED>

Channel corresponding to the alias file.

<!ATTLIST pcmaliasblockfile min CDATA #REQUIRED>

Minimum value, normalized from -1 and 1, of the audio segment corresponding to the summary file.

<!ATTLIST pcmaliasblockfile max CDATA #REQUIRED>

Maximum value, normalized from -1 and 1, of the audio segment corresponding to the summary file.

<!ATTLIST pcmaliasblockfile rms CDATA #REQUIRED>

RMS value, normalized from -1 and 1, of the audio segment corresponding to the summary file.

<!ELEMENTenvelope (controlpoint*)>

Each waveclip may have, besides the audio files forming it, an amplitude envelope, defined by 0 or more control points (nested child elements). The envelope is a (usually) slowly varying function between 0 and 1 that multiplies the successive waveclip samples.
It has a single attribute:

<!ATTLIST envelope numpoints CDATA #REQUIRED>

Number of control points of the envelope.

<!ELEMENT controlpoint EMPTY>

A control point is an ordered pair (time in s, normalized value in the range from −1 to 1) that belongs to the envelope. The envelope is completed by exponential interpolation between successive points.
It does not contain any data but its attributes.

<!ATTLIST controlpoint t CDATA #REQUIRED>

Time instant in seconds.

<!ATTLIST controlpoint val CDATA #REQUIRED>

Envelope value at the specified time.
Note: The described DTD, declared in the .aup files themselves, is out of date. Indeed, the current version of Audacity (2.1.2 in January 2017) generates project files .aup that do not comply with that DTD, which was valid for version 1.3 (Audacity team, n.d.). Two possible improved versions are the following ones:

<!ELEMENT tags (tag*)>
<!ELEMENT tag EMPTY>
<!ATTLIST tag name CDATA #REQUIRED>
<!ATTLIST tag value CDATA #REQUIRED>

<!ELEMENT tags (tag*)>
<!ELEMENT tag EMPTY>
<!ATTLIST tag (TITLE | ARTIST | ALBUM | TRACK | YEAR | GENRE | COMMENTS | ID3V2)>
<!ATTLIST tag value CDATA #REQUIRED>

The first one is more general, since the tag names are arbitrary; in the second one the tag names match those used in the recording industry.

References

Bray T, Paoli J, Sperberg-McQueen CM, Maler E, Yergeau F (eds) (2008) Extensible markup language (XML) 1.0, 5th edn. Disponible en http://www.w3.org/TR/REC-xml/

Miloslav N (no date) DTD Tutorial. http://www.zvon.org/xxl/DTDTutorial/General/book.html. Accessed 26 Feb 2016

Appendix M
Brief Description of Scilab

Scilab is a free software mathematical programming environment that works directly with matrices and vectors, both with classic operations such as matrix product, determinants and inversion of square matrices and mass operation on a data set, for instance applying a function or procedure to every component of a vector or matrix. This makes it particularly interesting for digital signal processing since using a single command it is possible to perform an iterative procedure.

It is developed by Scilab Enterprises (http://www.scilab-enterprises.com/) and can be freely downloaded from https://www.scilab.org/. It has a CeCILL license, which is not identical but is compatible with the General Public License (GPL) of the Free Software Foundation.[25] There are installers for Linux, Windows, and Mac OS X available at http://www.scilab.org/download.

First of all, there are two useful commands; **help** and **apropos** (from *à propos*, "about" in French). The first one, preceding any function, opens the manual page corresponding to that function. The second one looks for any relevant appearance of a character string in the manual. Usually one would use this command preceding an approximate or incomplete spelling of a function whose exact name is unknown or cannot be recalled. There is also a comprehensive downloadable manual (Scilab, n/d).

After installation, the working environment presents several panels, but the embedded text editor (SciNotes), where the programs (scripts) are written and tested, is not open by default. You can open it from the menu *Applications/SciNotes*. Once open, you can dock it at a convenient location by clicking on the bar immediately below the icons of the SciNotes window and dragging it to the desired position.

In order to apply any procedure or algorithm we need to start from a vector or matrix structure. There are many ways to create a matrix in Scilab. The simplest one is using the assignment operator, "=", enclosing between square brackets a list of

[25]Generally speaking, this implies the freedom to use the software for any purpose, either personal, academic or commercial, to modify, copy and share it, to share and publish whatever modification with the only condition of doing so under the same kind of license GPL. More details can be obtained from http://www.gnu.org/copyleft/gpl.html.

F. Miyara, *Software-Based Acoustical Measurements*, Modern Acoustics and Signal Processing, DOI 10.1007/978-3-319-55871-4

numbers separated by commas to delimit row components and semicolons to delimit rows:

```
a = [1, 2; 3, 4];
```

A blank space separating two numbers is equivalent to a comma. The ending semicolon is optional. If not used, Scilab may present the result on the screen, which is interesting at the debugging stage but may be annoying in the case of very large matrices, since much time is wasted in displaying a lot of numbers whose visualization is irrelevant.

The comma (or space) and semicolon may also be used as concatenation operators. If **b** and **c** are row vectors, then

```
a = [b, c];
```

is a row vector that has all the components of **b** followed by those of **c**; and if **b** and **c** are column vectors,

```
a = [b; c];
```

is a column vector that includes all the components of **b** followed by those of **c**. The colon operator "**:**" allows to create a vector with a succession of numbers that increase a unit at a time. For instance,

```
a = [1:m];
```

is a row vector with components **1, 2, ..., m**. Inserting a third number **h** in between (not necessarily an integer) as follows

```
a = [n:h:m];
```

we create a vector of numbers separated by **h**, i.e., **n, n + h, n + 2 * h**, ..., **n + r * h**, where **r** es such that **n + r * h** is the maximum number smaller than or equal to **m**.

It is possible to create matrices of **n** rows and **m** columns formed by zeros by means of **zeros(n, m)**, and formed by ones by means of **ones(n, m)**.

The size (number of rows and columns) can be found invoking the function **size(a)**, and the number of components of a vector can be found by means of **length(a)**. It can be also found in the variable browser window (usually docked at the upper right corner of the screen).

The value of the component **n** of a vector **x** is invoked writing **x(n)** and if we need to create a subvector **y** formed by components **n, n + 1, ..., m**, we may write

```
y = x(n:m);
```

In the case of matrices, the component in row **n** and column **m** is invoked with **x (n, m)**. It is also possible to invoke a complete row **n** or column **m** writing, respectively,

```
y = a(n, :);
y = a(:, m);
```

In both cases "**:**" may be replaced by an interval:

```
y = a(n, m1:m2);
y = a(n1:n2, m);
y = a(n1:n2, m1:m2);
```

There is a wildcard character, **$**, that represents the index of the last component of a vector or the last column or row of a matrix. For instance

```
y = x(1:$-1);
```

may be used to represent a vector that has all the components of **x** except the last one.

Note: If **x** is a matrix and an element is invoked using a single index, such as **x(n)**, then the **n**-th component is retrieved, counting column-wise.

A prime or apostrophe "'" at the right of the symbol of a matrix acts as a transposition operator.[26] Hence,

```
y = y';
```

converts a column vector into a row vector.

Product, division and power operators, *****, **/**, **^**, work as matrix operators. If they are preceded by a dot, **.***, **./**, **.^**, they perform component-by-component operations. Functions such as the exponential (**exp**), natural logarithm (**log**), decimal logarithm (**log10**), sine (**sin**), cosine (**cos**) or tangent (**tan**) work component-by-component, while **expm** and **logm** compute the exponential and logarithm in a matrix sense.

You can create a sine wave signal with the following code:

```
T = 1;
Fs = 44100;
f = 100;
t = [0:1/Fs:T];
y = sin(2*%pi*f*t);
```

[26]Actually, it is a conjugate transpose. For real matrices or vectors it is just a plain transpose.

where **T** is the duration in s, **Fs** the sample rate in Hz, **f** the frequency in Hz, **t**, a vector containing the time instants associated to the signal samples, and **y**, the signal. Note that the constant π is represented as **%pi** in Scilab. Built-in constants are preceded by the percent symbol.

We can create a random noise signal using the function rand with the following syntax:

```
r = rand(n1, n2, key);
```

where **n1** is the number of rows and **n2** is the number of columns, while **key** may be either **'uniform'** or **'normal'**, yielding respectively a uniformly or normally distributed noise. Note that the random number generator is actually a pseudorandom algorithm whose initial value depends on a seed. Type **help rand** to get more information.

In the case of audio signal processing we can create a meaningful column vector (an $n \times 1$ matrix) from a mono sound file by means of the procedure **wavread**. For instance,

```
[x, Fs, Nbits] = wavread('c:\sounds\sound.wav');
```

creates a column vector **x** that contains the audio samples of a file called **sound.wav** located at the directory **c:\sounds**, normalized between −1 and +1 (i.e., −1 is assigned to the minimum binary value and +1 to the maximum). The parameters **Fs** and **Nbits** retrieve respectively the sample rate in Hz and the number of bits (typically 8 or 16). The function **sound** allows to play an audio signal vector using the sound device:

```
sound(x, Fs, Nbits)
```

Conversely, we can save a vector or a matrix formed by two column vectors as a wav file by means of the procedure **wavwrite**:

```
wavwrite(y, Fs, file_name)
```

where **y** is the vector or matrix, **Fs** is the sample rate in Hz and **file_name** is a string with the name of the file (if ".wav" is lacking at the end of the name it is appended).

Scilab can also operate symbolically with polynomials and rational functions. Indeed, the function **poly** defines a polynomial from its roots, using one of the forms,

```
p = poly([r1, r2], 'x', 'roots');
```

```
p = poly([r1, r2], 'x');
```

or from its coefficients,

```
p = poly([a0, a1, a2], 'x', 'coeff');
```

The function **roots** solves for the zeros of a polynomial:

```
z = roots(p);
```

The result is a complex column vector. Similarly, the function **coeff** gives the coefficients as a vector:

```
c = coeff(p);
```

It is possible to operate with polynomials as if they where numbers. A division of polynomials yields a rational function. Scilab reduces the numerator and denominator removing common factors whenever they are identical. A convenient way of generating a polynomial is to create an elementary polynomial.

```
x = poly(0, 'x');
```

equivalent to variable x and then operate. For example,

```
p = 1 - 2*x + x^2;
```

generates the polynomial $1 - 2x + x^2$.

In order to evaluate a polynomial (or rational function) **p** at a given argument value the function **horner** is used.[27] For instance,

```
q = horner(p, 5);
```

evaluates the polynomial (or rational function) **p** at **x = 5**. The function **horner** also works if the second argument is a vector or matrix, in which case the function is evaluated at each component. It may be also applied to another polynomial or rational function, yielding the composed function.

It is possible to create a transfer function **H** from two polynomials **p** (numerator) and **q** (denominator) by means of the function **syslin**:

[27]The name of this function stands for William George Horner, who introduced a recursive method for evaluating polynomials saving additions and multiplications. For instance,

$$ax^3 + bx^2 + cx^1 + d = ((ax + b)x + c)x + d$$

This is the method implemented in Scilab.

```
H = syslin('d', p, q);
```

where **'d'** indicates that it is a discrete-time transfer function (it depends on the z variable). For continuous-time transfer functions, use **'c'** instead.

The discrete-time transfer function can be applied to a signal **x** by means of the function **flts**:

```
y = flts(x, H);
```

The frequency response of a transfer function **H** may be obtained with the function **frmag**:

```
[Hm, f] = frmag(H, n)
```

where **n** is the desired number of points, **f** a vector containing the normalized frequencies (0.5 corresponds to $F_s/2$) where the magnitude of the response is to be computed and **Hm** a vector with the corresponding magnitude values of the response.

It is possible to perform a line spectrum analysis by means of the discrete Fourier transform, implemented as a fast Fourier transform, FFT. If **x** is a vector,

```
y = fft(x, -1);
```

is its discrete Fourier transform. The parameter -1 indicates the sign of the exponent of the exponentials that appear in the transform. If 1 is used instead, the inverse FFT is computed. By default, the FFT is computed on the total length of the signal. If one needs to limit the computation to the samples between **N1** and **N2**, both included, then

```
y = fft(x(N1:N2), -1);
```

Sometimes it is necessary to change the number of digits of a result to be shown. This can be done with the function **format**. It has several syntactic options, being the most simple

```
format(m)
```

where **m** is the number of digits, including eventually the sign and the decimal point, of all results to be presented henceforth. Bear in mind, however, that variables have still full precision no matter the number of digits shown.

It is possible to plot a signal with the command **plot2d**. Although it has many options, a simple plot can be drawn with

```
plot2d(x, y)
```

where **x** and **y** are equal-length vectors that contain the abscissa and ordinate values. **plot2d** produces piecewise linear plots, but starting from several dozen points the plot looks like a smooth curve. If **y1** and **y2** are column vectors of the same length as **x**, it is possible to plot **y1** and **y2** simultaneously versus **x** by means of

```
plot2d(x, [y1, y2])
```

Sometimes it is necessary to use logarithmic axis. This can be achieved with an optional argument preceding the abscissa. It is a two-character string argument formed by any combination of the characters **l** (logarithmic) and **n** (normal). For instance,

```
plot2d('ln', f, 20*log10(Hm))
```

plots the frequency response (see above) in dB with a logarithmic frequency axis.

In order to easily identify a figure and its content, it is often desirable to add some text such as a title and an axis description. This can be done with the command:

```
xtitle(title, xlabel, ylabel)
```

where **title**, **xlabel** and **ylabel** are character strings containing the title, the horizontal axes label and the vertical axis label to be added to the figure. There are many other properties that can be set in a figure using the command **set**, which can be investigated using the **help** command. The properties can be also set interactively navigating the *Edit* menu of the figure window. Then the **get** command can be used to see the property names and values. For instance,

```
get('current_figure')
```

will present a list of all the parameters for the current figure and its respective values.

If new calls to the **plot** function are made, the graphs will be overlapped. If this is not the desired behavior but several graphic windows are needed, they can be opened with

```
xset('window',i);
```

where **i** is the number of graphic window where the next plot will be drawn. If no window has been previously open before the first call of **plot2d**, window 0 is assumed. Inside a single graphic window it is possible to include several plots by means of the command **subplot**. Hence,

```
subplot(m,n,p)
```

splits the window into a matrix of sub-windows with **m** rows and **n** columns, and locates the next plot at the **p**-th, cell counting first by rows and then by columns.

It is possible to close the graphic window **n** using

```
xdel(n)
```

In order to close all graphic windows, the argument **winsid()**, which returns the list of graphic windows, can be used:

```
xdel(winsid())
```

As in all programming languages, Scilab allows the use of conditional and control structures. The **for** structure allows to iterate a group of statements a specified number of times. For instance,

```
a = [];
for i=1:5
  a = [a, i*ones(1,i)];
end
```

creates the vector $a = [1, 2, 2, 3, 3, 3, 4, 4, 4, 4, 5, 5, 5, 5, 5]$.

The **if** structure allows to perform or not a group of operations depending on a given condition:

```
if i < 5 then
  a = 1;
elseif i < 10 then
  a = 2;
else
  a = 3;
end
```

Notice that the first time a condition is fulfilled, the conditional ends.

The **while** structure iterates a series of operations while a specified logical condition keeps being true. The code

```
i = 6;
a = [];
while (5 < i) & (i < 10)
  a = [a, i];
  i = i + 1;
end
```

creates the vector [6, 7, 8, 9]. If the initial value of **i** had been ≤ 5 or ≥ 10, the loop would not have been executed.

The **break** command inside a conditional (**if**) which is in turn inside a **for** or a **while** loop allows to quit the loop if certain condition turns to be true.

Finally, the **select** structure allows to carry out different groups of operations according to the value of a specified variable. In the following structure, for instance,

```
select a
case 1
  b = 1;
case 2
  b = 3;
case 3
  b = 0;
end
```

if **a = 2,b** is assigned a value of **3**.

Sometimes during debugging one is interested in measuring the speed of a given algorithm. The instructions **tic** and **toc** are very useful for that purpose. They are used as follows:

```
tic
instruction 1
...
instruction n
toc
```

giving the time in seconds taken by the execution of instructions 1 to n.

The intrinsic functions provided by Scilab cover a wide range of applications. However, it is frequently necessary to introduce specific functions that have not been implemented in the latest available version (or they are so specific that will never be included). It is possible to create new functions in Scilab by means of the syntax

```
function [y1, ..., yn] = name(x1, ..., xm)
  instructions
endfunction
```

where **x1,...,xm** are input arguments and **y1,...,yn** are output arguments, **name** is the name of the function and **instructions** is a sequence of instructions or statements which define or compute the values of the output arguments **y1,...,yn** from the input arguments.

The preceding syntax may be written directly on the console (the main Scilab window) or in any text editor (for instance, SciNotes, the internal editor bundled with Scilab). In the first case, once the function has been defined, it can be immediately be invoked writing

```
[y1, ..., yn] = name(x1, ..., xm);
```

where the input arguments **x1, ..., xm** must have been previously assigned. The syntax

```
name(x1, ..., xm);
```

yields only the first output argument. This syntax is best suited for functions that produce an action rather than a numeric result (such as **plot** or **sound**)

In the second case, once the code has been saved with extension .sci, it is necessary to load it by means of the statement

```
exec path
```

where **path** is the complete path (including file name) of the .sci file. Once the function has been loaded it can be invoked as in the previous case.

If it is desired to have the function permanently available in Scilab without having to load it over and over again, it will be necessary to include it in a library using, for example, the command

```
genlib('worklib', 'SCI/macros/work/', %f, %t)
```

which generates the library called **worklib** (including the necessary structure)[28] from the functions located as **.sci** files in the directory **SCI/macros/work/**.[29] The function **genlib** compiles each function as a binary file with extension **.bin** and creates two files, **names** and **lib**. **names** contain the function names and **lib** is a binary file associated with the library.

In order that the so generated library be available each time Scilab starts the statement

```
worklib = lib('SCI/macros/work/')
```

[28]The symbols **%f** y **%t** are the logical constants *false* and *true*.

[29]SCI is an environment variable that contains a character string indicating the directory where Scilab has been installed.

must be located in a file **.scilab** located in the directory **SCIHOME**.[30] This file is executed during the initialization process and the preceding line of code declares the library.

Maintenance, upgrade and readability of any piece of code are easier if comments are inserted now and then explaining the code. In Scilab these can be inserted by means of a double slash **//**. For instance:

```
// This is a comment
x = y  // x is made equal to y
```

A group of consecutive comment lines before the first instruction or the first empty line is interpreted as the standard help for the function.

Sometimes it is necessary to load into a variable the content of a file (not necessarily a sound file). In this case we must first open the file with the function **mopen** using the syntax

```
[fd, err] = mopen(path_and_name, mode, swap)
```

Here **fd** is a numeric identifier of the file, **err** is an error code (0 means that there have been no errors), and **mode** is a numeric string that indicates how the file is open:

‘r’ read
‘rb’ read binary data
‘rt’ read text
‘w’ write
‘r+’ read and write.

If **swap** is 1 (the default value), when little-endian numbers have been detected the bytes are automatically reordered to recover the normal order. If it is 0 it does nothing.

After opening the file it is possible to read it by means of the functions **mget**, **mgetl** or **mgeti**. The function **mget** reads bytes or numeric words:

```
x = mget(n, type, fd)
```

Here **n** is the number of elements to read, **type** indicates how are interpreted the data:

‘d’ double
‘f’ float
‘l’ long

[30]Te environment variable SCIHOME contains a character string indicating where are located some of the user’s initialization files.

'i' integer
's' short
'c' character

All of these can be followed by

'u' unsigned
'l' little endian
'b' big endian

fd is the file identifier. If it is −1, it refers to the last open file (the default behavior if it is omitted). For instance:

```
x = mget(10, 'ulf', 2)
```

will read 10 unsigned little-endian float values from the file opened with identifier 2.

The function **mgetl** reads text lines:

```
text = mgetl(file_desc, m)
```

where **text** is the string variable that will contain the text, **file_desc** may be the path and file name or the identifier **fd** returned by **mopen**, and **m** is the number of lines to read.

Reference

Scilab (no date) Scilab manual. http://www.scilab.org/download/5.1.1/manual_scilab-5.1.1_en_US.pdf. Accessed 27 Feb 2016

Appendix N
Fast Fourier Transform (FFT)

The discrete Fourier transform is given by

$$X(n) = \sum_{k=0}^{N-1} x(k) \mathrm{e}^{-j\frac{2\pi kn}{N}} \tag{N.1}$$

We will assume that the kernel of the transform $\mathrm{e}^{-j2\pi kn/N}$ can be previously computed and stored in a table (there are only N different values, since the complex exponential is a periodic function with period N). Then the calculation of the N DFT spectral components, i.e., $\{X(n)\}_{n=0,\ \ldots,\ N-1}$, requires N^2 products and N^2 sums, i.e., $2N^2$ floating point operations. For N large, this process is computationally costly. For instance, if $N = 4096$ (an appropriate resolution to represent virtually all the spectral features that the human ear can perceive), we need 33 million operations for each transform. This computational load is prohibitive in many systems.

In 1965, Cooley and Tuckey introduced an algorithm that reduces the order N^2 of the problem to an order $N \log_2 N$ (Cooley and Tuckey 1965).[31] The basic idea is to split the problem into the calculation of two transforms of $N/2$ points, which require $2 \times 2(N/2)^2 = N^2$ floating point operations instead of $2N^2$. If N is a power of 2, it is possible to apply this procedure recursively, accomplishing a substantial reduction of the total number of operations.

To see this, let us call $W_N = \mathrm{e}^{-j2\pi/N}$. Then we can write

$$X(n) = \sum_{k=0}^{N-1} x(k) W_N^{kn} \tag{N.2}$$

Now consider separately the even and odd samples of the signal:

[31]The algorithm had been introduced by Gauss in 1805 and posthumously published in 1866, to simplify the computation of the trajectories of asteroids (Gauss 1866; Wikipedia 2015; Heideman et al. 1984).

F. Miyara, *Software-Based Acoustical Measurements*, Modern Acoustics and Signal Processing, DOI 10.1007/978-3-319-55871-4

$$\begin{aligned} X(n) &= \sum_{h=0}^{\frac{N}{2}-1} \left(x(2h) W_N^{2hn} + x(2h+1) W_N^{(2h+1)n} \right) \\ &= \sum_{h=0}^{\frac{N}{2}-1} \left(x(2h) W_N^{2hn} + W_N^n x(2h+1) W_N^{2hn} \right) \end{aligned} \tag{N.3}$$

Observe that

$$W_N^{2hn} = \left(W_N^2\right)^{hn}$$

and

$$W_N^2 = \left(e^{-j2\pi/N}\right)^2 = e^{-j2\pi/(N/2)} = W_{N/2},$$

from where

$$X(n) = \underbrace{\sum_{h=0}^{\frac{N}{2}-1} x(2h)\, W_{N/2}^{hn}}_{Y(n)} + W_N^n \underbrace{\sum_{h=0}^{\frac{N}{2}-1} x(2h+1) W_{N/2}^{hn}}_{Z(n)}. \tag{N.4}$$

$Y(n)$ and $Z(n)$ are, respectively, the discrete Fourier transforms of the even and odd components of $x(k)$. We can split the calculation of (N.4) into the values of n smaller than $N/2$ and the values greater than or equal to $N/2$. In the first case,

$$X(n) = Y(n) + W_N^n Z(n).$$

In the second case, with a small variable change we have

$$X(n+N/2) = Y(n+N/2) + W_N^{n+N/2} Z(n+N/2).$$

Now, since $\{x(2h)\}$ and $\{x(2h+1)\}$ have only $N/2$ points, their transforms are periodic of period $N/2$, i.e.,

$$\begin{aligned} Y(n+N/2) &= Y(n) \\ Z(n+N/2) &= Z(n) \end{aligned}$$

so

$$X(n+N/2) = Y(n) + W_N^{n+N/2} Z(n),$$

expression valid for $n = 1, \ldots, N/2 - 1$. But it is also easy to see that $W_N^{N/2} = -1$, so we get the following espressions

$$\begin{cases} X(n) = Y(n) + W_N^n Z(n) \\ X(n+N/2) = Y(n) - W_N^n Z(n) \end{cases} \tag{N.5}$$

It suffices to compute $Y(n)$ and $Z(n)$ for $n = 0, \ldots, N/2 - 1$ to be able to get $X(n)$ for $n = 0, \ldots, N - 1$. This computation requires N sums and $N/2$ products.

To get $Y(n)$ and $Z(n)$ we proceed in the same way, splitting each one into the even and odd terms. For each one we need to perform $N/2$ sums and $N/4$ products, but as we have now to compute two transforms, we will need N sums and $N/2$ products.

The process is iterated and at each new step we have to compute twice as many transforms of half the number of points. This continues until after $\log_2 N$ steps we must compute N transforms of 1 point each. But the transform of a signal of length 1 is the signal itself, so there is no need to compute anything else. The whole process has involved $N \log_2 N$ sums and $N/2 \log_2 N$ products. In the example mentioned above where $N = 4096$, we will need 73 728 floating point operations, i.e., about 455 times less operations that if we had applied the discrete Fourier transform formula.

Why is it that we can save so many operations? Conceptually, the direct application of (N.1) implies to compute many times the same values, since although the exponent adopts N^2 values, the complex exponential is periodic and takes only N different values. The described procedure rearranges the calculations in such a way that we can take advantage of this situation.

The only inconvenience is that at each step it is necessary to interleave the even and odd components, which complicates component addressing. It can be shown that it is equivalent to counting from 1 to N reversing the bit position with respect to normal binary numbering (the MSB comes to be the LSB and vice versa). For instance, for $N = 16$ we would have

0000
1000
0100
1100
0010
1010
...
0111
1111

The algorithm just described is but an example of a broad class of algorithms known collectivelly as *Fast Fourier Transform*, FFT. All mathematical software packages as well as digital signal processing systems (DSP) feature highly optimized FFT algorithms, including not only the preceding algorithm but also cases where N is not a power of 2 but a multiple of prime factors, where even if the reduction in the computational load is not so impressive, significant reductions are accomplished. For this reason it is not frequent to have to implement an FFT algorithm from scratch. However, even without the objective of attaining the best code, writing a piece of code that implements the FFT is an excellent programming exercise.

References

Cooley JW, Tuckey JW (1965) An algorithm for the machine computation of complex Fourier Series. Math Comput 19:297–301

Gauss CF (1866) Theoria interpolationis methodo nova tractata. Werke, Band 3:265–327 (Königliche Gesellschaft der Wissenschaften, Göttingen, 1866)

Heideman MT, Johnson DH, Burrus CS (1984) Gauss and the history of the fast Fourier transform. IEEE ASSP MAGAZINE. Oct 1984, pp 14–21. http://www.cis.rit.edu/class/simg320/Gauss_History_FFT.pdf. Accessed 26 Feb 2016

Wikipedia (2015) Cooley–Tukey FFT algorithm. Wikipedia: The Free Encyclopedia. Wikimedia Foundation, Inc., 23 Dec 2015. https://en.wikipedia.org/wiki/Cooley%E2%80%93Tukey_FFT_algorithm. Accessed 27 Feb 2016

Appendix O
Parseval's Identity and Symmetry of the DFT

The discrete Fourier transform of a discrete signal $x(k)$ is given by

$$X(n) = \sum_{k=0}^{N-1} x(k) e^{-\frac{j2\pi kn}{N}} \tag{O.1}$$

while the inverse discrete Fourier transform, which allows to retrieve the signal from its transform, is

$$x(k) = \frac{1}{N} \sum_{n=0}^{N-1} X(n) e^{\frac{j2\pi kn}{N}}. \tag{O.2}$$

Frequently, the versions implemented by the mathematical software packages start from index 1 instead of 0. Some authors place the factor $1/N$ in the direct transform instead of the inverse one.

We can get a relationship with the RMS value of the signal given by

$$X_{\mathrm{ef}} = \sqrt{\frac{1}{N} \sum_{k=0}^{N-1} x^2(k)}, \tag{O.3}$$

in the following way:

$$\begin{aligned} X_{\mathrm{ef}}^2 &= \frac{1}{N} \sum_{k=0}^{N-1} x^2(k) = \frac{1}{N} \sum_{k=0}^{N-1} \left(x(k) \frac{1}{N} \sum_{n=0}^{N-1} X(n) e^{\frac{j2\pi\, kn}{N}} \right) \\ &= \frac{1}{N} \sum_{k=0}^{N-1} \left(\frac{1}{N} \sum_{n=0}^{N-1} x(k) X(n) e^{\frac{j2\pi kn}{N}} \right) \\ &= \frac{1}{N} \sum_{n=0}^{N-1} \left(\frac{1}{N} \sum_{k=0}^{N-1} x(k) X(n) e^{\frac{j2\pi kn}{N}} \right) \\ &= \frac{1}{N} \sum_{n=0}^{N-1} X(n) \left(\frac{1}{N} \sum_{k=0}^{N-1} x(k) e^{\frac{j2\pi kn}{N}} \right) \end{aligned}$$

F. Miyara, *Software-Based Acoustical Measurements*, Modern Acoustics and Signal Processing, DOI 10.1007/978-3-319-55871-4

If the signal $x(n)$ is real, the conjugate of each term is equivalent to change the exponent sign in the complex exponentials, so

$$X_{\mathrm{ef}}^2 = \frac{1}{N}\sum_{n=0}^{N-1} X(n)\left(\frac{1}{N}\overline{\sum_{k=0}^{N-1} x(k)\mathrm{e}^{\frac{-j2\pi kn}{N}}}\right) = \frac{1}{N^2}\sum_{n=0}^{N-1} X(n)\overline{X(n)}.$$

Finally,

$$X_{\mathrm{ef}}^2 = \sum_{n=0}^{N-1}\left|\frac{X(n)}{N}\right|^2. \tag{O.4}$$

This is known as Parseval's identity.

It is also possible to get some conjugate symmetry conditions for the Fourier coefficients for real signals. Starting from the transform definition, we conjugate and operate,

$$\begin{aligned}\overline{X(n)} &= \sum_{k=0}^{N-1} x(k)\overline{\mathrm{e}^{-\frac{j2\pi kn}{N}}} = \sum_{k=0}^{N-1} x(k)\mathrm{e}^{-\frac{j2\pi(-n)k}{N}} \\ &= \sum_{k=0}^{N-1} x(k)\mathrm{e}^{-\frac{j2\pi(N-n)k}{N}}\end{aligned}$$

The last step is possible because $\mathrm{e}^{j2\pi K} = 1$ for any integer K. The last member can be interpreted as the transform at $N - n$, so

$$X(N-n) = \overline{X(n)}. \tag{O.5}$$

In particular, this implies

$$|X(N-n)| = |X(n)|,$$

therefore we can write

$$X_{\mathrm{ef}}^2 = \left|\frac{X(0)}{N}\right|^2 + \sum_{n=1}^{N/2-1} 2\left|\frac{X(n)}{N}\right|^2 + \left|\frac{X(N/2)}{N}\right|^2. \tag{O.6}$$

We can reinterpret the terms in the sum if we group together the conjugate terms in the transform definition:

$$x(k) = \frac{X(0)}{N} + \sum_{n=1}^{N/2-1} \left(\frac{X(n)}{N} e^{\frac{j2\pi nk}{N}} + \frac{\overline{X(n)}}{N} e^{\frac{j2\pi(N-n)k}{N}} \right)$$
$$+ \frac{X(N/2)}{N} e^{\frac{j2\pi(N/2)k}{N}}.$$

If we now express $X(k)$ as

$$X(n) = |X(n)|e^{j\varphi},$$

the terms between parenthesis can in turn be rewritten as:

$$\frac{|X(n)|}{N}\left(e^{j\left(\frac{2\pi nk}{N}+\varphi\right)} + e^{-j\left(\frac{2\pi nk}{N}+\varphi\right)}\right)$$
$$= \frac{2|X(n)|}{N} \cos\left(\frac{2\pi nk}{N} + \varphi\right),$$

which, replacing, yields

$$x(k) = \frac{X(0)}{N} + \sum_{n=1}^{N/2-1} \frac{2|X(n)|}{N} \cos\left(\frac{2\pi nk}{N} + \varphi\right) + \frac{X(N/2)}{N} e^{\frac{j2\pi(N/2)k}{N}}.$$

Comparing with (O.6) we see that $2|X(k)/N|^2$ is the squared RMS value of a real sine wave signal. This allows introducing the real spectrum:

$$X_r(n) = \begin{cases} \frac{X(0)}{N} & n = 0 \\ \frac{2|X(n)|}{N} & n = 1,\ldots,N/2-1 \\ \frac{X(N/2)}{N} & n = N/2 \end{cases}. \tag{O.7}$$

The first term is a DC component whose RMS value is equal to its absolute value. The next terms have an RMS value that is equal to the amplitude divided by $\sqrt{2}$. Even if last term ($n = N/2$) is written as a complex exponential, it is a component that alternates between values $\pm X(N/2)/N$ so its RMS value is equal to its absolute value, which is constant. This term corresponds to the Nyquist frequency, i.e., the maximum one that does not create aliasing.

Appendix P
Spectrum of Window Functions

When applying a window function, such as Hann or Blackman, they multiply in the time domain the signal to analyze,

$$x_w(t) = w(t)x(t), \tag{P.1}$$

so the spectrum of the windowed signal is the convolution of the Fourier spectra of the signal and the window:

$$X_w(\omega) = W(\omega) * X(\omega). \tag{P.2}$$

To understand what this implies, suppose the signal is the spectrally most simple one, a complex exponential, $x(t) = e^{j\omega t}$, whose Fourier transform has a single nonzero line. According to the definition of convolution, the spectrum will be formed by the spectrum of the window, $W(\omega)$, shifted to the position of that spectral line. In other words, the original spectrum of the signal will be replaced by a spectrum that involves the whole frequency spectrum, creating distortion.

This is known as *spectral leakage*. The consequence of spectral leakage is the presence of spurious components similar to noise, even though they actually constitute a nonlinear distortion of the signal.

If we now add other spectral components to the signal, each one will contribute a corresponding copy of the window spectrum (weighted by the component amplitude and frequency-shifted), causing an increase in spectral noise.

The ideal window, i.e., that one which would produce no change in the spectrum, would be one that contributes no spectral components other than the original signal ones. It is easily seen that its spectrum should be a Dirac delta, which actually corresponds to a constant window over the whole time axis. Such a window would not be useful in practice since it would imply sampling the signal from $t = -\infty$ to $t = \infty$, so it would not satisfy its intended purpose. However, the shape of the spectrum of the ideal window is useful as a guide to an acceptable tradeoff: it should concentrate its spectral energy around 0 so that the inevitable spectral leakage falls rapidly at high frequency.

That is why it is important to compare the spectrum of different windows, since it provides a criterion of acceptability and shows their limitations. Consider the

F. Miyara, *Software-Based Acoustical Measurements*, Modern Acoustics and Signal Processing, DOI 10.1007/978-3-319-55871-4

Hann window of duration T in the continuous-time domain. To simplify the analysis, we will suppose it is centered at 0:

$$w_{\mathrm{hann}}(t) = 0.5\left(1 - \cos\frac{2\pi}{T}\left(t - \frac{T}{2}\right)\right)r(t), \tag{P.3}$$

where $r(t)$ is the rectangular window between $-T/2$ and $T/2$. To compute the spectrum, recall that the cosine can be written as the sum of two complex exponentials:

$$w_{\mathrm{hann}}(t) = \left(0.5 - 0.25e^{j\frac{2\pi}{T}\left(t-\frac{T}{2}\right)} - 0.25e^{-j\frac{2\pi}{T}\left(t-\frac{T}{2}\right)}\right)r(t), \tag{P.4}$$

i.e.,

$$w_{\mathrm{hann}}(t) = \left(0.5 + 0.25e^{j\frac{2\pi}{T}t} + 0.25e^{-j\frac{2\pi}{T}t}\right)r(t). \tag{P.5}$$

Each exponential corresponds to a single spectral line so the spectrum of the rectangular window,

$$R(\omega) = T\frac{\mathrm{sen}(\omega T/2)}{\omega T/2}. \tag{P.6}$$

is simply shifted towards the frequency of the exponential. Therefore, the complete spectrum of the Hann window is,

$$W_{\mathrm{hann}}(\omega) = 0.5R(\omega) + 0.25R(\omega - 2\pi/T) + 0.25R(\omega + 2\pi/T), \tag{P.7}$$

In Fig. P.1 the three components and their sum are shown. The shifted spectra are called *lateral bands*.

The effectiveness of the Hann window is due to the fact that for high frequency the sinc function (i.e., $\sin(\pi x)/\pi x$) falls down very slowly, so one pseudocycle differs only slightly from the next one. Therefore, the superposition of two negative pseudocycles and a positive one of double amplitude yields almost complete cancellation.

Figure P.2 compares the spectra of the rectangular and Hann windows in a plot with amplitude scale in dB.

We can see a widening of the low-frequency lobe (also called *central lobe* in the bilateral spectrum). This implies that it will be more difficult to separate very close tonal components than in the case of the rectangular window. This is the price to pay for the reduction of the first lateral lobe from −13 dB (which corresponds to the rectangular window) to −33 dB. Anyway, this lobe should be actually compared with the second lateral lobe of the rectangular window, which reaches −18 dB.

With a similar reasoning we get the spectrum of the Blackman window, expressed in this case by a rectangular window and two pairs of lateral bands:

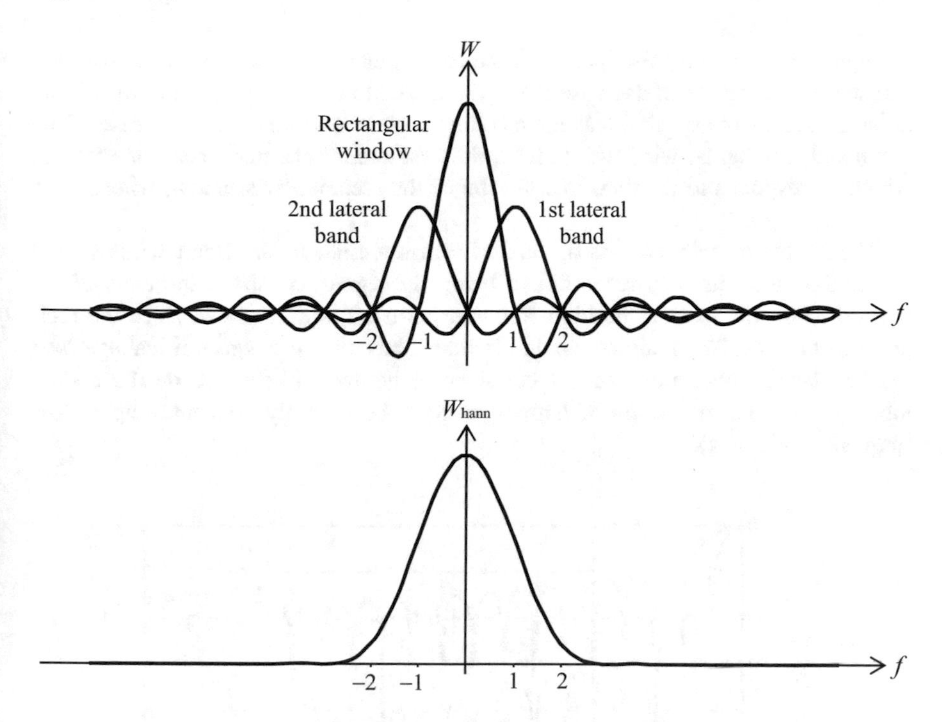

Fig. P.1 *Top*, spectral components of the Hann window spectrum for $T = 1$ s. *Bottom*, the complete bilateral spectrum of the Hann window (linear amplitude scale)

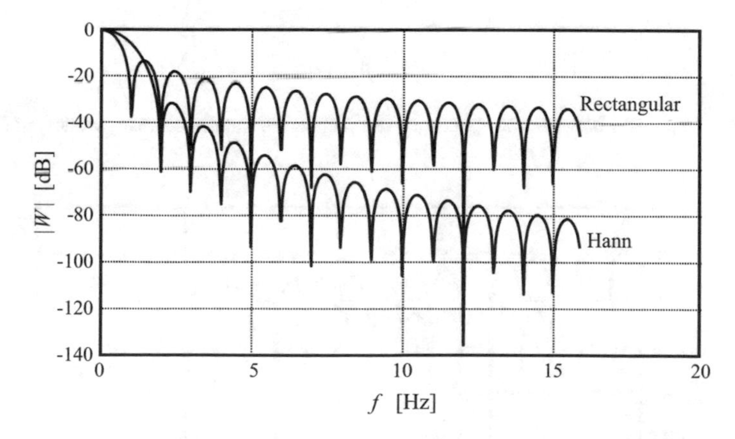

Fig. P.2 Comparison between the spectra of the rectangular and Hann window functions ($T = 1$ s)

$$\begin{aligned} W_{\text{blackman}}(\omega) = {} & 0.42R(\omega) + 0.25R(\omega - 2\pi/T) + 0.25R(\omega + 2\pi/T) \\ & + 0.04R(\omega - 4\pi/T) + 0.04R(\omega + 2\pi/T) \end{aligned} \tag{P.8}$$

Figure P.3 compares the spectra of the rectangular and Blackman windows. We can see a very important decrease at the intermediate frequencies, so the wide band noise caused by the spectral leakage has much less relevance than in the case of the Hann and rectangular windows. Indeed, the first lateral lobe has a peak of −58 dB, which corresponds to the third lateral lobe of the rectangular window, whose peak is 21 dB.

The Hamming window has an analytic form similar to the Hann window, but unlike the latter, its purpose is not to bring the signal to zero at the ends of the window. For this reason, the high frequency roll-off is quite slow, as in the rectangular window. The main advantage is that it has very low spectral leakage near the first lateral lobe, attaining a spectral noise below −40 dB outside the central lobe. It has a general slope of Brownian type, i.e., mainly concentrating at low frequency (Fig. P.4).

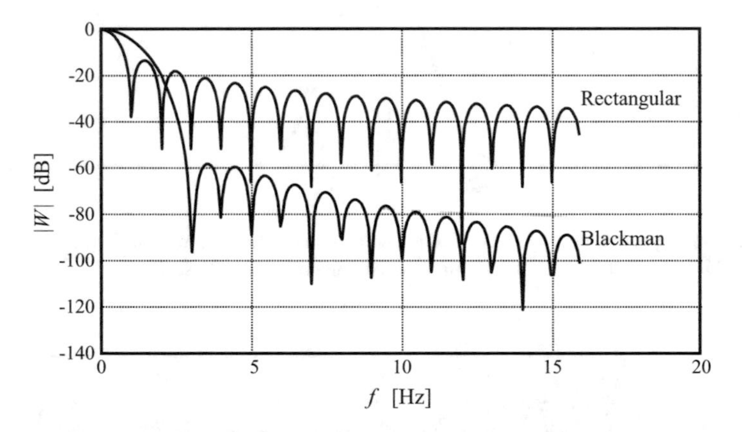

Fig. P.3 Comparison between the spectra of the rectangular and Blackman windows (T = 1 s)

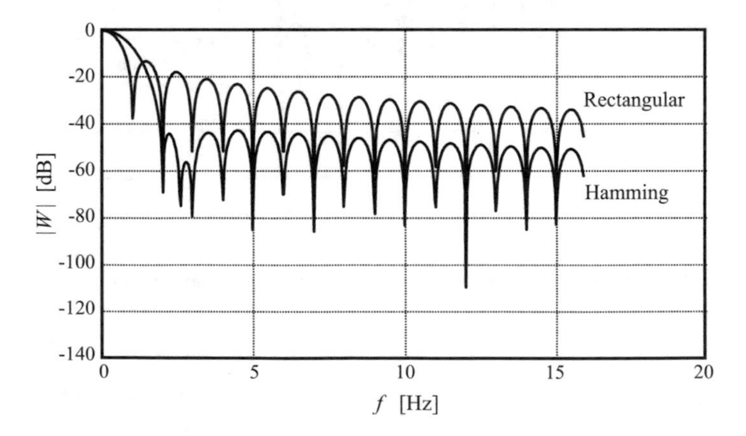

Fig. P.4 Comparison between the spectra of the rectangular and Hamming windows (T = 1 s)

P.1 Equivalent Bandwidth of Window Functions

The *noise effective bandwidth* or *equivalent bandwidth* of a window $w(t)$ is defined as (Gade et al. 1987a, b)

$$B_{\mathrm{ef}} = \frac{\int_{-\infty}^{\infty} |W(2\pi f)|^2 \mathrm{d}f}{\max|W(2\pi f)|^2}. \quad \text{(P.9)}$$

We can apply the Parseval's identity (see Chap. 1 and Appendix O) to transform the spectrum domain integral in a time domain integral:

$$\int_{-\infty}^{\infty} |W(2\pi f)|^2 \mathrm{d}f = \frac{1}{2\pi} \int_{-\infty}^{\infty} |W(\omega)|^2 \mathrm{d}\omega = \int_0^T w^2(t) \mathrm{d}t$$

Considering that the lateral band shift is an integer number of times the frequency $1/T$ (see Fig. P.1 and Eqs. (P.7) and (P.8)), the maximum happens to be at the origin and is equal to the coefficient of the independent term. We have, therefore, for the Hann window,

$$B_{\mathrm{ef,Hann}} = \frac{\int_0^T w^2(t)\mathrm{d}f}{0.5^2 T^2} = \frac{1}{T^2} \int_0^T \left(1 - \cos\frac{2\pi}{T}t\right)^2 \mathrm{d}t = \frac{3}{2T} \quad \text{(P.10)}$$

i.e.,

$$B_{\mathrm{ef,Hann}} = \frac{3}{2}\frac{F_{\mathrm{s}}}{N}. \quad \text{(P.11)}$$

This widening with respect to the analysis resolution of the DFT is due to the spectral leakage. Similarly, for a Blackman window,

$$\begin{aligned} B_{\mathrm{ef,Blackman}} &= \frac{\int_0^T w^2(t)\mathrm{d}f}{0.42^2 T^2} \\ &= \frac{1}{0.42^2 T^2} \int_0^T \left(0.42 - 0.5\cos\frac{2\pi}{T}t + 0.08\cos\frac{4\pi}{T}t\right)^2 \mathrm{d}t \\ &= \frac{1}{T}\left(1 + \frac{1}{2}\left(\frac{0.5}{0.42}\right)^2 + \frac{1}{2}\left(\frac{0.08}{0.42}\right)^2\right) = \frac{1.727}{T} \end{aligned}$$

$$B_{\mathrm{ef,Blackman}} = 1.727\frac{F_{\mathrm{s}}}{N} \quad \text{(P.12)}$$

References

Gade S, Herlufsen H (1987a) Use of weighting functions in DFT/FFT analysis (Part I). Brüel & Kjær. Technical Review 1987 N° 3. http://www.bksv.com/doc/bv0031.pdf. Accessed 26 Feb 2016

Gade S, Herlufsen H (1987b) Use of weighting functions in DFT/FFT analysis (Part II). Brüel & Kjær. Technical Review. 1987 N° 4. http://www.bksv.com/doc/bv0032.pdf. Accessed 26 Feb 2016

Index

A
Absolute positioning, 134
Accuracy, 45, 310
Acoustical measurement, 1
Acoustic coupler, 202
Acoustic resistance, 198
Adaptive Transform Audio Coding (ATRAC), 107, 149
Adobe Audition, 168
A filter, 178
Alias frequency, 80, 366, 368
Aliasing, 80, 366
Allocation file, 158
Allocation units, 153
Analog/digital, 83
Analog/digital conversion, 92
Analog/digital converter, 77, 83
Analysis protocol, 174
Android, 157
Anechoic chamber, 191
Anti-aliasing filter, 82, 106, 367, 368
Application, 374
apropos, 395
Ardour, 168
ASCII, 115
Associated data list (ADTL), 114
Attributes, 124, 385
Audacity, 168, 231, 284
    recording, 171
    select, 173
    dependency, 172
    labels, 174
Audacity project (AUP), 124, 125, 131, 172
Audacity project (AU) file, 130, 385
Audio data structure, 371
Audiometric earphones, 201
Audit, 156
AUF format, 131
AU, format, 130
AUP format, 124
Average, 323
A-weighted level C, 12
A-weighted sound level, 59
A-weighting, 12, 236
Azo, 139, 147

B
B*-Tree, 158
B + tree, 156
Background noise, 16
Back plate, 187
Backup, 77
Band, 18
Bandpass filter, 248, 351
Band spectrum, 264
BD-R, 151
BD-RE, 151
Beamforming, 137, 157, 200
Big endian, 370
Bilinear transformation, 228
Binary logic, 83
Binary system, 83
Bit, 83
Bitmap, 158
Blackman–Harris window, 261, 262
Blackman
    main lobe, 263
    window, 261, 262, 289, 416
Block, 135, 371
Bloque streaminfo, 377
Blu-ray Disc (BD), 151
Boltzmann's constant, 198
Brownian motion, 198
Buffer, 136
Burst error, 142, 143
Butterworth approximation, 287
Butterworth filter, 247
B weighting, 12

F. Miyara, *Software-Based Acoustical Measurements*, Modern Acoustics and Signal Processing, DOI 10.1007/978-3-319-55871-4

C
Calibration, 62, 233, 319
Calibration constant, 176
Calibration signal, 110
Calibration tone, 111, 175–177
Catalog file, 158
Cauer filter, 247, 248
CD-DA, 138, 139
CD-R, 139, 147
CD-ROM, 138
CD-RW, 139
CD-WO, 147
CD-WR, 147
CeCILL license, 231, 395
Center frequency, 19, 20
Central limit theorem, 61, 328
C filter, 180
Chain uncertainty, 72
Characteristic eq., 225
Characteristic polynomial, 223
Chebychev filter, 82, 247, 248
Child elements, 124, 386
Chi-square, 334
Chunk, 108
Circular convolution, 275, 276
Circumaural headset, 202
C language, 168
Clusters, 153
Codec, 115
Collinear holography, 152
Collision, 374
Combined standard uncertainty, 53, 316
Combined uncertainty, 353
Compact disc (CD), 138, 139
Compact Flash (CF), 105, 137
Compressed audio formats, 106
Computational load, 277
Condenser microphone, 187
Confidence interval, 45, 332, 333
Confidence level, 6
Constant band analyzer, 18
Constant bandwidth analyzer, 245
Constant linear velocity (CLV), 140
Constant model, 379
Constant percentage analyzer, 18, 246
Constant percentage band spectrum, 264
Constant predictor, 120
Continuous random variable, 325
Contrast of algorithms, 280
Control loop, 134
Convolution, 222, 275, 415
Convolution product, 213
Cool Edit, 168
Cooley and Tuckey, 253, 407
Correlation, 278
Covariance, 53, 314
Covariance matrix, 59, 317
Coverage factor, 52, 318
Coverage probability, 52
CRC-8, 380
CRC-16, 384
Creative Audigy, 286
Critical band, 273
Critical band filter, 279
Critical bandwidth, 273
Cross Interleaved Reed–Solomon Code (CIRC), 143, 145
Cubase, 168
Cue, 108, 111, 112
Cue chunk, 112
Cue sheet, 376
Cumulative
  distribution function, 326
  frequency distribution, 322
  percent distribution, 322
  probability function, 241, 326
Curie temperature, 148
C-weighted level, 12
C weighting, 12
Cyanine, 139, 147
Cyclic redundancy code, 144, 380
Cylinder, 135

D
Data chunk, 109
Data fork, 158
Data stream, 369
dB, 4
dBFS, 200
decibel, 4
Decimation, 101
Delayed allocation, 157
Delta modulation (DM), 98
Dependency, Audacity, 172
Diaphragm, 187
Difference equation, 215, 216
Difference equation, solution, 223
Differential pulse code modulation (DPCM), 95
Diffuse field, 191
Diffuse field corrección, 196
Diffuse-field microphone, 192, 195
Diffuse-field sensitivity, 204
Digital/analog conversion, 91
Digital audio, 77
Digital audio editor, 168
Digital audio tape (DAT), 104, 132, 152
Digital audio tape recorder, 104

Digital audio workstation, 168
Digital filter, 20, 235, 246
Digital recorder, 77, 297
Digital signal, 77
Digital signature, 373
Digital Versatile Disc (DVD), 138, 151
Digital word, 83
Digitization, 83
Dirac's delta, 36, 211
Directivity index, 207
Direct measurement method, 46
Directory entries, 154
Directory tree, 154
Discrete
impulse, 211
random variable, 325
signal, 83, 211
time, 77
Discrete dynamic systems, 213
Discrete Fourier transform (DFT), 150, 251, 253, 267, 407, 411
Discrete system frequency response, 217
Disk Operating System (DOS), 153
Dither, 86, 171
Dithering, 86
Document type definition (DTD), 124, 385
Downsampling, 101, 249
DSP, 409
DVD-R, 139, 151
DVD-RW, 151
D weighting, 8, 12

E
ECMA-167, 159
ECMA-377, 152
Edition events, 123
Effective bandwidth, 273, 419
Effective value, 2, 232
Eight-to-fourteen modulation (EFM), 139, 145
Eight-to-ten modulation (ETM), 152
Electret, 189
Elements, 124, 385
Elliptic filter, 82, 247
Energy average, 232
Energy envelope, 2, 17, 234
Entropy, 119
Envelope, 347, 392
Equivalent bandwidth, 419
Equivalent continuous sound level, 4
Equivalent level, 4, 56, 114, 275
Error bound, 313
Error-correcting code, 136, 142
Error correction, 78
Error-detecting code, 136
Error detection, 384
Error function, 16, 345
Euler, 231
Exceedance levels, 13, 239
exFA, 153
Expanded uncertainty, 52, 318
Expected value, 313
Exponential time average, 3
Export, 124
Expstudio Audio Editor, 168
EXT3, 153
EXT4, 153, 157
Extended File System, 157
Extensible markup language, 385
Extent, 153, 156, 157

F
Fast Fourier Transform (FFT), 157, 253–255, 259, 260, 267, 273, 274, 279, 400, 409, 410
Fast time response, 4
FAT 32, 153
FFT filter, 178, 195, 268–272, 285, 286
FFT spectrum analysis, 264, 289
FGMOS, 137, 138
Field effect transistor, 137
File allocation table (FAT), 153, 154
File system, 152, 153
Filter
Chebychev, 82
elliptic, 82
FIR, 246, 268
IIR, 246, 268
linear phase, 268
bank, 20
Filtered impulse, 213
Filtering window, 271
Finite impulse response (FIR), 214, 217, 246, 268
FIR filter, 237, 250, 268
FLAC file structure, 123, 371
FLAC partition order, 383
FLAC reference codec, 371
Flash converter, 93
Flash memory, 105, 132, 136
Floating-gate transistor, 137
`flts`, 400
Forced response, 31
Format chunk, 108
Fourier coefficients, 251
Fourier transform, 363
Fragmentation, 136
Frame, 143, 371

Frame header, 377
Free Audio Editor, 168
Free field, 191, 193–195
Free-field microphone, 192, 195, 197
Free-field sensitivity, 203
Free Lossless Audio Codec (FLAC), 107, 115
Free response, 31
Free Software Foundation (FSF), 105, 168, 231
Frequency distribution, 322
Frequency response, 217, 222, 298
Frequency weighting, 9, 10
Frmag, 400
Frontal incidence, 193–196

G
Gamma function, 331
Gaussian distribution, 328
Gaussian dither, 88
General Public License (GPL), 157, 168, 231, 395
Goldwave, 168
Golomb–Rice code, 117, 122, 382
GPL License, 168, 395
GPS, 162, 174, 374
Granularity, 100, 158
Group channel code, 144
Guide for the Expression of Uncertainty in Mesurement, 310
GUM, 44, 310
G weighting, 8, 9, 11, 12

H
Hamming window, 261, 262, 418
Hanning, *see* Hann, 261
Hann, overlapping, 264, 266
Hann window, 261–264, 417
  overlapping, 269
Hard disk, 132
Hard disk drive, 133
Hard disk geometry, 135
Hash algorithm, 373
HDA 200 headset, 201
Head, 135
Header, 371, 372
Head related transfer functions (HRTF), 205
Hearing threshold, 4
Hexadecimal, 108
HFS+, 153, 158
Hierarchical file System, 158
High capacity Secure Digital, 105
Histogram, 239, 322
Holographic Versatile Disc (HVD), 152
Homogeneous solution, 223, 225
Horner, 399
Hot-electron injection, 138

I
ID3, 388, 393
Identifier, 108
IEC 61260, 245, 249, 281, 286
IEC 60537, 10, 12
IEC 60651, 10, 12
IEC 61183, 195
IEC 61260, 27
IEC 61672, 10, 12, 105, 297
IIR, 214, 235, 246, 247, 268
IIR filter, 237, 268, 287
Impedance tube, 192
Import, 124
Impulse response, 4, 213, 221, 222, 275
Index points, 376
Indirect measurement method, 46
Inference, 331
Infinite frequency response, 246
Infinite-impulse-response, 214
Integration, 341
Integration time, 337
Interchannel decorrelation, 122
Internal noise, 300
Interoperability, 162
Inverse discrete Fourier transform (IDFT), 253
Inverse error function, 345
Inverse z transform, 225
ISO 7196, 11, 12
ISO 1996-2, 105, 273, 297
ISO-GUM, 310

J
Jitter, 90
Joint probability density function, 53, 314

K
Kaiser window, 262
Kerr effect, 148
Kundt tube, 192

L
L_{A90}, 16
Label, 108, 111
Label chunk, 115
Labeled text chunk, 114
Labeling, 78

Labels, 115
Audacity, 174
Label track, 174
Labeltrack element, 124
labl, 114, 115
Laboratory standard, 319
Laplace distribution, 116, 117
Laplace transform, 351
Laser-trimming, 92
Latency, 93, 132
Lateral band, 416
Lateral lobe, 263, 264
Least significant bit (LSB), 84
Level adjustment, 171
Levinson–Durbin algorithm, 122
Linear convolution, 278
Linearity error, 62, 68, 92
Linear-phase filter, 268
Linear predictive model, 119
Linear predictor coding (LPC), 119, 122
Linear predictor model, 122
Linear system, 213
Linear velocity, 140
Line spectrum, 23, 255
Line spectrum analyzer, 18
Linux, 124, 157
LIST, 114
Little endian, 108
Lobe, 263
Long distance code, 151
Lossless compression, 107
Lossy audio data compression, 149
Lossy compression, 107
Low-level distortion, 85
L_p, 4
LPC model, 382
LPM, 119
LTI, 213
ltxt, 114

M
Mac OS X, 124
Magnetic selection, 175
Main lobe, 264
Masking, 149
Masking threshold, 275
Master File Table (MFT), 156
Mathematical expectancy, 314, 324
MATLAB, 231, 284
MD5, 373
MD, 148
Mean, 313, 323
Mean value, 323
Measurand, 43, 309
Measurement, 309
error, 312
method, 45, 309
model, 45, 46, 309
principle, 45, 309
procedure, 44, 45, 309
protocol, 167
report, 46
standard, 318
Measures of central tendency, 322
Measures of dispersion, 322
Measures of variability, 322
Median, 323
Membrane, 187
MEMS, 200
MEMS microphones, 200
sensitivity, 200
frequency response, 201
Metadata, 111, 371, 372
Microphone, 1, 187
damping, 198
directional response, 196, 197
distortion, 199
frequency response, 193
intrinsic noise, 198
noise, 199
polarization, 189
protective grid, 195
response correction, 193
sensitivity, 233
Minidisc, 104, 132, 148
Mode, 323
Modified discrete cosine transform (MDCT), 150
Monel, 189
MOSFET, 137
MP3, 107
MPEG Layer 3, 107
Music Editor Free, 168
Mute, 125

N
NAND, flash memory, 138
Narrow band spectrum, 273
Natural angular frequency, 30
Natural response, 31
Nested elements, 124
New Technology File System (NTFS), 153, 155, 156
Newton–Raphson methodo, 280
Nodes, 158
Noise profile, 183
Noise reduction, 78
Noise removal, 78, 183

Noise shaping, 103, 200
Nominal frequencies, 20
Nonlinear systematic error, 357
Nonresident attributes, 156
Non-stationary noise, 341
NOR, flash memory, 138
Normal distribution, 6, 13, 15, 328
Normalized frequency response, 31
Normal vibration modes, 23
Note, 108, 111, 115, 116
Note chunk, 115
NTFS-3G, 157
Nuendo, 168
Nyquist condition, 80, 82, 366
Nyquist frequency, 80, 219, 256, 267, 366, 368
Nyquist language, 168

O
Observation error, 312
Octave, 231, 284
Octave band, 19, 27, 178, 264
Octave band analyzer, 18
Octave fraction, 18
Offset, 374
OGG Vorbis, 107
Omnidirectional source, 206
One-third octave analyzer, 18
One-third octave band, 20, 27, 178, 250, 264
Optical disc, 138
Optimal power calibration (OPC), 147
Oscillogram, 130, 169
OS X, 157, 158
Overlap-add, 270
Oversampling, 82, 101, 250
Oversampling factor, 101

P
Padding, 374
Paralax error, 62
Parseval's identity, 36, 411, 412, 419
Partial fraction expansion, 351
Particular solution, 223
Partition order, 383
PCM format, 106
Percent distribution, 322
Percentile, 13, 15, 239, 345, 346
Perceptual compression, 149
Phtalocyanine, 139, 147
Pink noise, 199
Platters, 133
`plot2d`, 400
Polarization voltage, 188
Poles, 225
Poly, 398
Polynomial approximation, 119
Polynomial model, 382
Polynomial predictor, 120
Population combined uncertainty, 353
Population mean, 314, 325
Population standard deviation, 50, 325, 330
Population variance, 335
Power calibration area (PCA), 147
Power Sound Editor, 168
Power spectral density, 24, 267
Preamplifier, 190
Preamplifier distortion, 200
Precision, 44, 310
Predictor, 115, 119
  constant, 120
  polynomial, 120
  verbatim, 119
Pre-echo, 270
Pressure field, 191
Pressure-field microphones, 192
Pressure-gradient microphone, 192
Primary measurement standard, 318
Primary standard, 318
Probability, 326
Probability density, 326
Probability density function, 16, 313, 326
Program memory area (PMA), 147
Project, 123
  Audacity, 172
  element, 124
  sample rate, 124
Pro Tools, 168
Pulse code modulation (PCM), 95
Pulse density modulation (PDM), 200

Q
Quadrature mirror filter, 149
Quality factor, 30
Quantity, 309
Quantization noise, 101, 106

R
R-2R ladder network, 91
Radiation acoustic impedance, 192
Random error, 313
Random field, 191
Random signal, 24
Random variable, 13, 321
RAW format, 106
Rayleigh distribution, 348
Read–write head, 132, 134
Realization of the measurand, 44, 309
Reconstruction formula, 366
Recordable CD's, 147

Recording, 132
  Audacity, 171
  layer, 148
Redundancy, 122
Reed–Solomon code, 143
Reference
  conditions, 47
  pressure, 4
  sensitivity, 203
  standard, 319
Relative bandwidth, 20
Relative positioning, 134
Relative systematic error, 64, 65
Renard series, 20, 319
Repeatability, 311
Repeatability conditions, 311
Repetition, 311
Replication, 311
Reproducibility, 312
Reproducibility conditions, 312
Requantization, 101
Resident attributes, 156
Residuals, 116, 383
Resolution, 44, 61, 83, 310
Resource fork, 158
Rewritable compact disc, 147
Rice parameter, 119, 383
RMS, 2, 232
RMS value, 337, 341
Roots, 399
Run-length compression, 120

S
Sample, 321, 325
Sample and hold, 78, 94
Sample combined uncertainty, 355
Sample covariance, 54, 315, 356
Sampled signal, 363
Sample mean, 315, 325, 329, 331
Sample rate, 8, 14, 16, 78, 363, 367, 373
Sample standard deviation, 50, 315, 325, 330
Sample variance, 334, 335
Sampling, 78
Sampling frequency, 78
Sampling period, 78
Sampling theorem, 80, 257, 363, 366, 369
Scale error, 64
Scilab, 231, 233, 284, 395
SD card, 154
Secondary measurement standard, 319
Secondary standard, 319
Sector, 135, 153
Secure Digital (SD), 105, 137
Secure Digital Extended Capacity (SDXC), 105
Secure Digital High Capacity (SDHC), 105, 137, 167
Seek point, 374
Seek table, 374
Select, Audacity, 173
Selection, 173
Self-generated noise, 300
Sensitivity, 65, 176, 187
Servo system, 134
Sigma-delta modulation (SDM), 100
Sigma-delta modulator, 200
Signal-to-nose ratio, 84
Sinc interpolation, 364
Slack, 153
Slow time response, 4
Snap guide, 175
Software libre, 115
Solo, 125
Sound, 398
  card, 283
  Engine, 168
  field, 190
  file formats, 106
  pressure, 2
  pressure level, 4
  transmission loss, 60
Sound forge, 168
Sparse files, 157
Spectral analysis, 245
Spectral density, 24, 36
Spectral leakage, 23, 37, 259, 415
Spectral lines, 274
Spectral masking, 150
Spectral resolution, 255, 267
Spectrogram, 181
Spectrum analysis, 18, 181, 182, 245, 254
  FFT, 264
Spectrum analyzer, 18
Stabilization time, 6–8, 15–17, 343
Stable system, 225
Standard deviation, 6, 15, 45, 50, 53, 117, 314, 324
Standard reference, 64
Standard uncertainty, 50, 316
Stationariy phenomenon, 14
Stationary field, 191
Stationary noise, 337
Statistical dispersion, 15, 3374, 338, 341, 345
Statistical distribution, 15, 321

Statistical inference, 321
Steady state, 31
Stereo channels, 122
Storage, 132
Streaminfo, 372
Structure of a FLAC file, 371
Student *t* distribution, 330
Sub-bands, 149
Subblock, 371
Subcode, 144
Subframe, 371
Subframe header, 380
Subharmonic frequency, 258
Successive approximation converter, 93
Successive approximation register, 93
Summary, 130, 169
Summary files, 131
Synchronization word, 144
`syslin`, 399
Systematic error, 62, 312
  additive, 62
  multiplicative, 62, 64
  nonlinear, 68

T
Table of contents (TOC), 147
Tag element, 124
Tags, 385
Temporal leakage, 271
Temporal masking, 150
Thermal noise, 6, 198
Time-frequency uncertainty, 23
Time-frequency uncertainty principle, 263
Timetrack element, 124
Time weighting, 3, 17
Tolerance, 27, 281, 311
Tonal auudibility, 275
Tonality adjustment, 275
Tonal noise, 273
Total harmonic distortion (THD), 106, 301
Traceability, 318
Track, 134, 135, 376
Traffic flow, 7, 17, 341
Traffic noise, 341
Transducer, 65, 187
Transducer sensitivity, 110
Transfer function, 280
Transient, 31
Transient response, 302, 351
Traverso DAW, 168
True value, 43, 309
Truncation errors, 248
Two's complement, 110
Type A uncertainty, 50, 316, 356
Type B uncertainty, 50, 51, 56, 59, 316, 355

U
Unary code, 117
Unbiased estimate, 315, 325, 334
Uncertainty, 1, 44, 49, 316
Uncertainty and resolution, 61
Uncertainty due to recorder, 303
Uncertainty in environmental conditions, 56
Uncompressed audio formats, 106
Undersampling, 368
Unicode, 379
Uniform distribution, 84, 328
Unit pulse, 211
Universal Disk Format (UDF), 153, 159
Unstability, 249
USB, 132
UTF-8, 375, 379

V
Variable length code, 116
Variance, 314, 324
Verbatim, 119
Verbatim model, 381
Verbatim predictor, 119
Vorbis comment, 375

W
Waveblock, 390
Waveclip, 389
Waveclip element, 125
Wave Editor, 168
Wave Pad, 168
Wavesurfer, 168
Wavetrack, 389
Wavetrack element, 124, 125
WAV file, 112
WAV format, 107, 115
Wavosaur, 168
`wavread`, 132, 232, 234, 398
Weighted levels, 8
White noise, 6, 17, 337
Window, 124, 260, 415
  Blackman, 262, 416
  Blackman-Harris, 262
  functions, 261
  Hamming, 262, 418
  Hann, 262
  Kaiser, 262
  lateral lobes, 263

main lobe, 263
spectrum, 263
Windowing, 23
Working standard, 319
Write-once CD, 147

X

XML, 124, 174, 385

Z

Zero-padding, 250, 270
Zoned bit recording (ZBR), 136
Zoom H4, 297
Z transfer function, 221, 225
Z transform, 219
Z transform of a delayed signal, 221
Z weighting, 283

Printed in the United States
By Bookmasters